Microencapsulation and Related Drug Processes

DRUGS AND THE PHARMACEUTICAL SCIENCES

A Series of Textbooks and Monographs

Edited by

James Swarbrick
School of Pharmacy
University of North Carolina
Chapel Hill, North Carolina

Volume 11. PHARMACEUTICAL ANALYSIS: MODERN METHODS (in two parts), *edited by James W. Munson*

Volume 12. TECHNIQUES OF SOLUBILIZATION OF DRUGS, *edited by Samuel H. Yalkowsky*

Volume 13. ORPHAN DRUGS, *edited by Fred E. Karch*

Volume 14. NOVEL DRUG DELIVERY SYSTEMS: FUNDAMENTALS, DEVELOPMENTAL CONCEPTS, BIOMEDICAL ASSESSMENTS, *Yie W. Chien*

Volume 15. PHARMACOKINETICS, Second Edition, Revised and Expanded, *Milo Gibaldi and Donald Perrier*

Volume 16. GOOD MANUFACTURING PRACTICES FOR PHARMACEUTICALS: A PLAN FOR TOTAL QUALITY CONTROL, Second Edition, Revised and Expanded, *Sidney H. Willig, Murray M. Tuckerman, and William S. Hitchings IV*

Volume 17. FORMULATION OF VETERINARY DOSAGE FORMS, *edited by Jack Blodinger*

Volume 18. DERMATOLOGICAL FORMULATIONS: PERCUTANEOUS ABSORPTION, *Brian W. Barry*

Volume 19. THE CLINICAL RESEARCH PROCESS IN THE PHARMACEUTICAL INDUSTRY, *edited by Gary M. Matoren*

Volume 20. MICROENCAPSULATION AND RELATED DRUG PROCESSES, *Patrick B. Deasy*

Other Volumes in Preparation

Microencapsulation and Related Drug Processes

Patrick B. Deasy
School of Pharmacy
University of Dublin
Dublin, Ireland

Marcel Dekker, Inc.   New York and Basel

Library of Congress Cataloging in Publication Data

Deasy, P. B.
Microencapsulation and related drug processes.

(Drugs and pharmaceutical sciences ; v. 20)
Includes bibliographical references and indexes.
1. Microencapsulation. 2. Drugs--Controlled release.
I. Title. II. Series. [DNLM: 1. Capsules. 2. Delayed-action preparations. 3. Technology, Pharmaceutical.
Wl DR893B v. 20 / QV 785 D285m]
RS201.C3D43 1984 615'.191 83-26267
ISBN 0-8247-7162-1

MARCEL DEKKER, INC.
270 Madison Avenue, New York, New York 10016

Current printing (last digit):
10 9 8 7 6 5 4 3 2 1

PRINTED IN THE UNITED STATES OF AMERICA

To my wife, Philda,
and my children,
Patricia, Deirdre, and Nicola
for their patience and understanding
during the extensive period devoted by
me to the preparation of this text.

Preface

Microencapsulation is now the most frequently employed method of producing controlled-release dosage forms. Over the past two decades enormous progress has been made in developing the technology and in applying it to a diversity of medical and other uses. This book is concerned mainly with defining criteria for drug selection for microencapsulation, with a detailed review of the many techniques used and with a discussion of the mechanism of drug release from such products.

To this end more than 800 references have been cited in the various chapters. This represents a careful selection of the vast amount of published material in this area. In this regard I am most grateful to the many research workers whose findings I have abridged and blended together to present a balanced and comprehensive overview of the subject. I am also grateful to the many publishers and authors for permission to reproduce figures and tables. The extensive references will enable those interested in acquiring greater detail about a particular aspect of the subject to refer easily to primary literature sources.

Many of the references cited are patents. Also, many of the research papers discussed are based on approaches that are the subject of patents. The potential user of the information in this book is hereby advised not to infringe any patent referred to directly or indirectly. However, the reader should be aware that the text contains many useful opportunities for technology transfer between processes without breaching patents. A number of these approaches are discussed in the text.

The physicochemical and pharmacological properties of drugs and the intended use of their products will tend to dictate how they should be microencapsulated. The index will help locate some microencapsulation processes that have been applied to particular drugs. However, more detailed perusal of the text may indicate that a different process, or a modification thereof, may be more suitable. The reader should also be aware that many processes, such as pan coating, are simple to describe, but their practical application requires skills and experience not easily mastered. Also, scaling-up problems may be encountered. For these reasons it may be necessary to examine a number of microencapsulation processes experimentally before selecting one that readily achieves the specifications required of the end product.

A number of other related processes, such as microparticles, nanocapsules, and nanoparticles, are also referred to in the text. The book should be of primary interest to those in the pharmaceutical industry and schools of pharmacy concerned with the formulation of drugs as microcapsules or related dosage forms. The text should be suitable as a reference for use in conjunction with an undergraduate or postgraduate pharmacy course in this subject area. The book should also be useful to anyone contemplating initiating research in the area so as to avoid the duplication of approach that is often obvious in the existing literature. It should be of interest to those concerned with the development of this type of product in the many other industries, such as foods, cosmetics, photography, and printing, which also use microencapsulation technology.

Research in microencapsulation is being actively pursued by many groups and the next 20 years should continue to yield many useful and innovative ideas. Much work has yet to be done in areas such as the use of novel polymers and other additives for coatings, the construction of laminated films, and the elimination of difficulties such as clumping and lack of coat uniformity associated with existing technologies. Judging by the enormous interest shown in the technology to date, it is likely that many of these problems will be resolved in the future.

I would like to thank Joan Barnes, Joan Byrne, Helen Chambers, Philda Deasy, Dorothy O'Brien, Dolores O'Higgins, and Elizabeth Sherlock for typing the original manuscript. I am grateful to the staff of Marcel Dekker, Inc., for their cooperation in the publication of this book.

Patrick B. Deasy

Contents

Preface v

Chapter 1 General Introduction 1
1.1 Some Historical and Other Considerations 1
1.2 Reasons for Microencapsulation 3
1.3 Pharmacological and Physicochemical Considerations 8
1.4 General References 13
References 14

Chapter 2 Core and Coating Properties 21
2.1 Core Properties 21
2.2 Coating Properties 23
2.3 Desolvation and Gelation of the Coating 44
2.4 Mechanical Properties of Films 45
2.5 Permeability to Oxygen, Carbon Dioxide, and Water Vapor—Photostability 47
2.6 Miscellaneous Other Properties Relating to Microcapsules and Similar Dosage Forms 49
References 53

Chapter 3 Coacervation-Phase Separation Procedures Using Aqueous Vehicles 61
3.1 Introduction 61
3.2 Simple and Complex Coacervation 64
3.3 Microencapsulation of Drugs by Simple Gelatin Coacervation 70
3.4 Microencapsulation of Drugs by Complex Gelatin—Acacia Coacervation 71

	3.5	Other Aspects of Microencapsulation by Simple or Complex Coacervation Involving Gelatin and Acacia	75
	3.6	Other Wall-Forming Polymers	82
	3.7	Solvent Evaporation Process	85
	3.8	Gelatin Nanoparticles	86
		References	89
Chapter 4	Coacervation-Phase Separation Procedures Using Nonaqueous Vehicles		97
	4.1	Ethylcellulose	97
	4.2	Cellulose Acetate Phthalate	107
	4.3	Cellulose Acetate Butyrate	108
	4.4	Hydroxypropylmethylcellulose Phthalate	108
	4.5	Carboxymethylethylcellulose and Polylactic Acid	109
	4.6	Cellulose Nitrate and Polystyrene	109
	4.7	Acrylate	110
	4.8	Poly(Ethylene-Vinyl Acetate) and Chlorinated Rubber	110
	4.9	Hardened Oils and Fats	111
	4.10	Miscellaneous Polymers	112
		References	114
Chapter 5	Interfacial Polycondensation		119
	5.1	Polyamide	121
	5.2	Polyester	137
	5.3	Polyurethane	138
	5.4	Miscellaneous	139
		References	139
Chapter 6	Pan Coating		145
	6.1	Introduction	145
	6.2	The Process	146
	6.3	Side-Vented Coating Pan Process	153
	6.4	Some Further Examples of the Microencapsulation of Drugs by Pan Coating	154
		References	159
Chapter 7	Air Suspension Coating		161
	7.1	Introduction	161
	7.2	Some Process Considerations	164
	7.3	Mathematical Determination of Parameters Associated with Air Suspension Coating	168
	7.4	Air Suspension Equipment and Its Operation	172
	7.5	Some Applications of Air Suspension Coating	173
		References	177

Chapter 8 Spray Drying, Spray Congealing, Spray Embedding, and Spray Polycondensation 181
8.1 Some Basic Principles of Spray Drying and Spray Congealing 181
8.2 Some Examples of the Microencapsulation of Excipients and Drugs by Spray Drying 184
8.3 Some Examples of the Microencapsulation of Excipients and Drugs by Spray Congealing 185
8.4 Spraying into Chilled Organic Solvent, Dehydrating Liquid, or Sorptive Solid Particles 187
8.5 Spray Embedding 188
8.6 Spray Polycondensation 191
References 191

Chapter 9 Polymerization Procedures for Nonbiodegradable Micro- and Nanocapsules and Particles 195
9.1 Introduction 195
9.2 Acrylic Products 199
9.3 Polystyrene Products 213
9.4 Polysiloxane Products 214
References 215

Chapter 10 Polymerization Procedures for Biodegradable Micro- and Nanocapsules and Particles 219
10.1 Polymers and Copolymers of Lactic/Glycolic Acids—Other Aliphatic Polyesters 219
10.2 Albumin Products 225
10.3 Polyalkyl Cyanoacrylate Products 231
10.4 Epoxy Products 233
10.5 Miscellaneous Products 235
References 236

Chapter 11 Ion-Exchange Resins 241
11.1 Introduction 241
11.2 Drug Release from Ion-Exchange Resins 245
11.3 Drug Release from Coated Ion-Exchange Resins 247
11.4 Miscellaneous 250
References 250

Chapter 12 Congealable Disperse-Phase Encapsulation Procedures 253
12.1 Introduction 253
12.2 Extrusion Devices 253
12.3 Hydrophilic Congealable Systems 256

	12.4	Hydrophobic Congealable Systems—Waxes, Fats, and Oils	261
		References	262
Chapter 13	Miscellaneous Other Methods of Encapsulation and Entrapment		265
	13.1	Physical Methods	265
	13.2	Dip Coating	272
	13.3	In Situ Polymerization	272
	13.4	Liposomes	273
	13.5	Spherical Matrices from Liquid Suspension	278
	13.6	Granulation Processes	278
	13.7	Spheronization	279
	13.8	Molecular-Scale Entrapment	281
	13.9	Other Approaches and Considerations	284
		References	286
Chapter 14	Release of Drug from Microcapsules and Microparticles		289
	14.1	Introduction	289
	14.2	Permeation Considerations	289
	14.3	Diffusion—Some Initial Mathematical and Other Considerations	293
	14.4	Diffusion Coefficient	295
	14.5	Partition Coefficient	299
	14.6	Drug Solubility and Solubility Gradient	300
	14.7	Coating Area and Thickness	302
	14.8	Kinetics of Drug Release from Microcapsules and Microparticles	302
	14.9	Microcapsules Conforming to Release from Reservoir-Type Devices	303
	14.10	Lag Time and Burst Effects	307
	14.11	Microcapsules and Microparticles Conforming to Release From Monolithic Devices	308
	14.12	Pore Effects	312
	14.13	Boundary-Layer Effects	312
	14.14	Moving Boundaries	314
	14.15	Summary	315
		References	316
Appendix			321
Author Index			327
Subject Index			345

1

General Introduction

Microcapsules developed for use in medicine consist of a solid or liquid core material containing one or more drugs enclosed in coating as shown in Fig. 1.1. The core may also be referred to as the nucleus or fill and the coating as the wall or shell. Depending on the manufacturing process, various types of microcapsule structure can be obtained as illustrated. The most common type is the mononuclear spherical. Microcapsules usually have a particle size range between 1 and 2000 μm. Products smaller than 1 μm are referred to as nanocapsules, because their dimensions are measured in nanometers. When no distinct coating and core regions are distinguishable, the analogous terms used are microparticles and nanoparticles.

Microcapsules are often described by other terms, such as coated granules, pellets or seeds, microsperules, and spansules. These products differ from larger conventional hard or soft gelatin capsules in a number of important ways apart from size. They differ most notably in the greater variety of coating materials and procedures used, in the comparative thinness of the coating formed, in their unique release properties, and in their greater diversity of application in medicine.

1.1 SOME HISTORICAL AND OTHER CONSIDERATIONS

The first research leading to the development of microencapsulation procedures for pharmaceuticals was published by Bungenburg de Jong and Kaas [1] in 1931 and dealt with the preparation of gelatin spheres and the use of a gelatin coacervation process for coating. In the late 1930s and 1940s, Green and co-workers of The National Cash Register Co., Dayton, Ohio, developed the gelatin coacervation process, which eventually lead to several patents for carbonless carbon paper. This product used a gelatin microencapsulated oil phase usually containing a colorless dye precursor. The microcapsules were afixed to the undersurface of the top page and released the dye precursor upon rupture

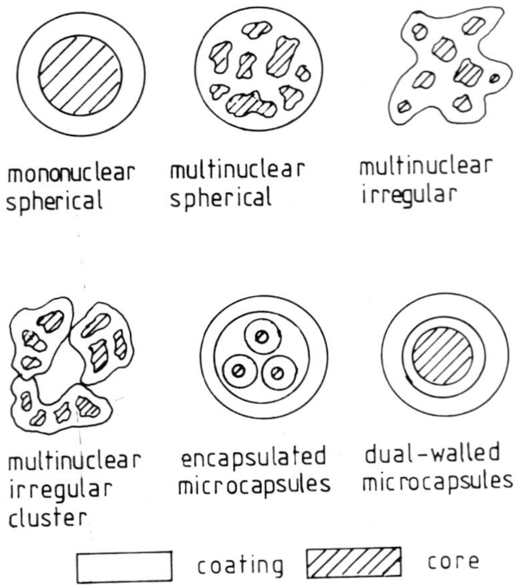

Figure 1.1 Some typical structures of microcapsules.

by pressure from the tip of the writing instrument. The liberated dye precursor then reacted with an acid clay coating on the top surface of the underlying page to form the copy image.

Since then many other coating materials and processes of application have been developed by the pharmaceutical industry for the microencapsulation of medicines. Drug companies have been quick to realize and exploit the enormous potential of the technology for overcoming formulation and delivery problems in many dosage forms such as capsules, tablets, powders, topicals, and injectables. Over the last 25 years numerous patents have been taken out by pharmaceutical companies for microencapsulated drugs. Manufacturers of medical equipment, diagnostic agents, and other health care-related products have also been interested in the commercial possibilities of microencapsulation. Other industries, such as the food, cosmetic, horticultural, paint, print, photographic, computer, fertilizer, adhesives, cleaning, and aerospace industries, have been concerned with microencapsulated products. At the same time the plastics industry has been continually involved in the production and evaluation of new polymers with potential application in microencapsulation. Also during this period research groups in various academic, governmental, and other bodies have become increasingly interested in the technology and have contributed extensively to the published literature.

Despite the intensity of research into the new technology, there are still many difficulties to be resolved. Problems frequently encountered include incomplete or uneven coating deposition, clumping of microcapsules, unsatisfactory or nonreproducible core release, and scale-up difficulties. Every microencapsulated product requires an individual design approach, and there is no one methodology that is suitable in all cases. Any newcomer to the science of microencapsulation will very quickly discover how difficult it is to select and apply the optimum encapsulation procedure for a particular product because of patent restriction and lack of adequate information in the literature.

Another problem hindering the development of microencapsulation procedures for medicines is the reservation of persons in industry and drug regulatory agencies to accept new dosage forms, particularly when they involve the use of novel adjuvants and technologies. Obviously, satisfactory toxicological data on polymers and other materials for use in microencapsulated medicines for use in humans must be available before being authorized for clinical trials and marketing. However, in many potential applications of microencapsulation the coating material is not absorbed, and so long as it is nonreactive with body surfaces, it may be safely used in products whose internal absorption into the body would be contraindicated. It is probable that in the future, as the use of microencapsulated products containing nonbiodegradable coating material increases, greater attention will have to be given to environmental pollution considerations associated with such products.

1.2 REASONS FOR MICROENCAPSULATION

Table 1.1 lists some of the many drugs that have been microencapsulated. Drugs from many different pharmacological classes have been microencapsulated, in particular analgesics, antibiotics, antihistamines, cardiovascular agents, iron salts, tranquilizers, and vitamins. It should be

Table 1.1 Some Examples of Drugs That Have Been Microencapsulated

Drug	Principal reasons for microencapsulation[a]	Primary pharmacological class
Acetazolamide	S.R.	Diuretic
Aminophylline	T.M.	Smooth muscle relaxant
Amitriptyline	S.R.	Antidepressant
Ampicillin (Na and trihydrate)	T.M.	Antibiotic
Aspirin	E.P./S.I./S.R./T.M.	Analgesic

Table 1.1 (Continued)

Drug	Principal reasons for microencapsulation[a]	Primary pharmacological class
Attapulgite	S.I.	Adsorbent
Beclamide	S.I./T.M.	Anticonvulsant
Butobarbitone	T.M.	Hypnotic and sedative
Camphor	S.I.	Counterirritant
Castor oil	L.S.C./O.M./T.M.	Laxative
Chloramphenicol	S.I./S.R.	Antibiotic
Chlorpheniramine maleate	S.R.	Antihistamine
Chlorpromazine HCl	S.R.	Tranquilizer
Citric acid	E.P./S.I./T.M.	Excipient
Clofibrate	L.S.C.	Cholesterol-reducing agent
Cloxacillin	T.M.	Antibiotic
Codeine phosphate	S.R.	Analgesic
Cod liver oil	L.S.C./O.M./T.M.	Vitamin oil
Cyclandelate	T.M.	Peripheral vasodilator
Cysteine	O.M./S.I./T.M.	Amino acid
Diazepam	S.R.	Tranquilizer
Dicloxacillin	T.M.	Antibiotic
Dimethicone fluid	L.S.C.	Silicone
Diphenhydramine HCl	S.R.	Antihistamine
Disulfiram	T.M.	Enzyme inhibitor
Doxycycline HCl	T.M.	Antibiotic
Eprazinone	L.S.C.	Antitussive and expectorant
Fenfluramine	S.R.	Anorectic agent
Ferrous citrate	E.P.	Iron supplement
Ferrous fumarate	S.I./S.R./T.M.	Iron supplement
Ferrous sulfate	E.P./G.I.R./S.I./S.R./T.M.	Iron supplement

Table 1.1 (Continued)

Drug	Principal reasons for microencapsulation[a]	Primary pharmacological class
Glyceryl trinitrate	S.R.	Vasodilator
Hyoscine methonitrate	S.I./T.M.	Anticholinergic agent
Indomethacin	G.I.R.	Analgesic
Intrinsic factor	T.M.	For vitamin B_{12} absorption
Levodopa	E.P.	Antiparkinsonian agent
Lithium carbonate	S.R./T.M.	Tranquilizer
Meclofenoxate HCl	E.P./T.M.	Central and respiratory stimulant
Meprobamate	S.R./T.M.	Tranquilizer
Methaqualone	T.M.	Hypnotic and sedative
Methionine	O.M./T.M.	Amino acid
Methylamphetamine HCl	S.R./T.M.	Central and respiratory stimulant
Nitrofurantoin	G.I.R./S.R./T.M.	Antimicrobial agent
Nortriptyline	S.R./T.M.	Antidepressant
Noscapine	S.R.	Cough suppressant
Oxytetracycline base	T.M.	Antibiotic
Papaverine HCl	S.R.	Smooth muscle relaxant
Paracetamol	G.I.R.	Analgesic
Pentaerythritol tetranitrate	S.R.	Vasodilator
Phenacetin	E.P./S.I.	Analgesic
Phenformin HCl	S.R./T.M.	Antidiabetic agent
Phenobarbitone Na	S.R./T.M.	Hypnotic and sedative
Phenylbutazone	G.I.R./T.M.	Analgesic
Phenylephrine HCl	S.R./T.M.	Sympathomimetic
Phenylpropanolamine HCl	S.R./T.M.	Sympathomimetic
Prednisolone	T.M.	Corticosteroid
Procainamide HCl	E.P./S.R.	Cardiac depressant

Table 1.1 (Continued)

Drug	Principal reasons for microencapsulation[a]	Primary pharmacological class
Propantheline Br	S.R./T.M.	Anticholinergic agent
Propoxyphene (HCl and napsylate	E.P./S.I./T.M.	Analgesic
Propranolol	S.R.	β-blocker
Quinidine sulfate	T.M.	Cardiac depressant
Reserpine	E.P.	Antihypertensive agent
Sodium bicarbonate	E.P.	Excipient
Succinimide	T.M.	Antioxaluric agent
Sulfamethoxydiazine	S.I./T.M.	Sulfonamide
Tartaric acid	S.I.	Excipient
Tetracycline (base, HCL, phosphate)	E.P./T.M.	Antibiotic
Trifluoperazine embonate	S.R.	Tranquilizer
Trimeprazine tartrate	S.R./T.M.	Antihistamine
Vitamin B_1 (thiamine HCl)	E.P./T.M.	Vitamin
Vitamin B_2 (riboflavin)	E.P./T.M.	Vitamin
Vitamin B_6 (pyridoxine HCl)	E.P./T.M.	Vitamin
Vitamin B_{12} (cyanocobalamin)	E.P.	Vitamin
Vitamin B factor (nicotinamide)	E.P./T.M.	Vitamin
Vitamin C (ascorbic acid)	E.P.	Vitamin

[a]E.P. = environmental protection, G.I.R. = gastric irritation reduction, L.S.C. = liquid solid conversion, O.M. = odor masking, S.I. = separation of incompatibilities, S.R. = sustained release, T.M. = taste masking.

appreciated that a number of these drugs have been chosen for microencapsulation studies more as model compounds whose physical characteristics or assay is convenient rather than for sound clinical reasons.

There are many reasons why drugs and related chemicals have been microencapsulated. A liquid such as eprazinone [2] may be converted to a pseudo-solid by the process as an aid to handling or storage. Toxic chemicals such as insecticides may be microencapsulated to reduce hazards to operators [3]. Ampicillin trihydrate has been microencapsulated to reduce the possibility of sensitization of factory personnel [4]. Also the hygroscopic properties of many core materials such as sodium chloride may be reduced by microencapsulation [5]. National Cash Register [6] reported that the flow properties of the vitamins thiamine hydrochloride, riboflavin, and niacin with iron phosphate could be improved by microencapsulation prior to compression into tablets. The choice of suitable coating material should improve the compaction and subsequent disintegration of encapsulated drugs from tablets.

Microencapsulation has been employed to provide protection to the core material against atmospheric effects. Bakan and Anderson [7] reported that microencapsulated vitamin A palmitate had enhanced stability compared to the unencapsulated control. Klaui et al. [8] described the enhancement of stability of fat-soluble vitamins by microencapsulation. The process has been used to reduce the volatility of several substances such as methyl salicylate and peppermint oil [7]. Also, incompatibilities between drugs such as aspirin and chlorpheniramine maleate can be prevented by microencapsulation [7]. Pharmaceutical eutetics have been microencapsulated to effect separation.

Microencapsulation has been used to disguise the unpleasant taste of a number of drugs as indicated in Table 1.1. In many of these examples, however, taste masking has been a subsidiary consideration to the provision of other properties. Macroencapsulation and sugar or film coating conventional tablets are usually much cheaper ways of masking an unpleasant taste associated with a drug. Carbon tetrachlroide and a number of other substances have been microencapsulated to reduce their odor and volatility.

Many drugs have been microencapsulated to reduce gastric and other gastrointestinal (GI) tract irritation, including ferrous sulfate [9] and potassium chloride [10]. Likewise, sustained-release aspirin preparations have been reported to cause significantly less gastric bleeding than conventional aspirin preparations [11, 12], although Hoom [13] has questioned the validity of this claim. Also, the local irritation and release properties of a number of topically applied products can be altered by microencapsulation [14, 15]. Liquid crystals have been microencapsulated for use in thermography of peripheral vascular disorders [16]. Microcapsules have also been proposed as an intrauterine contraceptive device [17].

Chang and his co-workers [18–20] have prepared a series of microcapsules known as artificial cells containing charcoal and a number of enzymes such as urease and asparaginase. Because of their lack of antigenicity and their selective permeability, they have been investigated for the treatment of chronic hepatic or renal failure, acute intoxication, and for their use in erythrocyte and enzyme replacement therapy.

Despite the many potential applications for microencapsulated drug products outlined above, comparatively few of these have been exploited commercially to any major extent. However, the use of microencapsulation for the production of sustained-release dosage forms has been widely employed in the last 30 years since the successful introduction by Smith, Kline and French in the early 1950s of their Spansule range of products. These dosage forms are often described by other terms, such as delayed-action, long-acting, retard, slow-release, sustained-action, or timed-release. Sometimes the term "controlled release" is used in the same sense or where, for example, an enteric coating is applied to localize the core release in the small intestine rather than the stomach. These products usually consist of a large number of microcapsules having variable release rates because of the composition or amount of the coatings applied, filled into outer hard gelatin capsule shells. Upon ingestion the outer shell quickly disintegrates in the stomach to liberate up to 3000 microcapsules which spread over the GI tract, thus ensuring more reproducible drug absorption with less local irritation than occurs with many nondisintegrating tablets designed for sustained release. Less frequently, microcapsules are tableted because of the risk of capsule rupture during compression. Luzzi et al. [21] have shown that tablet hardness inversely affects drug release from compressed microcapsules. Microcapsules have also been formulated into suspensions having controlled drug-release properties [22].

The potential usefulness of microencapsulated products for parenteral depot therapy is limited by the fact that the particle size should be small so as to limit injury at the site of injection. Many microencapsulation procedures are not capable of producing an adequately fine product. Also, the adjuvants chosen must be biodegradable and nonantigenic, which restricts their choice considerably. Sterlization procedures may affect stability. Products intended for intravenous (IV) therapy require a particle size less than 5 μm and preferably in the nanometer size range in order to prevent obstruction of the vascular system. However, these minute particles tend to be extensively phagocytized by cells of the reticuloendothelial system, which may cause unwanted accumulation in organs such as the liver, spleen, and lungs. Various attempts to produce suitable parenteral products will be reviewed later in the book.

1.3 PHARMACOLOGICAL AND PHYSICOCHEMICAL CONSIDERATIONS

Figure 1.2 shows the typical plasma drug concentration versus time profile following oral administration for a single dose of an idealized

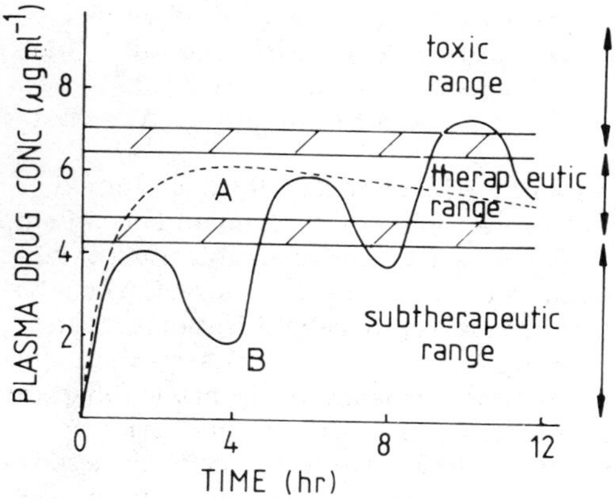

Figure 1.2 Typical plasma drug concentration versus time profiles for an idealized sustained-release microencapsulated product (A) and a conventional dosage form (B) repeatedly given orally.

sustained-release microencapsulated product (A) and for conventional dosage forms given repeatedly (B). With conventional dosage forms there is a peak-and-valley effect that tends toward a plateau, because most drugs exhibit exponential accumulation when given repeatedly at equal intervals. The initial doses may not be adequate to reach the therapeutic range, which could be a serious disadvantage with drugs such as antibiotics and analgesics, where prompt onset of action is desirable. Also, the spikes associated with later doses may well account for undesirable side effects. The more frequent the dosing interval with dosages of appropriate size, the smaller will be the oscillations in the plasma drug profile and the faster the plateau will be reached. This in fact is the underlying principle of microencapsulated oral dosage forms, which often contain a portion of the dosage in a non-sustained-release form so as quickly to establish the drug level in the therapeutic range, which is then maintained over an extended period by the progressive release of drug from the various microcapsule fractions present. By combining microcapsules of different release rates in the one dosage form, a pseudo-zero-order or steady-state release of drug can be obtained over an extended period. The aim of these products is to make the rate-limiting step the release of drug from the sustained-release dosage form rather that its rate of absorption. The result is improved management of the disease state with the use of less total drug and with fewer undesirable side effects. Pines et al. [23] reported that a microencapsulated sustained-release product containing tetracycline hydrochloride reduced the incidence of undesirable side effects compared to conventional capsules. The effect of variable clearance time from the stomach associated with enteric-coated tablets designed for controlled release with consequent variable onset in drug absorption is minimized by fractionating the dose into many enteric-coated microcapsules that are exponentially cleared from the stomach. Bogentoft et al. [24] reported a gastric emptying time (t90%) of 2 to 5 hr for enterically coated microcapsules.

Because of the reduced frequency of administration, sustained-release products are often claimed to be more convenient and to enhance ambulatory patient compliance, which is a particular problem associated with psychiatric and antihypertensive drugs [25–27]. Apart from the general economy associated with the more rapid rehabilitation of the patient, there is a considerable saving in nurses' and pharmacists' time when patients are maintained on sustained-release products in hospitals.

Other nonmicroencapsulated sustained-release dosage forms for oral use, such as plastic or waxy compressed matrices, seek to achieve the same effect. However, their drug release profiles are usually exponential, not as easily adjusted, and more adversely affected by physiological factors such as gastric emptying time. Accordingly, the majority of sustained-release oral dosage forms available are based nowadays on the microencapsulation principle. With the decline in the number of new drugs being introduced, greater attention is being paid to improving the efficacy of existing drugs. Microencapsulation will continue to be a major formulation approach to achieving this goal.

There are, however, a number of potential disadvantages associated with microencapsulated products for sustained release. The bulk of many of the products may cause difficulties in swallowing the dosage form. Patients may chew such dosage forms to assist their ingestion, with resultant alteration in the release rate from damaged microcapsules. Accidental or intentional overdosage with sustained-release products causes special problems [28, 29]. Fatal poisonings have been reported with sustained-release capsules containing microcapsules of fenfluramine [30] and phenformin [31].

Considerable controversy surrounds the clinical merits, if any, of many microencapsulated products such as amitriptyline, aspirin, chlorpromazine hydrochloride, ferrous sulfate, meprobamate, and nortriptyline over their conventional dosage form-equivalents of lower cost. Care should be taken in the interpretation of studies where the specificity of the drug assay procedure and pharmacokinetic model is questionable and where results obtained in normal subjects have been extrapolated to patients with disease states. In most cases where insignificant clinical improvement has been reported, this has been due to poor understanding on the part of the manufacturer at the time of production of those factors that influence the suitability of particular drugs as candidates for microencapsulated sustained-release products. With the growth in knowledge of physicochemical and pharmacokinetic properties of drugs, the unsuitability of many drugs for such products has been better understood, and a number of these products have now been withdrawn from the market.

The transit time of up to approximately 12 hr limits the extent of prolongation possible with oral microencapsulated products. Roentgenographic studies have shown that within this time granules released from an outer gelatin capsule shell were well dispersed in the GI tract [32, 33]. In contrast, Pemberton [34] reported that a bulky potassium chloride tablet formulated to have prolonged release properties had lodged in the esophagus, producing local ulceration. Levy and Hollister [35] showed how variation in gastric emptying time and intestinal peristaltic activity affected the absorption of aspirin from microcapsules. Environmental conditions vary enormously along the GI tract, and certain drugs are preferentially absorbed at particular regions. Vitamin B_2 (riboflavin) is absorbed high up in the GI tract, particularly in the upper duodenal area, by an active transport mechanism that is saturatable so that little is absorbed in the lower intestine [36]. Accordingly, it was not suprising that Morrison et al. [37] found that a number of sustained-release vitamin B_2-containing preparations designed to release their drug throughout the intestinal tract showed no sustained urinary excretion, nor gave full availability compared to conventional therapy. Likewise, as the maximum amount of iron is absorbed in the duodenum and upper jejunum [38. 39], Bothwell et al. [40] showed that the greatest fraction of iron was absorbed from the microcapsules in the Spansule with the fastest release rate and that there was a marked reduction in the amount of iron absorbed from the slowest-releasing microcapsules. The clinical efficiency of such sustained-release iron preparations over conventional dosage forms is questionable. Other drugs that are selectively or erratically absorbed at particular regions of the GI tract and so are poor candidates for sustained-release dosage forms include dicumarol [41] and hydrocortisone [42]. Compounds such as quaternary ammonium salts, which bind extensively to mucin in the intestine [43], tend to have inherent sustained-release properties. The absorption of anionic and cationic agents can be affected by many coadministered agents [44].

Many drugs are extensively metabolized during absorption from the gut or by first-pass effects. Levodopa can be extensively metabolized by high concentrations of dopa decarboxylase in the gut wall, so that prolonged-action products gradually releasing drug along the GI tract produce no apparent sustained-release effect [45, 46]. Likewise, more than 50% of propantheline is metabolized in the small intestine [47], and accordingly it would be a poor choice for a sustained-release product. Nitroglycerin is extensively metabolized by a first-pass effect [48, 49]. Accordingly, Pilkington and Purves [50] questioned the usefulness of a sustained-release product of nitroglycerin as an effective prophylactic against attacks of angina. Similar considerations apply to the β-lactam antibiotics, which, depending on chemical structure, are susceptible to acid degradation and attack by β-lactamase [51].

The biological half-life of the drug is very important. Drugs such as ampicillin, cloxacillin, furosemide, levodopa, and propylthiouracil have very short half-lives and therefore require frequent dosing to maintain an adequate therapeutic level. If such drugs require a large quantity of drug for a conventional unit dosage form, they may well give rise to unacceptably bulky products if they are to contain sufficient drug for a sustained-release effect. Eriksen [52] has stated that if the conventional dosage form exceeds 0.2 g, it is most probable that its sustained-release formulation will be excessively bulky to be consumed as a single dosage unit. On the other hand, as the biological half-life of a drug progressively exceeds 4 hr, it becomes essentially prolonged-acting and gains comparatively little by additional formulation to render it sustained-release. For example, aspirin has a biological half-life of about 6 hr, and a sustained-release product, which because of problems of bulk requires a usual dosage of two tablets every 12 hr, is reported to have little clinical advantage over conventional therapy other than maintaining a more uniform therapeutic plasma level and reducing the incidence of morning stiffness following dosing the previous evening [53]. Some other examples of drugs with long biological half-lives include amitriptyline, dicumarol, chlordiazepoxide, chlorphentermine, chlorpropamide, diazepam, guanethidine, meprobamate, paraminosalicylic acid, phenobarbitone, phenothiazines, phenytoin, and warfarin. Also, the duration of action of drugs such as anticholinesterases, corticosteroids, and monoamine oxidase inhibitors is longer than is suggested by their biological half-lives. It is not suprising that microencapsulated sustained-release products of such drugs with intrinsic extended duration of action have been reported in the cases of amitriptyline [54–56], meprobamate [57], and phenothiazines [58–59] to offer little advantage over conventional dosage forms.

For drugs with narrow therapeutic indices, such as barbiturates, procainamide, or theophylline, the design of sustained-release microencapsulated products can be difficult because of the danger of fatal overdosage if the sustained release mechanism should fail, causing the immediate release of all of the drug contained in the product. However, enteric-coated microcapsules of such drugs may be used to flatten the plasma profile.

Consideration of physiocochemical properties of drug molecules is also very important in the design of their effective sustained-release microencapsulated products. The size of the drug molecules is important, particularly as their molecular weight exceeds 500, because large molecules find it more difficult to diffuse through biological and other polymeric membranes. Drugs with very high water solubility are often difficult to formulate into sustained-release microencapsulated products. Their release properties are frequently controlled by a diffusion process across the coating whose driving force drops off rapidly as the core concentration of drug becomes depleted. On the other hand, drugs such as diazoxide, digoxin, griseofulvin, hydrochlorthiazide, prednisolone, salicylamide, tolbutamide, and warfarin are dissolution rate-limited in their absorption because of their low water solubility. Such drugs have inherent sustained-release properties, and further reduction in their rates of release by microencapsulation procedures may serve only to decrease their availabilities.

Certain drugs show marked changes in solubility with variation in pH. Tetracycline, for example, is more soluble by a factor of approximately 100 at the low pH of 1 to 3 in the stomach than at the pH of 5 to 6 in the upper small intestine [60]. Accordingly, the use of enteric coating alone would significantly reduce the overall dissolution of the drug, which in turn would limit its absorption in the small intestine. Also, as drugs are better absorbed in their nonionized form, consideration of the pK_a of the drug is important. Niazi [61] lists pK_a values for many drugs and discusses how weakly acid or basic compounds will be altered in their absorption by the varying pH along the GI tract. Finally, drugs with high partition coefficients (high lipid solubility), such as phenothiazines, tend to accumulate in fatty tissues with slow elimination and are accordingly poor candidates for sustained-release formulations.

For a fuller discussion of these and other pharmacological and physiocochemical factors affecting the design of sustained-release products, including the importance of intra- or intersubject variability or coadministered food or drugs, the interested reader is advised to consult Refs. 62–67.

1.4 GENERAL REFERENCES

Many other medical applications of microencapsulated products are discussed later in this book. Ways of designing microencapsulated products to overcome certain pharmacological and physicochemical problems are considered. Properties of core and coating, methods of microencapsulation, and drug release from such products are described. Other related dosage forms, such as microparticles, nanoparticles, and drug-loading ion-exchange resin beads, are also reviewed. Subsequent chapters contain very extensive bibliographies covering all aspects of microencapsulation and related processes for drugs and collectively comprise the most complete coverage of the literature relating to these topics yet published.

There have been many excellent short reviews on the microencapsulation of medicines and related products published in various journals and textbooks [5, 7, 68–77]. A number of textbooks bearing the word "microencapsulation" or "microcapsule" in their titles have been published that contain the edited proceedings of various symposia on the subject [78–80] or reviews of the patent literature [81–83]. One other short textbook on microencapsulation has been published [84] that does not cover the many important advances made in the area since 1970 and that is not particularly pharmaceutical in orientation.

REFERENCES

1. H. G. Bungenburg de Jong and A. J. Kaas, Zur kenntuis der komplex-koazeration, V. Mitteilung: relative verschiebungen im elektrischen gleichstromfelde von flussigkeits—einschliebungen in komplex—koazervat—tropfehen, *Biochem. Z. 232*:338–345 (1931).
2. L. Si-Nang, P. F. Carlier, P. Delort, J. Gazzola, and D. Lafont, Determination of coating thickness of microcapsules and influence upon diffusion, *J. Pharm. Sci. 62*:452–455 (1973).
3. E. S. Raun and R. D. Jackson, Encapsulation as a technique for formulating microbial and chemical insecticides, *J. Econ. Entomol. 59*:620–622 (1966).
4. H. Seager, U.S. Patent 4,016,254 (April 5, 1977).
5. J. A. Bakan, Microencapsulation of foods and related products, *Food Technol. 27*(11):34–44 (1973).
6. Microencapsulation of pharmaceuticals and related materials, The National Cash Register Co., Dayton, Ohio (1966).
7. J. A. Bakan and J. L. Anderson, Microencapsulation, in *The Theory and Practice of Industrial Pharmacy*, 2nd ed. (L. Lachman, H. A. Lieberman, and J. L. Kanig, eds.), Lea & Febiger, Philadelphia, 1976, pp. 420–438.
8. H. M. Klaui, W. Hausheer, and G. Huschke, Technological aspects of the use of fat-soluble vitamins and carotenoids and of the development of stabilized marketable forms, in *Fat-Soluble Vitamins* (R. A. Morton, ed.), Pergamon, London, 1970, pp. 113–159.
9. P. C. Elwood and G. Williams, A comparative trial of slow-release and conventional iron preparations, *Practitioner 204*: 812–815 (1970).
10. J. Arnold, J. T. Jacob, and B. Riley, Bioavailability and pharmacokinetics of a new slow-release potassium chloride capsule, *J. Pharm. Sci. 69*:1416–1418 (1980).
11. E. P. Frenkel, M. S. McCall, C. C. Douglass, and S. Eisenberg, Fecal blood loss following aspirin and coated aspirin microspherule administration, *J. Clin. Pharmacol. 8*:347–351 (1968).
12. B. L. J. Treadwell, D. G. Carroll, and E. W. Pomare, Gastrointestinal blood loss with sustained release aspirin, *N. Z. Med. J.* 78:435–437 (1973).
13. J. R. Hoon, Bleeding gastritis induced by long-term release aspirin, *J. Amer. Med. Assoc. 229*:841–842 (1974).
14. R. H. Sudekum, Microcapsules for topical and other applications, in *Microencapsulation* (J. R. Nixon, ed.), Dekker, New York, 1976, pp. 119–128.
15. T. R. Tice, W. E. Meyers, D. H. Lewis, and D. R. Cowsar, Controlled release of ampicillin and gentamicin from biodegradable microcapsules, in *Proceedings of the 8th International Symposium*

on Controlled Release of Bioactive Materials, Ft. Lauderdale, July 26–29, 1981, pp. 108–111.
16. G. C. Maggi and F. M. Di Roberto, Encapsulated liquid crystals (ELC) thermography in the diagnosis and monitoring of peripheral vascular disorders, in *Microencapsulation* (J. R. Nixon, ed.), Dekker, New York, 1976, pp. 103–111.
17. D. L. Gardner, D. J. Fink, and C. R. Hassler, Potential delivery of contraceptive agents to the female reproductive tract, in *Controlled Release of Pesticides and Pharmaceuticals* (D. H. Lewis, ed.), Plenum, New York, 1981, pp. 99–109.
18. T. M. S. Chang, *Artificial Cells*, Thomas, Springfield, Ill., (1972).
19. T. M. S. Chang and N. Malave, The development and first clinical use of semipermeable microcapsules (artificial cells) as a compact artificial kidney, *Trans. Amer. Soc. Artif. Int. Organs 16*:141–148 (1970).
20. E. D. Siu Chong and T. M. S. Chang, In vivo effects of intraperitoneally injected L-asparaginase solution and L-asparaginase immobilized within semipermeable nylon microcapsules with emphasis on blood L-asparaginase, "body" L-asparaginase and plasma L-asparaginase levels, *Enzyme 18*: 218–239 (1974).
21. L. A. Luzzi, M. A. Zoglio, and H. V. Maulding, Preparation and evaluation of the prolonged release properties of nylon microcapsules, *J. Pharm. Sci. 59*:338–341 (1970).
22. Y. Raghunathan, L. Amsel, O. Hinsvark, and W. Bryant, Sustained-release drug delivery system 1: coated ion-exchange resin system for phenylpropanolamine and other drugs, *J. Pharm. Sci. 70*:379–384 (1981).
23. A. Pines, G. Khaja, J. S. B. Greenfield, H. Raafat, K. S. Sreedharan, and W. D. Linsell, A double-blind comparison of slow-release tetracycline and tetracycline hydrochloride in purulent exacerbations of chronic bronchitis, *Brit. J. Clin. Pract. 26*:475–476 (1972).
24. C. Bogentoft, G. Ekenved, and U. E. Jonsson, Controlled release of drugs to the small intestine, in *Proceedings of the 8th International Symposium on Controlled Release of Bioactive Materials, Ft. Lauderdale, July 26–29*, 1981, pp. 42–45.
25. A. M. W. Porter, Drug defaulting in a general practice, *Brit. Med. J. i*:218–222 (1969).
26. C. J. Latiolais and C. C. Berry, Misuse of prescription medications by outpatients, *Drug Intell. 3*:270–277 (1969).
27. R. B. Haynes, The effect of the therapeutic regimen on patient compliance and the possible influence of controlled-release dosage forms, in *Controlled Release Pharmaceuticals* (J. Urquhart, ed.), American Pharmaceutical Association, Washington, D.C., 1981, pp. 121–139.

28. C. L. Winek, W. D. Collom, and C. H. Wecht, Sustained-release-barbiturate-risk, *Lancet* 2:155-156 (1967).
29. S. R. Meadow and G. A. Leeson, Poisoning with delayed-release tablets: treatment of debendox poisoning with purgation and dialysis, *Arch. Dis. Childhood* 49:310–312 (1974).
30. H. Simpson and I. McKinlay, Poisoning with slow-release fenfluramine, *Brit. Med. J.* 4:462–463 (1975).
31. J. P. Bingle, G. W. Storey, and J. M. Winter, Fatal self-poisoning with phenformin, *Brit. Med. J.* 3:752 (1970).
32. T. M. Feinblatt and E. A. Ferguson, Timed-disintegration capsules. An in vivo roentgenographic study, *New Engl. J. Med.* 254:940–943 (1956).
33. T. M. Feinblatt and E. A. Ferguson, Timed-disintegration capsules (tymcaps)—a further study. An in vivo roentgenographic study, blood-level study and relief of anginal pain with pentaerythritol tetranitrate, *New Engl. J. Med.* 256:331–335 (1957).
34. J. Pemberton, Oesophageal obstruction and ulceration caused by oral potassium therapy, *Brit. Heart J.* 32:267–268 (1970).
35. G. Levy and L. E. Hollister, Dissolution rate limited absorption in man. Factors influencing drug absorption from prolonged-release dosage form, *J. Pharm. Sci.* 54:1121–1125 (1965).
36. G. Levy and W. J. Jusko, Factors affecting the absorption of riboflavin in man, *J. Pharm. Sci.* 55:285–289 (1966).
37. A. B. Morrison, C. B. Perusse, and J. A. Campbell, Physiologic availability and in vitro release of riboflavin in sustained-release vitamin preparations, *New Engl. J. Med.* 263:115–119 (1960).
38. E. B. Brown and B. W. Justus, In vitro absorption of radioiron by everted pouches of rat intestine, *Amer. J. Physiol.* 194: 319–326 (1958).
39. M. S. Wheby, Site of iron absorption in man, *Scand. J. Haemat.* 7:56–62 (1970).
40. T. H. Bothwell, G. Pirzio-Biroli, and C. A. Finch, Iron absorption 1. Factors influencing absorption, *J. Lab. Clin. Med.* 51: 24–36 (1958).
41. M. Weiner, S. Shapiro, J. Axelrod, J. R. Cooper, and B. B. Brodie, The physiological disposition of dicoumarol in man, *J. Pharmacol. Exp. Ther.* 99:409–420 (1950).
42. I. Komiya, J. Y. Parks, A. Kamani, N. F. H. Ho, and W. I. Higuchi, Quantitative mechanistic studies in simultaneous fluid flow and intestinal absorption using steroids as model solutes, *Int. J. Pharm.* 4:249–262 (1980).
43. R. M. Levine, M. R. Blair, and B. B. Clark, Factors influencing the intestinal absorption of certain monoquarternary anticholinergic compounds with special reference to benzomethamine

[N-diethylaminoethyl-N'-methyl-benzilamide methobromide (MC-3199)], *J. Pharmacol. Exp. Ther. 114*:78–86 (1955).
44. R. R. Levine, The influence of the intraluminal intestinal millieu on absorption of an organic cation and an anionic agent, *J. Pharmacol. Exp. Ther. 131*:328–333 (1961).
45. G. Curzon, J. Friedel, L. Grier, C. D. Marsden, J. D. Parks, M. Shipley, and K. J. Zilkha, Sustained-release levodopa in parkinsonism, *Lancet 1*:781 (1973).
46. A. C. Woods, G. A. Glaubiger, and T. N. Chase, Sustained-release levodopa, *Lancet 1*:1391 (1973).
47. B. Beerman, K. Hellstrom, and A. Rosen, On the metabolism of propantheline in man, *Clin. Pharmacol. Ther. 13*:212–220 (1972).
48. P. Needleman, S. Lang, and E. M. Johnson, Organic nitrates: relationship between biotransformation and rational angina pectoris therapy, *J. Pharmacol. Exp. Ther. 181*:489–497 (1972).
49. E. M. Johnson, A. B. Harkey, D. J. Blehm, and P. Needleman, Clearance and metabolism of organic nitrates, *J. Pharmacol. Exp. Ther. 182*:56–62 (1972).
50. J. R. E. Pilkington and M. J. Purves, Long-acting glyceryl trinitrate in angina pectoris of out patients, *Brit. Med. J. i*:38 (1960).
51. A. Tsuji, E. Miyamoto, I. Kagami, H. Sakaguchi, and Y. Tsukinaka, GI absorption of β-lactam antibiotics 1: Kinetic assessment of competing absorption and degradation in GI tract, *J. Pharm. Sci. 67*:1701–1704 (1978).
52. A. Eriksen, Sustained action dosage forms, in *The Theory and Practice of Industrial Pharmacy* (L. Lachman, H. A. Lieberman, and J. L. Kanig, eds.), Lea & Febiger, Philadelphia, 1970, pp. 408–436.
53. R. Harris and R. G. Regalado, A clinical trial of a sustained-release aspirin in rheumatoid arthritis, *Ann. Phys. Med. 9*:8–18 (1967).
54. Anon., Two new antidepressive preparations, *Drug Ther. Bull. 9*:99–100 (1971).
55. J. R. Gomez and G. Gomez, Depression in general practice: comparison between thrice daily amitriptyline tablets and a single sustained-release capsule at night, *Brit. J. Clin. Pract. 26*:33–34 (1972).
56. I. Haider, A single daily dose of a new form of amitriptyline in depressive illness, *Brit. J. Psychiat. 120*:521–522 (1972).
57. L. E. Hollister, Studies of delayed-action medication. 1. Meprobamate administered as compressed tablets and as two delayed-action capsules, *New Engl. J. Med. 266*:281–283 (1962).
58. L. E. Hollister, S. H. Curry, J. E. Derr, and S. L. Kanter, Studies of delayed-action medication V. Plasma levels and urinary excretion of four different dosage forms of chlorpromazine, *Clin. Pharmacol. Ther. 11*:49–59 (1970).

59. C. Bartlett, Comparison of thioridazine tablets to chlorpromazine spansules in the maintenance care of chronic schizophrenics. *Curr. Ther. Res. 13*:100–106 (1971).
60. W. H. Barr, J. Adir, and L. Garrettson, Decrease of tetracycline absorption in man by sodium bicarbonate, *Clin. Pharmacol. Ther. 12*:779–784 (1971).
61. S. Niazi, *Textbook of Biopharmaceutics and Clinical Pharmacokinetics*, Appleton-Century-Crofts, New York, pp. 18–23, 1979.
62. P. Speiser, Dosage form availability and biological action, *Pharma. Int. 3*:5–16 (1971).
63. M. D. Rawlins, Sustained and delayed release oral preparations, *Prescriber's J. 15*:145–150 (1975).
64. B. E. Ballard, An overview of prolonged action drug dosage forms, in *Sustained and Controlled Release Drug Delivery Systems* (J. R. Robinson, ed.), Dekker, New York, 1978, pp. 1-69.
65. V. H. Lee and J. R. Robinson, Drug properties influencing the design of sustained or controlled release drug delivery systems, in *Sustained and Controlled Release Drug Delivery Systems* (J. R. Robinson, ed.), Dekker, New York, 1978, pp. 71–121.
66. V. H. Lee and J. R. Robinson, Methods to achieve sustained drug delivery. The physical approach: oral and parenteral dosage forms, in *Sustained and Controlled Release Drug Delivery Systems* (J. R. Robinson, ed.) Dekker, New York, 1978, pp. 123–209.
67. H. G. Boxenbaum, Physiological and pharmacokinetic factors affecting performance of sustained release dosage forms, *Drug Develop. Indust. Pharm. 8*:1–25 (1982).
68. H. W. Mattson, Miniature capsules, *Int. Sci. Technol. 40*:66–76 (1965).
69. J. E. Finn and H. Nack, What is happening in microencapsulation, *Chem. Eng.*, Dec. 4:171–178 (1967).
70. J. A. Herbig, Microencapsulation, in *Encyclopedia of Chemical Technology*, 2nd ed., Interscience, New York, 1967, pp. 436–456.
71. H. Nack, Microencapsulation techniques, applications and problems, *J. Soc. Cosmet. Chem. 21*:85–97 (1970).
72. L. A. Luzzi, Microencapsulation, *J. Pharm. Sci. 59*:1367–1376 (1970).
73. G. O. Fanger, Microencapsulation: a brief history and introduction, *Amer. Chem. Soc., Div. Org. Coatings Plastics, Chem. Papers 33*:533–544 (1973).
74. G. O. Fanger, What good are microcapsules, *Chemtech. 4*:397–405 (1974).
75. W. Sliwka, Microencapsulation, *Angew. Chem. Int. Ed. 14*:539–550 (1975).

76. J. A. Bakan, Microcapsule drug delivery systems, in *Polymers in Medicine and Surgery* (R. L. Kronenthal, Z. Oser, and E. Martin, eds.) Plenum, New York, 1975, pp. 213–235.
77. Fourth International Symposium on Microencapsulation, published in *J. Pharm. Sci. 70*:351–394 (1981).
78. J. E. Vandegaer, ed., *Microencapsulation Processes and Applications*, Plenum, New York, 1974.
79. J. R. Nixon, ed., *Microencapsulation*, Dekker, New York, 1976.
80. T. Kondo, ed., *Microencapsulation: New Techniques and Applications*, Techno, Tokyo, 1979.
81. M. W. Ranney, *Microencapsulation Technology*, Noyes Development Corporation, Park Ridge, New Jersey, 1969.
82. M. H. Gutcho, *Microcapsules and Microencapsulation Techniques*, Noyes Data Corporation, Park Ridge, New Jersey, 1976.
83. *Microcapsules and Other Capsules: Advances since 1975*, Noyes Data Corporation, Park Ridge, New Jersey, 1979.
84. A. Kondo, *Microcapsule Processing and Technology* (J. Wade Van Valkenburg, ed.), Dekker, New York, (1979).

2

Core and Coating Properties

Various properties of the core and coating materials used in the formulation of micro- and nanocapsules are considered in this chapter, together with aspects of their final products. Such is the diversity of adjuvants used in their production that only selected examples from the various classes of materials employed can be considered. It should be appreciated that particular production procedures will involve only limited use of the range of materials discussed. Because of the structural similarity between microcapsules and coated tablets, appropriate examples involving the latter will be considered. A number of studies on the use of free polymeric films will also be considered so as to understand better permeation through polymers used as coatings. It is hoped that these observations will collectively provide a fundamental understanding of those factors of importance in the design and production of micro- and nanocapsules, though their relevance to the related dosage forms of micro- and nanoparticles should be easily appreciated.

2.1 CORE PROPERTIES

The core of microcapsules used in medicine contains one or more drugs either alone or in combination with suitable additives to form a liquid or solid phase. Liquid cores may be composed of polar or nonpolar substances that comprise the active ingredient or that act as vehicles for dissolved or suspended drugs. The solvent properties of such liquids will critically influence the rate of drug release and the selection of coating materials, which obviously should not be significantly affected by the vehicle over the shelf life of the product. For drugs of low water solubility with known bioavailibility problems associated with low rates of dissolution, decrease in particle size of suspended drugs may be important in enhancing in vivo absorption. There is, of course, a general tendency for smaller microcapsules to have faster release rates because of their increased surface area per unit volume or weight of core material. Addition of dispersed linear polymer may be necessary to afford adequate mechanical support to liquid cores by increasing apparent viscosity when fragile coatings are being deposited on their surface. McGinity et al. [1], for example, used gelatin or other materials to form rigid aqueous core matrices onto which were deposited thin nylon coatings. The concept of including physiologically acceptable buffers in the core containing a drug that is a weak acid or base, so as to render its release rate less dependent on pH variation in different regions of the gastrointestinal (GI) tract, has potential.

Solid cores are used more frequently than liquid ones. Apart from control of size for the reasons mentioned above, very small core particles tend to give rise to troublesome aggregation problems during production because of the relative importance of their surface attractive forces. Large particles can cause problems because of their rapid sedimentation. The shape of these cores is also very important. It is much easier to deposit uniform coatings on regular spherical particles of narrow size range that are devoid of sharp edges. Accordingly, choice of particular polymorphs of drugs, reduction of irregularly shaped crystalline materials by grinding, or the use of spray-dried forms may aid the ease of deposition of uniform coatings onto their surfaces. To provide a regular spherical core of very narrow size range, which is necessary for the success of pan coating, commercially available sucrose pellets or granules are used onto whose surface is applied micronized drug particles using a suitable binder such as polyvinylpyrrolidone prior to coating.

The density of the core is very important in controlling the transit time in the GI tract. Originally Marston [2] showed that increasing density was the most important factor promoting retention of pellets in the ramenoreticular sac of sheep. More recently, Bechgaard and Antonsen [3] and Bechgaard and Ladefoged [4] reported that coated heavy pellets containing barium sulfate of density 1.6 significantly increased the average transit time in ileostomy subjects compared to coated light pellets containing hard paraffin of density 1.0. The average transit times were 7 and 25 hr for light and heavy pellets, respectively. Variation in diameter had little effect on transit time. The results indicated that by enteric-coating cores of variable density, the dosage frequency of microencapsulated products taken orally could easily be reduced to once daily. Subsequently Bechgaard and Ladefoged [5] reported that the gastrointestinal transit time for unmedicated, nondisintegrating, hard paraffin tablets was far less reproducible than that for the microcapsules.

An alternative approach to increasing density was proposed by Umezawa [6] in a patent assigned to Zaidan Hojin Biseibutsu Kagaku Kenkyu Kai, Japan. Sodium bicarbonate was incorporated into cores containing the drug pepstatin and coated with hydroxypropylmethylcellulose. Upon administration immediately after a meal, the microcapsules floated on the stomach content as a result of the formation of carbon dioxide by reaction with gastric acid. This caused microcapsules to remain for up to 3 to 5 hr in the stomach, where they released adequate drug to suppress pepsin activity. Inclusion of waxy or other material might also increase buoyancy.

Swelling of the core with disruption of the coating and uncontrolled drug release can be a troublesome problem. For example, Raghunathan et al. [7] found it necessary to pretreat ion-exchange resin-drug complex particles with agents such as polyethylene glycol, which increased water uptake time, prior to air suspension coating with ethylcellulose, in order to retain core geometry during coating and dissolution. Obviously, moisture uptake leading to core swelling as a result of faulty storage must be avoided. Likewise, Rowe [8, 9] suggested that differences in the thermal expansion and contraction of the core and film coating of tablets could well cause a higher incidence of film defects such as cracking. Accordingly, in microencapsulation processes involving heating, it is desirable to minimize the difference in coefficients of expansion of the core and coating as occurs with sugar-based cores having cellulose coatings, both being predominantly carbohydrates.

2.2 COATING PROPERTIES

Depending on the microencapsulation procedure employed, coatings for microcapsules may contain several different additives such as film formers, plasticizers, and fillers and may be applied from various solvent systems. Each of these components will be considered in the following sections, stressing their mutual interdependence in determining the final properties of the deposited films. It should be understood that many materials have properties that enable them to be classified under a number of these headings, although for simplicity they have been arbitrarily ascribed to one section. Emphasis is placed on additives for pan and air suspension coating, which are the two most widely used commercial procedures for the production of microcapsules for medical use.

2.2.1 Film Formers

By far the most important material governing the properties of the coating is the film former. One or more of these materials, which are usually high-molecular-weight polymers, may be used alone or in combination with other additives to form the coating. Table 2.1 lists some of the more commonly used film formers employed for microencapsulation and related uses. Obviously, an enormous variety of polymers are employed, often in various grades.

Table 2.2, modified from Lehmann [10], shows the principal release mechanisms of a number of these film formers.

Table 2.1 Some Commonly Used Film Formers for Microencapsulation and Related Use

Acacia

Acrylic polymers and copolymers
- e.g., polyacrylamide
  - polyacryldextran
  - polyalkyl cyanoacrylate
  - polymethyl methacrylate

Agar and agarose

Albumin

Alginates
- e.g., calcium alginate
  - sodium alginate

Aluminium monostearate

Carboxyvinyl polymer

Cellulose derivatives
- e.g., cellulose acetate
  - cellulose acetate butyrate
  - cellulose acetate phthalate
  - cellulose nitrate
  - ethylcellulose
  - hydroxypropylcellulose
  - hydroxypropylmethylcellulose
  - hydroxypropylmethylcellulose phthalate
  - methylcellulose
  - sodium carboxymethylcellulose

Cetyl alcohol

Dextran

Gelatin

Hydrogenated beef tallow

Hydrogenated castor oil

12-Hydroxystearyl alcohol

Gluten

Glyceryl mono- or dipalmitate

Glyceryl mono-, di-, or tristearate

Myristyl alcohol

Table 2.1 (Continued)

Polyamide
e.g., nylon 6-10
poly(adipyl L-lysine)
polyterephthalamide
poly(terephthaloyl L-lysine)
Poly (ε-caprolactone)
Polydimethylsiloxane
Polyester
Polyethylene glycol
Poly(ethylene-vinyl acetate)
Polyglycolic acid, polylactic acid, and copolymers
Polyglutamic acid
Polylysine
Poly(methyl vinyl ether/maleic anhydride)
Polystyrene
Polyvinyl acetate phthalate
Polyvinyl alcohol
Polyvinylpyrrolidone
Shellac
Starch
Stearic acid
Stearyl alcohol
Waxes
e.g., beeswax
carnauba wax
Japanese synthetic wax
paraffin wax
spermaceti

Enteric film formers such as cellulose acetate phthalate (CAP), polyvinyl acetate phthalate (PVAP), hydroxypropylmethylcellulose phthalate (HPMCP), shellac, and various anionic acrylic polymers are polyelectrolytes containing ionizable carboxyl groups. At the low pH (1 to 3) of the stomach, these weak acids are practically nonionized,

Table 2.2 Principal Release Mechanisms of Some Film Formers

Film former	Chemical structure	Release mechanism
Paraffin	$\sim CH_2\text{-}CH_2\text{-}CH_2\sim$	Diffusion, disintegration
Glyceryl stearates	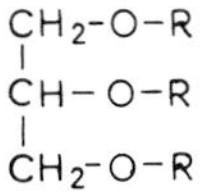 $R = CH_3(CH_2)_nCO$ R=H at one or two positions	Enzymatic breakdown by lipase of small intestine, diffusion
Cellulose acetate phthalate (CAP, cellacephate)	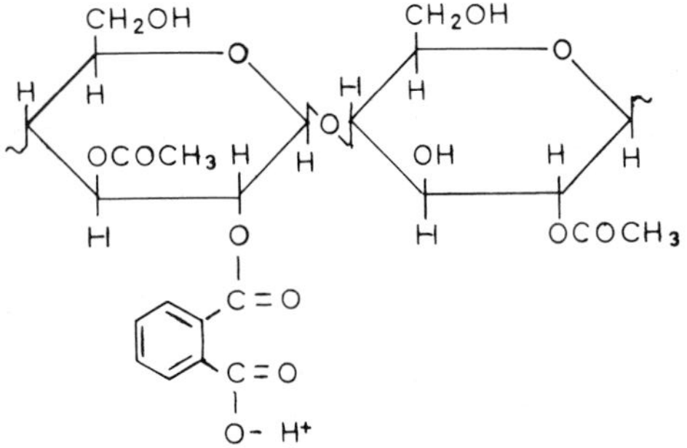	pH-dependent dissolution, diffusion, enzymatic breakdown by esterases of small intestine
Anionic acrylic polymers	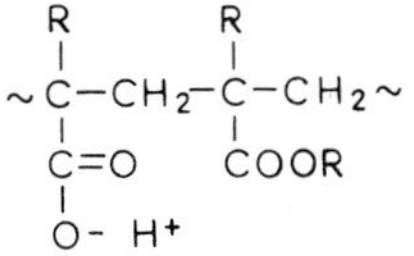	pH-dependent dissolution, diffusion
Cationic acrylic polymers	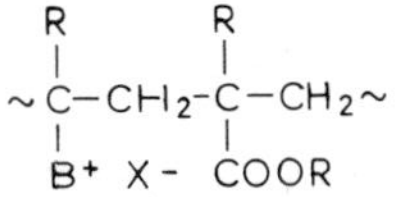 B=basic group	pH-dependent dissolution, swelling, diffusion

as they are some 2 to 4 pH units below their pK_a values and these film formers will not dissolve. However, upon passage of the enteric-coated dosage unit from the stomach through the pylorus, there is an abrupt change in pH of 5 to 6 in the upper duodenum to about 8 in the lower ileum. This rise in pH causes progressively more carboxyl groups to dissociate, as indicated in Table 2.2, and the polymer becomes more and more water-soluble. Figure 2.1 shows the solubility of some common enteric film formers in buffer solutions of various pH. The nature of the reagents used to acquire a particular pH, particularly the pK_a of the acidic component and the overall salts concentration, also affect the dissolution rate of enteric coating as reported by Spitael and Kinget [11]. Cationic acrylic polymers, on the other hand, would be more soluble at low pH, being slowly swellable, and so could be used for taste masking, as the pH of the oral cavity is 5 to 8. Recently Alhaique et al. [12] described a polymeric film composed of ethylene-vinyl-*NN*-diethylglycinate that responded in diffusional properties to environmental pH stimuli and that might be adaptable for use in microencapsulation. Polymers that are aliphatic such as paraffin, almost 100% crystalline such as cellulose, or that contain a predominance of nonionizable functional groups tend to be insoluble in water irrespective of pH.

A problem that frequently arises with enteric and other film formers is called after-hardening and leads to a progressive decline in solubility in digestive juices on storage. With CAP this effect is due to transesterification whereby acetic acid is formed and cellulose chains

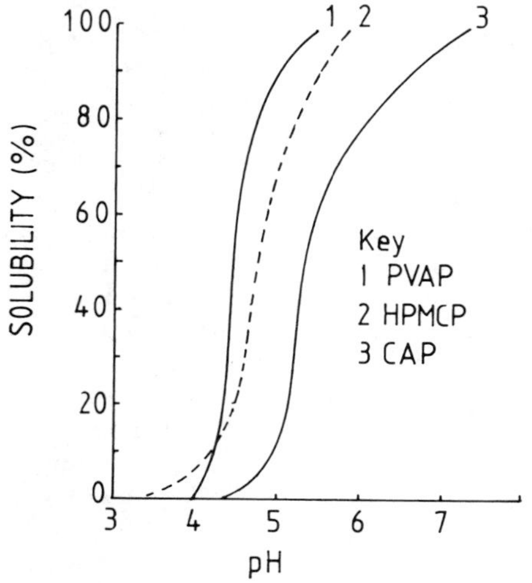

Figure 2.1 Solubility of some common enteric film formers in aqueous buffer of variable pH.

become crosslinked by phthalic acid bridges. Johnston et al. [13] avoided the hardening reaction associated with shellac by heating such film-coated products at 65 to 70°C for about 16 hr, whereby the shellac was polymerized to an equilibrium state that did not change further upon storage. An alternative way, more suited to cores containing thermolabile drugs, was described by Signorino [14, 15] to overcome the aging problem with shellac. Both patents were assigned to Colorcon, Inc. It involved mixing it with another polymeric material such as polyvinylpyrrolidone that contained basic groups, which by interaction prevented the carboxyl groups of the shellac from crosslinking. However, it was difficult to determine the stoichemistry of the reaction, because shellac is so variable in composition.

It is possible to increase the permeability of hydrophobic films by the use of a second, more hydrophilic film former. Donbrow and Friedman [16] found that the diffusion rate of aqueous solutions of caffeine through ethylcellulose films increased linearly with increasing polyethylene glycol 4000 content. This effect was due to the increased film porosity without the formation of continuous capillaries that resulted from the dissolution of the polyethylene glycol out of the film, while still retaining its barrier properties as evidenced by its permeability to an ionizable drug [17]. By mixing two acrylate-methacrylate copolymers with differing amounts of cation (quaternary ammonium) content, Okor [18] showed that it was possible to increase the permeability of the resulting films to urea by increasing their cation content. Apart from increasing the hydration of the film, greater mass transport of the penetrant may have been due to the formation of large pores at high cation content due to mutual repulsion of charged groups during film formation. Hunke and Matheson [19] reported that hydration and swelling increased in co(polyether) polyurethane membranes with increasing polyethylene glycol molecular weight in the copolymer. However, there have been numerous reports of interactions between macromolecules such as film formers and various drugs that may hinder drug release.

In a later paper by Donbrow and Samuelov [20] on laminated films, incorporation of hydroxypropylcellulose as a second film former with ethylcellulose to produce one layer of a double-layered film was also shown to enhance drug permeability. The hydroxypropylcellulose was largely retained in the film layer, so that enhancement was less than with polyethylene glycol and was ascribable to the formation of swollen hydrated channels. Borodkin and Tucker [21] reported a similar increase in drug permeability for laminated hydroxypropylcellulose-polyvinyl acetate films, where the hydroxypropylcellulose was apparently not leached out of the film [22].

Despite the importance of laminated films in the design of controlled-release products with constant rate-of-release characteristics [23], such composite films have rarely been applied to microcapsules. Two exceptions are the dual-walled microcapsules using a combination of complex coacervation and pan coating as described by Harris [24] and the triple-walled microcapsules of Morris and Warburton [25].

Obviously, the choice of film former for pharmaceutical application is restricted by the toxicity of the polymer, particularly if it is to be used parenterally. Orally consumed polymers are rarely absorbed because of their large molecular weight, and hazards associated with their use are often related to their residual monomer and catalyst content. Likewise, because of growing concern over environmental pollution caused by plastics, it is likely that increased attention will have to be given to the problem of the biochemical degradation of these polymers. Aspects of these topics have been reviewed by Baier [26] and Kulkarni [27]. Before contemplating the use of a particular film former, its various physicochemical properties and its regulatory status should be checked by reference to standard textbooks, manufacturers' trade literature, or if necessary by direct experimentation. Novel polymers are rarely used in formulation of products for use in humans because of the prohibitive cost of establishing their safety. It is outside the scope of this book to provide detailed information on the properties of all the commonly used film formers, though Refs. 28 to 30 should provide some useful information on some of the more frequently employed polymers. However, aspects of film formers will be considered below, citing properties of individual materials to illustrate particular points.

When a film former is applied to a core, two sets of forces are involved. These are the cohesional forces between the polymer molecules that comprise the film former and the adhesional forces between the coating and the core. High levels of cohesive force occur in high-molecular-weight polymers as a result of diffusion of individual macromolecules or segments thereof under favorable conditions such as elevated temperature (semisolid state) or suitable solvent (gelation). This results in the coalescence of individually applied layers to form a relatively homogeneous coating devoid of lamination. Increasing cohesiveness tends to increase film density and rigidity while reducing porosity and permeability, which properties and others may be modified by appropriate choice of film former, its grade, and the presence of other additives.

Polymer structure and chemistry have a major influence on the degree of cohesion attained in a film coating. The typical polymer molecule is a macromolecule formed from a sequence of repeating monomer units. Its size plays a very important role in determining the properties of the polymer, so it is important to assign a molecular weight to the material. However, unlike a pure chemical compound, polymers contain a range of molecular sizes and so some form of average value must be applied. Some methods for determining molecular weights of polymers such as light scattering give data that are a function of molecular mass and can be used to determine a weight-average molecular weight $\overline{M}_w$. Other methods of molecular weight determination, such as those based on colligative properties, effectively count the number of molecules of

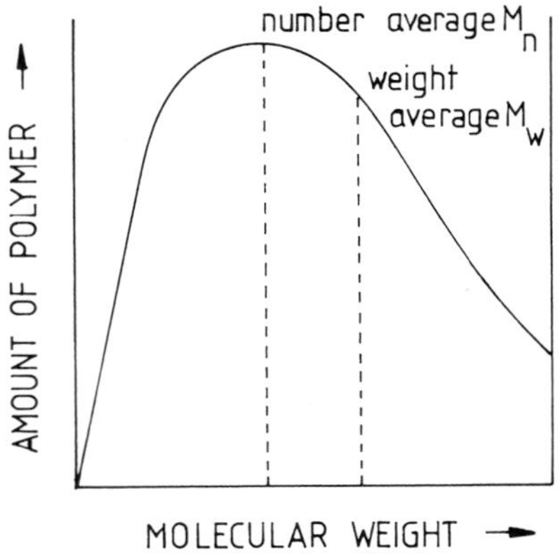

Figure 2.2 Distribution of molecular weight in a typical polymer.

a known mass in the polymer sample to give a number-average molecular weight $\overline{M}_n$, which is always equal or less than $\overline{M}_w$ (see Fig. 2.2). The molecular weight distribution may be expressed using the ratio $\overline{M}_w:\overline{M}_n$, the higher the ratio the wider the distribution. A more convenient but indirect method of approximating $\overline{M}_n$ is by viscosity determination based on the empirical relationship that

$$\eta_0 = K'\overline{M}_n^{\,a'}$$

where η_0 is the intrinsic viscosity, K' is a constant, and a' is an exponent or index.

Often, rather than use molecular weight data, apparent viscosities of polymer solutions are cited. For example, ethylcellulose is available in a range of grades such as 50 and 100 cP, where the apparent viscosity of a 5% w/w solution in 80:20 toluene:ethanol at 25°C is measured in centipoise units under specified test conditions. Increasing cP number should denote increasing $\overline{M}_n$, but any rheologist will recognize the unreliability of such determinations for characterizing the viscosity of these non-Newtonian systems.

Rowe [31] used gel permeation chromatography to determine the molecular weight and molecular weight distribution of nine grades of hydroxypropylmethylcellulose used in film coating. There tended to be a high-molecular-weight component ($>5.0 \times 10^5$) in most grades, the amount increasing in grades of higher nominal viscosity. However, there was a wide molecular weight distribution in all grades, with the presence of a high proportion of very low-molecular-weight ($<5 \times 10^3$) components in grades with low nominal viscosity designations. Batch-to-batch variation in molecular weight data of polymers with the same nominal viscosity was also considerable. The variability in molecular weight and distribution tended to affect mechanical properties of the films, such as increased hardness and resistance to abrasion, with decreased elasticity being associated with grades of higher nominal viscosity.

Rowe [32] also investigated the molecular weight of various grades of ethylcellulose used in film coating by intrinsic viscosity determination and by gel permeation chromatography. There was good agreement between the results obtained by the two methods. Tensile strength and elongation studies indicated that at a molecular weight of approximately $4-5 \times 10^4$ (30 cP) or above, mechanical properties should be similar. In practice, however, this was not the case, probably because of the presence of low-molecular-weight components in ethylcellulose samples of high nominal viscosity, which have a deleterious effect on mechanical properties disproportionate to their concentration.

Other rheological properties are sometimes used to characterize particular polymers. The melt index (MI) is defined as the number of grams of molten polymer that will flow through a standard orifice at a standard temperature and pressure. Gelatins are often graded by jelly strength, denoted by Bloom rating, which is the weight in grams that, when applied to a plunger 12.7 mm in diameter, will produce under controlled conditions a depression exactly 4 mm deep in a jelly matured at 10°C and containing 6.66% w/w of gelatin in water. Most gelatins used in pharmaceutical formulation have a minimum Bloom strength of 150. As an example of its importance, it was shown by Nixon et al. [33] that the rate of diffusion of methylene blue through a glycogelatin gel decreased as the Bloom strength of gelatin used was increased. However, it should be appreciated that conformity with such indices of molecular weight in different samples of polymers is not necessarily acceptable as indicative of equivalence in other physicochemical properties, particularly those affected by polymer structure and alinement.

Polymers tend to have two types of structures within the material, a crystalline region (A) and an amorphous region (B) as shown in Fig. 2.3.

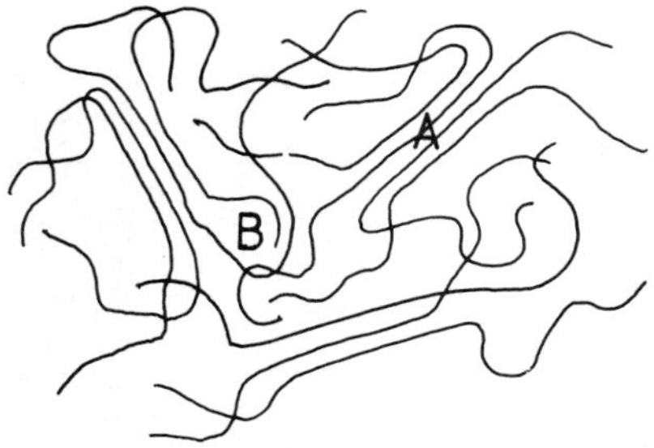

Figure 2.3 Plane structure of a typical film former showing ordered (crystalline A) and disordered (amorphous B) alignment of linear polymeric molecules.

Crystallization occurs when linear polymer molecules orientate themselves in whole or in part to form an ordered compact configuration known as a crystallite that lies scattered throughout the amorphous zone of the polymer material. Crystallites may in turn be ordered into an arrangement radiating outward from a central point to form a structure known as a spherulite or may form an ordered row structure. It may be necessary to control the growth of these superstructures, as they tend to weaken those regions of the polymer film, leading to cracks or even breaks. Highly crystalline polymers have a high cohesive strength and are characterized by compactness, toughness, rigidity, surface hardness and brittleness, and by lack of permeability and flexibility. The molecular order necessary for crystallization is favored by the bonding of interactant groups such as —OH and —COOH regularly distributed along the backbone of these molecules. It is also favored by the stereoconfiguration of neighboring chain segments, which is denoted by the tacticity of the polymer. Isotactic polymers contain molecules that are at least partially orientated in space in the same manner for each molecule of the polymer, which facilitates interlocking of two or more polymer segments to give rise to crystallinity. A polymer is rarely 100% crystalline because of the molecular weight distribution of its constituent molecules and their lack of perfectly ordered chemical and steric interactions. Atactic polymers tend to have no stearic order and are therefore more amorphous-like. The formation of amorphous polymers is thus favored in polymers that contain bulky side chains, irregular substitutions, or that have an irregular repeating sequence of dissimilar monomer units, which frequently occurs in copolymers. Elastomers ("rubbers") such as silicone rubber are typically amorphous polymers whose relatively weak interchain cohesive forces give them their property of elasticity, flexibility, and permeability. Cardarelli [34] has discussed the use of such materials to form matrices for controlled-release monolithic devices containing agents such as molluscicides and herbicides. Particular film formers used for microencapsulation can vary enormously in the relative amounts of crystalline and amorphous regions they contain, which obviously influence the fine structure of these materials.

An important property of a polymer is its glass transition temperature. Below this temperature there is a virtual cessation of molecular motion, causing the polymer to become crystalline-like in its properties. Table 2.3 shows the glass transition temperature (T_g) of some polymers used as film formers.

When a polymer is heated above its glass transition temperature, its physical properties may be modified by orientating polymer molecules with repeated pulling or stretching followed by a heat-setting process. Such annealing procedures may be used to evaluate the type and orientation of superstructures in polymers as discussed by Chu and Smith [37].

Table 2.3 Glass Transition Temperatures (T_g) of Some Polymers Used as Film Formers[a]

Polymer	T_g (°C)
Ethylcellulose	129
Hydroxypropylmethylcellulose	177
Nylon 6	47
Polyethylene	-110
Polymethyl acrylate	5
Polymethyl methacrylate	105
Polyvinyl acetate	30
Polyvinyl chloride	82
Silicone rubber	-123

[a]Average value from the literature [35, 36]; should be regarded as only approximate.

Above the glass transition temperature, increase in temperature causes the surfaces of applied layers of film former to become more cohesive as a result of the greater rate of diffusion of polymer molecules or fractions thereof at the interface caused by increased thermal motion. It is common practice to apply heated film coating solution to warm cores when using pan coating or air suspension techniques, which aids not only cohesion within the film but also adhesion of the coating to the core. Care must be taken not to cause the surface of partially coated cores to become too cohesive or tacky, as this gives rise to troublesome aggregation problems that may cause coating damage when the cores are subsequently prized apart. Obviously, the longer the coating and core are held at elevated temperature, the more marked these effects will be.

The adhesion of film former to the core surface is also favored by chemical bonding at the interface and the flow of film former into irregularities on the surface of the core. Various tests to measure film coating adhesiveness, including peel tests or preferably various adhesiveness testers, have been developed. Using one such tester, Fung and Parrott [38] showed that the force required to pull a hydroxypropylcellulose coating from the surface of a tablet was reduced to between a half and a quarter when an aqueous rather than an organic solvent-based coating solution was applied. Using another type of tester, Rowe [39] showed a decrease in adhesiveness with increasing film thickness up to

35 μm and then a gradual increase with thicknesses up to 140 μm for hydroxypropylmethylcellulose-coated tablets. Increasing porosity of the substrate increased adhesion. In a later paper relating to the film coating of tablets by the same author [40], the adhesive force of various film coats were all found to be much less than the cohesive force within the film, though this may have been due partly to the removal of some adhering substrate with particular film coats.

2.2.2 Solvents

Film formers are usually applied to cores using a suitable organic or inorganic solvent system. This dissolves the polymer by a process of union or solvation with solvent molecules, which disrupts the cohesive force between polymer molecules. The more crystalline a polymer is, the greater will be its cohesive force and consequently it will be more difficult to dissolve in a solvent system. Dissolution may be aided in polar solvents by dissociation of functional groups along linear polymer molecules, causing charge repulsion, leading to separation of neighboring molecules or uncoiling within molecules.

Dissolution of polymers is also affected by steric considerations. By substitution of bulky groups such as methyl into cellulose, which is highly crystalline and water-insoluble, the cellulose chains are wedged apart, allowing access by water to hydroxyl groups within the polymer. Substitution into cellulose of an even bulkier and more charged grouping such as sodium carboxymethyl produces even greater water solubility. Normal grades of ethylcellulose used for microencapsulation are water-insoluble because of the predominance of alkyl ether substituent groups over remaining hydroxyl groups, though grades of the polymer have been observed by Deasy et al. [41] to swell upon immersion in water, indicating some degree of hydration of the hydroxyl groups.

Increasing dissolution of a fixed overall concentration of a film former in a particular solvent system is normally accompanied by a progressive increase in apparent viscosity of the system as a result of the extended configuration of dispersed molecules, which increases the resistance to flow (see Fig. 2.4).

Polar or nonpolar solvents tend to dissolve polar or nonpolar film formers and other additives for film coating, respectively, though the prediction of solubility of polymeric molecules in particular organic solvent systems is difficult. Banker [42] claimed that a polymer was most soluble in solvents that had cohesive energy density (CED) values close to that of the polymer. The cohesive energy density is defined as:

$$\mathrm{CED} = \frac{\Delta H - RT}{V'}$$

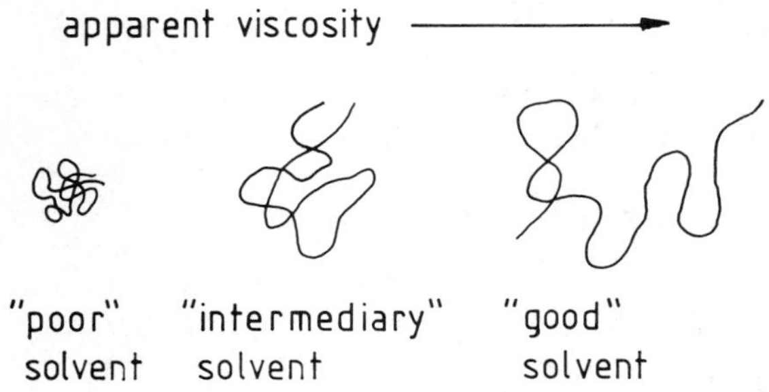

Figure 2.4 Uncoiling and extension of a linear polymer molecule with increasing efficiency of solvent used.

where ΔH is the molar heat of vaporization of the solvent and V' is its molar volume.

For solid solutes a hypothetical value for its cohesive energy density can be calculated from U/V', where U is the lattice energy of the crystal. The approach is based on the Hildebrand regular solution theory of nonelectrolytes, and CED values were tabulated for a range of common solvents and film formers, examples of which are shown in Table 2.4.

Table 2.4 Some Cohesive Energy Density (CED) Values for Common Solvents and Film Formers

Solvents CED (cal ml^{-1})	Film formers CED (cal g^{-1})
Water 551.1	Polyethylene 62
Methanol 210.3	Polystyrene 75
Ethanol 163.4	Acrylate 85
n-Propanol 147.5	Polyvinyl chloride 90
Acetone 98.5	Polyvinyl acetate 85–95
Ethyl acetate 83.0	Acrylics 95
Chloroform 85.4	Cellulose acetate 130
Ethyl ether 54.1	Cellulose nitrate 110
Carbon tetrachloride 73.6	
n-Hexane 52.4	

Source: (Reproduced with permission of the copyright owner, American Pharmaceutical Association; from Ref. 42.)

Crowley et al. [43] observed that use of values of solubility parameter (i.e., the square root of the CED value), hydrogen bonding, and dipole moment on a three-dimensional plot was more effective in predicting the solubility of cellulosic polymers in organic solvents. It has often been observed that the properties of coatings are dependent on the solvent used. Superior film formation is favored by solvents that promote better polymer solution, leading to greater molecular uncoiling and extension. Nadkarni et al. [44] have shown that film substrate adhesion is greatest when the film former is cast from a solvent with a solubility parameter close to that of itself. Organic solvents capable of only poor hydrogen bonding, such as toluene and chloroform, were observed by Kent and Rowe [45] to be the best solvents for the widely used film former ethylcellulose.

The solubility of copolymers is difficult to predict, as they are usually prepared from substantially polar and nonpolar monomers. The solubility of the copolymer is generally poor in "good" solvents for either homopolymer and is generally best in a mixture of these solvents that promotes greatest solvation and extension of the copolymer chains. Likewise, when a mixture of homopolymers or a homopolymer and another additive requiring dissolution is required, use of a mixed solvent is normally preferable to the use of a single solvent of intermediary polarity.

Mixed solvent systems frequently give rise to problems during evaporation. One or more of the solvents may evaporate more rapidly than the others, giving rise to major changes in solvent composition with premature separation of film components. This problem may be avoided if a suitable azeotrope is formed that retains satisfactory solubility characteristics. For example, Porter and Ridgway [46] used an azeotropic mixture of 93:7 by weight of dichloromethane:methanol, because 50:50 mixture caused premature precipitation of cellulose acetate phthalate when film coating. Separation problems may also be overcome with a film former such as ethylcellulose by using a mixture of an alcohol and a chlorinated hydrocarbon where the poor solvent, alcohol, evaporates at a faster rate than the other solvent, so that at the point of gelation the remaining mixed solvent is richer in the more effective solvent.

In recent years considerable interest has been shown in the use of aqueous-based coatings for microencapsulation, particularly by pan and air suspension technology. This is because organic solvents give rise to health hazards and environmental problems, may be potentially explosive and inflammable, require the product to be monitored for residual organic solvent, and are expensive. Water-soluble celluloses such as hydroxypropylcellulose and hydroxypropylmethylcellulose may be used [47]. Elegant coatings of both polymers were obtained on tablet cores by Banker et al. [48] using polyethylene glycol 4000 or 6000 as plasticizer and aqueous-based pigments. Delporte et al. [49] and Delporte [50] have also described the use in pan and air suspension coating of low-viscosity hydroxypropylmethylcelluloses for aqueous coating of pharmaceutical dosage forms. Aulton et al. [51] have reported the influence of various plasticizers on certain mechanical properties of cast films prepared from aqueous-based hydroxypropylmethylcellulose systems. Aqueous dispersions of enteric and nonenteric acrylic polymers have been prepared, and their uses were described by Lehmann and Dreher [52] and Lehmann [53] for coating pharmaceutical dosage forms. When the water evaporates, the latex particles form a dense spherical mass as shown in Fig. 2.5. With further evaporation of water, the droplets merge to form a continuous film whose formation requires a minimum film-forming temperature to avoid the appearance of cracks. This temperature may be as low as room temperature, depending on the polymer and other additives used. Cores that are highly water-soluble or that absorb water very rapidly tend to interfere with the fusing of these films by competing for water and need sealant treatment with hydrophobic precoatings. Hall et al. [54] described the use of an aqueous dispersion of ethylcellulose (Aquacoat, FMC Corporation) for microencapsulation. Because the polymer chain length does not contribute directly to the apparent viscosity, the dispersion can be used at a much higher solids content than the corresponding solution in an organic solvent, resulting in a saving in process time to apply the coating. It was found that a suitable plasticizer was necessary to fuse the film former into a continuous film upon drying.

Stafford [55] has described how a completely dissolved aqueous enteric film former system containing neutralized hydroxypropylmethylcellulose phthalate could be successfully applied by pan coating onto tablet cores. This approach could obviously be extended to other enteric film formers and to the coating of microcapsules.

Finally, it should be realized that whatever the solvent system used, it should not appreciably dissolve the core material, which is normally a low-molecular-weight drug. Factors that influence the solubility of drugs in liquids are outside the scope of this book but have been extensively reviewed by Florence and Attwood [56].

2.2.3 Plasticizers

A plasticizer has been defined by Mellan [57] as a substantially nonvolatile, high-boiling, nonseparating substance that when added to another changes certain physical and chemical properties of that material. Plasticizers are normally added to polymeric film formers, particularly those used in pan and air suspension coating procedures, in order to increase segmental mobility, impart flexibility, reduce brittleness, and increase resistance of the film coating to failure produced by mechanical stress.

Substances such as phthalate esters, fatty acid esters, and glycol derivatives are examples of common types of external plasticizers because

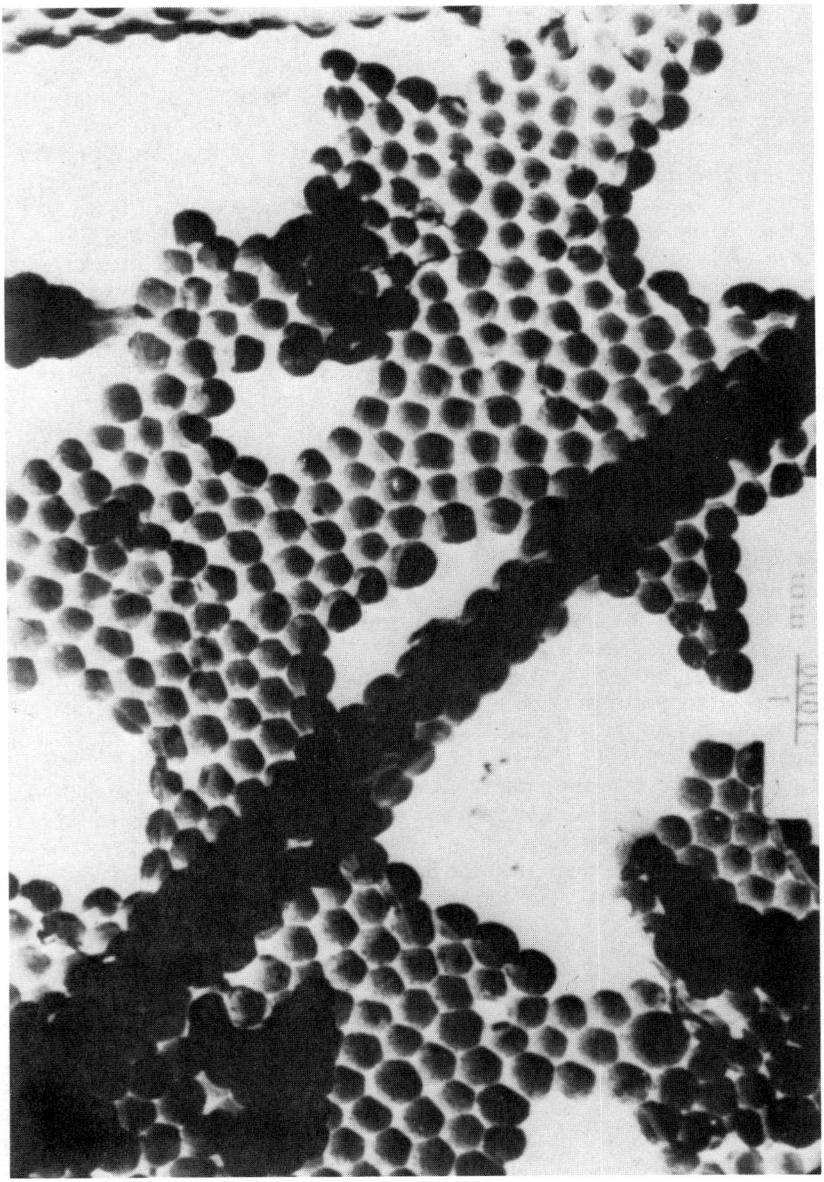

Figure 2.5 Electron micrograph of a drying acrylic resin dispersion with latex particles deposited close together. (Reproduced with permission from Ref. 52 and courtesy of Dr. K. Lehmann, Rohm Pharma Gmbh, Darmstadt, West Germany.)

they are added to film formers to influence their physicochemical properties. Internal plasticization, in contrast, involves the direct chemical modification of a polymer, usually by copolymerization to achieve similar effects. Examples of these latter types of polymers include various acrylic and vinyl copolymers, which normally require the addition of little or no external plasticizer. Internal plasticization will not be considered further.

The plasticizer (external) achieves its effects by being interposed between the polymer chains to reduce cohesive force between polymer molecules as a result of a chemical association by secondary valence forces of both species. Effective plasticizers tend to have a chemical structure similar to the polymers they plasticize. Thus cellulose ethers retaining a high percentage of hydroxyl groups are best plasticized by hydroxy-containing compounds such as polyhydric alcohols. Concentrations in the range 20 to 50% of plasticizer relative to film former are usually required for cellulosic polymers. Polymers with lower cohesive force will normally require 20% or less of plasticizer. Strongly crystalline polymers, unlike amorphous ones, are usually difficult to plasticize, as it is hard to disrupt their high intermolecular cohesive forces. Long cylindrical plasticizer molecules tend to be more effective than spherical molecules of the same molecular weight. Evidence of good plasticization may be observed from a decrease in the glass transition temperature of the polymer, as its chains are more mobile, which may alter a hard, brittle coating at room temperature to one that is flexible yet tough. Entwistle and Rowe [36] reported that the calculated glass transition temperature of hydroxypropylmethylcellulose (determined from differential scanning calorimetry data) appeared to be lowered by the use of various glycol plasticizers. Within homologous series of phthalates for ethylcellulose and glycols for hydroxypropylmethylcellulose there was a critical molecular weight of plasticizer that gave minimal resistance to mechanical failure of cast films. Vemba et al. [58] observed that ethyl phthalate increased the mechanical resistance of ethylcellulose by reducing film stresses.

Obviously, to be effective at the molecular level the plasticizer must dissolve in the solvent system used for the film former. This is normally facilitated by the fact that the polymer and plasticizer have common functional groups that prevent premature separation of either component as the solvent is evaporated off to deposit the film. Ideally the solubility parameter of the plasticizer should be close to that of the polymer, as observed by Entwistle and Rowe [59] in the effectiveness of diethyl phthalate for plasticizing ethylcellulose films. Where mixed solvents of different polarities are used with multiple film formers or with copolymers, plasticization may be more effectively achieved by the use of more than one plasticizer. Here unwanted separation of components caused by alteration in solvent blends during evaporation are more difficult to avoid.

Because of the loose association between the plasticizer and film former, problems of permanence and uniformity of distribution of the former in the coating may arise. Migration of low-molecular-weight plasticizer during storage can profoundly affect the physical and mechanical properties of the film and may be associated with the leaching of undesirably high levels of plasticizer into the surrounding medium. Autian [60] has discussed the problem of leaching of plasticizers from plastics. Greater permanence may be achieved by the use of the higher-molecular-weight, bulky compounds in an effective series of plasticizers that have low volatility during any heating process involved and that have a slow rate of diffusion from the coating.

As plasticizers fill the interstitices of the polymer, a secondary effect of these compounds is that they alter the permeability of the coating. Compounds of high water solubility will tend to make the coating more permeable to aqueous media, as they will rapidly dissolve out of the coating upon immersion into water. However, the effect can be more complex, as was observed by Okor and Anderson [61], who showed that films composed of a mixture of acrylic copolymers became more permeable to urea as the content of the more hydrophilic plasticizer, glyceryl triacetate, decreased by leaching when used in combination with glyceryl tributyrate. This effect was ascribed to greater copolymer chain flexibility upon sorption of water produced by the permanence of the glyceryl tributyrate, particularly at high concentration. On the other hand, poorly water-soluble plasticizers will limit water permeation through the film and are often necessary for use with enteric and slow-release coatings to control diffusion of low-molecular-weight drugs out of the core. Again the effect may not be simply predictable. For example, Porter and Ridgway [46] investigated the effect of diethyl phthalate on two enteric film formers, polyvinyl acetate phthalate (PVAP) and cellulose acetate phthalate (CAP). Whereas increasing concentration of plasticizer made apparently nonporous PVAP films more permeable to water vapor and simulated gastric juice, CAP films tended to decrease in permeability, presumably due to the reduction in the porous nature of such films, or the immobilization of water within the more expanded hydrophilic backbone of this polymer upon plasticization. Gurny et al. [62] have illustrated the complex influence of plasticizer content on salicylic acid penetration through a methacrylate membrane by means of an orthogonal composite design as shown in Fig. 2.6.

Thus plasticizers may be used to control drug release from and through polymeric films. Plasticizers may also have other functions, such as to act as antitack agents or waxy sealants, which for convenience will be discussed in later sections of this chapter. Finally, caution should be exercised in the choice of plasticizers, notably with phthalate derivatives, which apart from having a bitter taste may represent a toxicological hazard when taken internally [63].

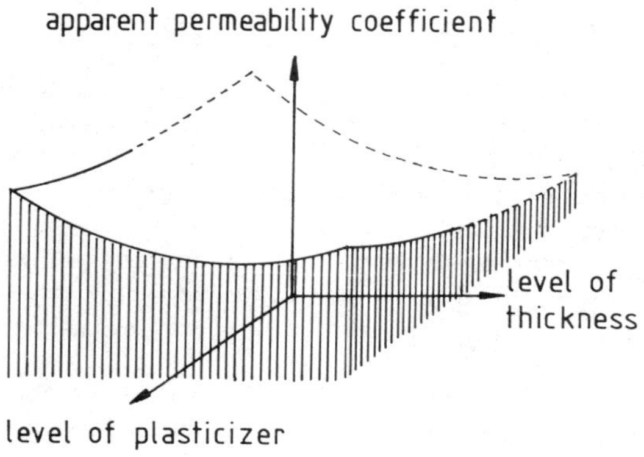

Figure 2.6 Orthogonal composite design illustrating a three-dimensional representation of the permeability of methacrylate film with respect of thickness and plasticizer content. (From Ref. 62.)

2.2.4 Surfactants

Various anionic, cationic, and nonionic surfactants are sometimes added to film coating formulations that are to be sprayed or poured onto the surface of cores in order to aid wetting and even spreading of the film coating solution. Likewise, interfacial polycondensation and emulsion polymerization methods involve the use of surfactants. Such agents may permit the use of otherwise immiscible or insoluble additives by solubilizing them. They also influence the permeability of coatings. Fites et al. [64], for example, reported that soluble poly (methyl vinyl ether/maleic anhydride) films could be made water-insoluble by addition of polysorbate 20, the permeability of which could be controlled by adjustment of the surfactant content. Surfactants also aid in the penetration and dissolution of the coating and core upon ingestion.

Dopper et al. [65] used a photographic procedure to measure the contact angle of drops of film former solution on the surface of tablet cores. In order to promote wetting of substrates on which contact angles of greater than 90° were initially observed, addition of 10 to 20% surfactant was needed.

Cationic surfactants have been recommended by Maierson [66] to inhibit the clustering of gelatin-acacia-coated microcapsules in an aqueous slurry. Addition of surfactant to the prepared microcapsules caused it to be adsorbed strongly on the microcapsule surface, which, because of the resultant steric and charge effects, kept the microcapsules apart and prevented them from aggregating. Obviously, this approach could be extended to the use of other types of surfactants that might be added to coating solutions or as an aftertreatment.

2.2.5 Channeling Agents, Waxy Sealants, and Crosslinking Agents

In order to enhance the permeability of a hydrophilic film to penetration by water and drugs in aqueous solution, a channeling agent may be included in the coating formulation. Donbrow and Friedman [67] incorporated up to 50% of polyethylene glycol 4000 into films of ethylcellulose, which aided drug penetration as the polyethylene glycol was leached out. The effect seemed to be due to the increased porosity produced in the polymer matrix, though the film still retained barrier properties. Sjökvist et al. [68] increased the permeability of nylon film by incorporating palmitic acid. At high pH the ionization of the fatty acid was greater, so aiding the penetration by ion-pair formation of oppositely charged quarternary ammonium compound.

However, the use of channeling agents is rarely necessary with microcapsules and related dosage forms, where because of their ultrathin coatings, control of unwanted overpermeability is the more usual problem. Excessive permeability may be reduced by the incorporation of waxy sealants such as paraffin wax, carnauba wax, beeswax, or spermaceti into the coating after casting. This approach was used by Deasy et al. [41] to reduce the in vitro release of sodium salicylate from ethylcellulose-coated microcapsules. Likewise, Motycka and Nairn [69] treated drug-loaded ion-exchange beads with various waxes to retard drug release. Of course, the approach of incorporating waxy materials into tablets to retard drug release as described by Cain and Federici [70] is easily adaptable to the formulation of the cores of microcapsules. Brophy and Deasy [71] used such an approach successfully when formulating the cores of drug-containing microcapsules.

Another approach for retarding drug release from microcapsules involves the use of crosslinking agents for the cast film former to reduce its segmental mobility and hence diffusion through the polymer. Thus various aldehydes such as formaldehyde and glutaraldehyde are used to reduce the aqueous solubility and permeability of gelatin-coated microcapsules. Calcium salts and other metallic chelating agents have also been used to reduce the permeability of sodium alginate-coated microcapsules. These and other examples are discussed in later chapters.

2.2.6 Antitack Agents

To reduce adhesion and friction between surfaces during coating, particularly in pan technology, various antitack agents may be used. Fatty acids such as stearic acid are most frequently used. These compounds act as boundary lubricants by having the polar region of the molecule adsorbed onto the substrate, leaving the nonpolar aliphatic region projecting from the surface of the coating. This causes a reduction in surface adhesion and also lowers the surface free energy. Excessive use of antitack agent is undesirable, as it may promote sliding and slippage of cores in the pan, despite the presence of baffles, rather than the good rolling action required for proper distribution of the coating material. Such slippage of cores is usually more noticeable during the initial stages of the coating buildup, where a degree of polymer orientation occurs on the core substrate rather than the more random orientation and greater surface friction associated with the application of later coats.

2.2.7 Dispersed Solids, Including Colorants, Opaquant-Extenders, and Fillers

The classes of materials previously discussed are normally dissolved in the solvent system used in order to achieve a molecular dispersion having an apparent viscosity suitable for application by spraying or pouring, and for the uniform suspension of a range of finely divided solids. Because coating systems tend to have pseudo-plastic flow properties, variable thixotrophy, and no absolute yield value, periodic agitation of the coating system during application is necessary to obtain a homogeneous distribution of these solids in the film formed. Sometimes these solids are added at intervals during the buildup of successive layers of coating to give a heterogeneous distribution in the final film. The microcapsules formed are not suitable for parenteral use.

Colorants are often added to the coating system prior to application by pan or air suspension technology to facilitate the visual observation of the buildup of coating, to reduce the risk of accidental contamination of one batch of prepared microcapsules with another, and to confirm the uniformity of mixing of batches of microcapsules having different release rates as denoted by color when filled into outer hard gelatin capsule shells. Approved, finely dispersed, insoluble lakes of color, often referred to as pigments, tend to be used. Soluble colors are infrequently used, as they are more liable to color migration and photodegradation problems during storage, giving rise to an unsightly, mottled product. Obviously, the inclusion of such insoluble pigment or other particles in the film will delay drug permeability, either by adsorbing it onto the extensive surface of these particles and/or by increasing the effective diffusional path length as the penetrant is forced to pass around these impermeable inclusions in the film. This effect was observed by Porter and Ridgway [46] for a cellulose acetate phthalate film formulation containing a red iron oxide pigment.

Opaquant extenders are occasionally used in coatings for microencapsulation. These are white inorganic particles employed primarily to pearlize the coating and render it nontransparent so that surface defects and the core are less obvious. Titanium dioxide is the most widely used material, but various silicates, carbonates, or other oxides are sometimes employed. Like the more expensive colorants with which they are often mixed to extend the effect of the colorant, these opaquant-extenders also tend to reduce film permeability.

A somewhat similar type of material that likewise affects permeability and is used in the formulation of film coats is a filler. For microcapsules, 10 to 100% of coating solids based on the weight of core material is frequently required. These coating solids can be composed mainly of film formers and may require large volumes of expensive solvents to achieve systems of suitable apparent viscosity for application. In order to reduce the cost of materials and the unduly long drying times, insoluble particulate fillers or extenders may be added to the coating system. Talc and other silicates are the most frequently used. These solid inclusions may immobilize film-former segments by adsorption, causing an increase in the glass transition temperature of the polymer. However, these inclusions may also decrease molecular order in films, producing a net reduction in the strength of films. Aulton et al. [72] noted that increasing solids content in hydroxypropylmethylcellulose films caused them to become more brittle and less tough.

Such filler-type materials may be used to lessen the tendency to surface adhesion, resulting in the coalescence frequently observed during various microencapsulation processes. If not rapidly corrected, this coalescence can easily lead to damage to the coating if partially fused cores subsequently have to be forced apart. Rowe [73], for example, recommended the use of talc with a particle size of less than 50 μm to lessen coalescence during the preparation of drug-containing microcapsules by coacervation procedures involving nonaqueous vehicles. Likewise, fillers are often periodically dusted onto cores during pan coating to ensure that individual cores continue to roll freely in the pan. The resultant alternate deposition of film coating and dusting powder on cores tends to lessen the cohesion between successive layers of film former, and this effect may cause a loss of film durability [42].

2.3 DESOLVATION AND GELATION OF THE COATING

As the solvent evaporates there is a tendency for the coating material to convert itself from a sol to a gel at a certain polymer concentration. This gelation of the film associated with increasing concentration of polymer gives rise to the formation of a network type of structure as individual molecules come into closer proximity. Gelation is also aided by decrease in temperature, which reduces further the mobility of polymer chains. With further loss of solvent the gel contracts more, finally forming a dry, viscoelastic film that unless adequately plasticized tends to be dimensionally constrained on the relatively immobile core substrate. This unwanted stress in the final film may be relieved by the formation of minute cracks in the coating. Accordingly, it is important to desolvate the coating slowly enough by proper choice of solvent and drying temperature so that polymer chains have adequate time to orientate themselves in order to dissipate as much stress as possible. Spitael and Kinget [74] noted that leaks occurred in a free film of cellulose acetate phthalate when cast by spraying from a solution in the very volatile solvent acetone, though none were observed when using the less volatile mixed azeotropic solvent ethylacetate:isopropanol 77:23. The requirement of allowing stress dissipation may involve an extension of coating time, as the rate of desolvation is influenced by the evaporation rate of the solvent. Overdelay in the removal of the solvent is undesirable in certain circumstances, as the gel state may be associated with a troublesome tacky condition.

When multiple applications of coating material are being used, it is important that the solvent associated with later-applied fractions should have adequate time to gel the surface of previously dried layers so as to favor the formation of a homogeneous polymer continuum. For such intermittent applications, hydraulic or airless atomizers are often preferred to avoid the possibility of applying prematurely gelled or dried coating material. Spitael and Kinget [74] reported that the permeability by hydrochloric acid and caffeine of air-sprayed free films of cellulose acetate phthalate was greater than if prepared by airless spraying or pouring. Air-sprayed films were observed to have a dropletlike structure that increased their porosity.

2.4 MECHANICAL PROPERTIES OF FILMS

Various mechanical properties of free films are often studied together with the effect of additives upon same in order to assess suitability for film coating. Whereas these studies usually relate to much thicker isolated films cast on glass, mercury, or other metallic substrates by methods such as those used by Kanig and Goodman [75] or Munden et al. [76], they do provide valuable insight into the durability of the thin film coats applied to microcapsules. The mechanical properties most commonly studied involve stress-strain testing (see Fig. 2.7).

The modulus of elasticity (or Young's modulus) is the proportionality constant of the relationship of the stress applied to the film to the strain (elongation) produced. It may be obtained from the slope of the initial straight-line part of the stress-strain plot as shown in Fig. 2.7a, and its value is affected by the rate of strain. In general, the higher the modulus of elasticity, the stronger will be the film in that it undergoes little elastic deformation in response to high stress. With increasing stress, however, the film may undergo irreversible strain; the minimum stress just causing this effect is termed the lower yield stress. Further increasing stress will eventually cause the film to break at the upper yield stress. The difference in strain values at the maximum and minimum yield points is a measure of the degree of plastic deformation

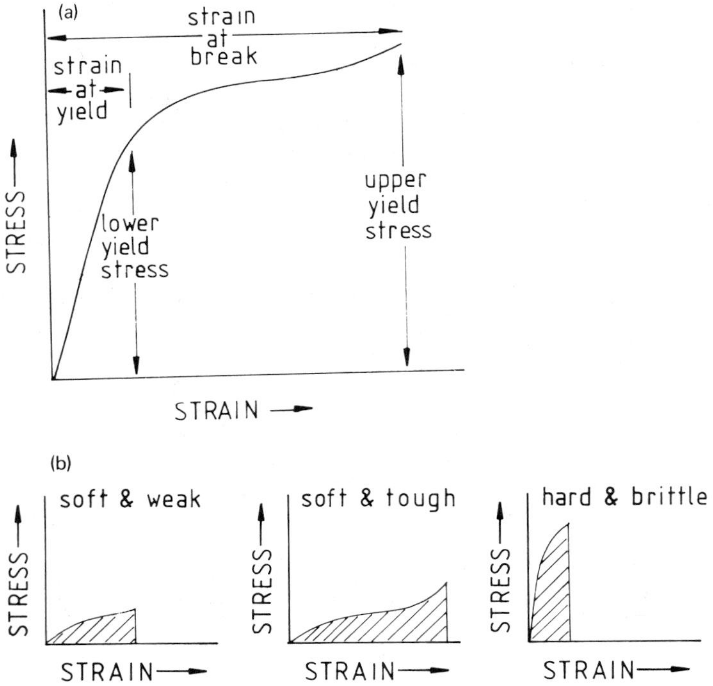

Figure 2.7 (a) Typical tensile stress-strain curve for a hard and tough film. (b) Typical tensile stress-strain curves for other types of film.

available in the film. A low or negligible value indicates that the film is brittle. The area under the curve may be taken as a measure of the toughness of the film. Such stress-strain curves are usually determined by measuring continuously the force developed as the film is elongated at constant rate of extension using an instrument such as an Instron tensile tester.

Some recent pharmaceutical examples of the usefulness of such mechanical testing for the evaluation of film performance will now be briefly discussed. Kellaway et al. [77] showed that the maximum modulus of elasticity of gelatin films was recorded when they contained 14 to 16% moisture. The effect of moisture content in hydrophilic polymers is similar to that of plasticization in that elongation begins at a lower stress and is greater for a given load. In a subsequent paper by the same authors [78], titanium dioxide and dyes were found to have little effect at the concentrations used on the modulus of elasticity and tensile strength, though the dyes did modify the strain-induced contraction of the gelatin film. Delporte [79], however, reported that titanium dioxide increased the modulus of elasticity and decreased the tensile strength of a hydroxypropylmethylcellulose film, causing it to become brittle and less flexible. Use of propylene glycol and polyethylene glycol 400 as plasticizers caused both the tensile strength and elongation at break of the film to be reduced, so giving a softer film.

Stanley et al. [80] have presented a theoretical analysis of the influence of polymeric film coating on the mechanical strength of tablets that may have implications for microcapsules. The theory predicted that the ratio of the tensile fracture strength to the modulus of elasticity of the core and coating, rather than coating thickness, determined which will fracture first. In practice, tablet cores will usually fracture before their coatings, as reported by Fell et al. [81].

2.5 PERMEABILITY TO OXYGEN, CARBON DIOXIDE, AND WATER VAPOR—PHOTOSTABILITY

Because of the importance of oxygen, carbon dioxide, and water vapor permeability from the atmosphere to the stability and shelf life of many core materials, these properties are often measured for free films using methods similar to those described in Refs. 74 and 75. Table 2.5 shows the permeability of various polymers to these three agents, from which it can be seen that there is a wide range between and within different polymers. Highly crystalline polymers tend to be less permeable, whereas elastomers are highly permeable. There often tends to be an inverse relationship between water vapor transmission and gas permeability, as observed by Munden et al. [76]. This is because water vapor permeability tends to be greater in more polar polymers, which usually have a more ordered and less porous structure that limits gas diffusion through them [83]. Obviously, suitable choice of polymeric coating material can minimize the penetration of these agents into microcapsules and related dosage forms. Because of their importance in packaging design, numerous publications have appeared on the subject, including a good review by Stannett et al. [84] and a textbook edited by Hopfenberg [85]. Some relevant pharmaceutical examples will now be discussed.

Nixon and Nouh [86] investigated the effect of microcapsule size on the oxidative decomposition of benzaldehyde as core material. A gelatin-acacia complex coacervation process of microencapsulation was used. Microcapsules with larger diameter had higher oxidative rates, as they contained proportionally larger amount of core material and were less affected by termination reactions. Banker [42] reported that plasticizers increased the water permeability of acrylic films by increasing polymer chain segmentation and possibly by directly sorbing water. Other additives for different films, such as surfactants, may also directly promote sorption of water. Insoluble fillers, which tend to increase

Table 2.5 Permeability of Various Polymers to Oxygen, Carbon Dioxide, and Water Vapor at 30°C

Polymer	P_O	P_{CO_2}	P_{H_2O}
Polyacrylonitrile	0.0002	0.0008	300
Polymethacrylonitrile	0.0012	0.0032	410
Polyethylene terephthalate	0.035	0.017	175
Nylon 6	0.038	0.016	275
Polyvinyl chloride	0.045	0.016	275
Polyethylene (density 0.964)	0.40	1.80	12
Cellulose acetate	0.080	2.40	6,800
Butyl rubber	1.30	5.18	120
Polycarbonate	1.40	8.0	1,400
Polypropylene	2.20	9.2	65
Polystyrene	2.63	10.5	1,200
Polyethylene (density 0.922)	6.90	28.0	90
Natural rubber	23.3	153	2,600
Polydimethyl siloxane	605	3,240	40,000

Units: CCS.(S.T.P.)/sq cm/cm/sec/cm Hg $\times 10^{10}$.

Source: Ref. 82.

the glass transition temperature, lower water permeability. Pickard et al. [87] reported that the water vapor permeability of poured films was less than that of sprayed films. Finally, increasing room humidity during drying has been reported by Gillard et al. [88] to be responsible for an observed marked increase in the permeability of ethylcellulose films.

It is important that coating films, apart from the core, should be stable to light during storage, and this property has been determined for various films [75, 76]. Most films have adequate photostability but may be affected slightly in their physiocochemical properties rather than their gross appearance. For example, Matsuda et al. [89] found that the rate of water vapor transmission decreased through gelatin film with increasing exposure time to ultraviolet irradiation, presumably due to greater polymer crosslinking.

2.6 MISCELLANEOUS OTHER PROPERTIES RELATING TO MICROCAPSULES AND SIMILAR DOSAGE FORMS

2.6.1 Coating Defects

The coating thickness of microcapsules is normally very thin, and for a given coating core ratio it decreases rapidly as the microcapsule size decreases. Figure 2.8 shows the calculated wall thickness of idealized spherical capsules as a function of capsule size, from which it can be seen that microcapsules having a diameter of 400 μm or less tend to have coating thicknesses of less than 40 μm. Unless it contains a liquid, the core surface may contain depressions in the form of fissures. Rowe and Forse [91] investigated the effect of film thickness on the incidence of bridging of intagliated or monogramed core tablets, which embossing likewise acts as surface depressions. They noted that when the film thickness of a plasticized hydroxypropylmethylcellulose coating exceeded approximately 16 μm, the proportion of tablets showing this defect increased markedly. The effect was considered to be due to the rise in cohesive force upon evaporation of the solvent within the coating upon increase in coating thickness, which eventually exceeded the constant film-core adhesional force. By analogy, it is reasonable to assume that in certain irregular microcapsules, increasing coating thickness

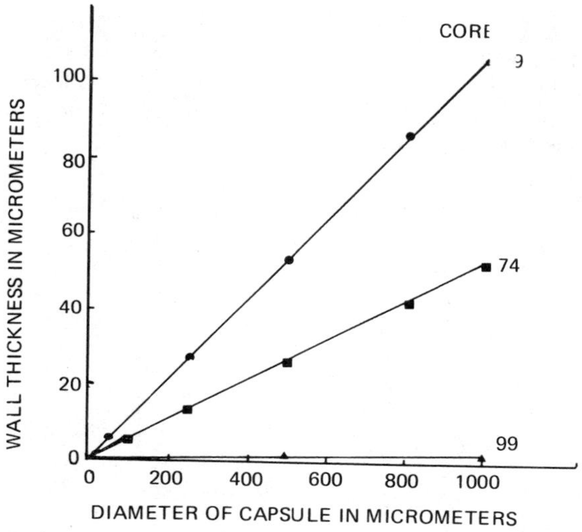

Figure 2.8 Calculated capsule wall thickness as a function of capsule diameter and core content assuming ideal spherical capsules. Core, water; coating, polyethylene of uniform thickness. (From Ref. 90.)

eventually leads to loss of core coating adhesion. Likewise, use of a suitable plasticizer in adequate concentration should decrease cohesive force within the film and decrease the incidence of this defect, as reported by Rowe and Forse [92] for film-coated tablets. A certain amount of foam infilling of instagliations in tablets, air suspension-coated with an aqueous hydroxypropylmethylcellulose formulation, was reported by Down [93].

Rowe [94] also studied the effect of some factors on the surface roughness of tablets film coated with hydroxypropylmethylcellulose. The polymer grade with intermediary molecular weight gave coated tablets with minimal surface roughness. Increasing polymer concentration in the solution applied above 2% and greater coating thickness achieved above 20 μm gave increased roughness. Coatings of 140 μm thickness have a flaky appearance, as though the film had cracked upon drying. Increasing addition of titanium dioxide or pigment caused a marked increase in roughness. There was a tendency for the surface roughness to increase with increasing particle size of the added dispersed solid [95], though the effect was concentration- and substrate roughness-dependent.

Rowe [9] has recently presented a theoretical approach to the formation of cracks and other defects in film coatings for tablets and has reviewed the formulation factors affecting them.

Various different microencapsulation processes give rise to the formation of microcapsules with characteristic size ranges. Table 2.6 shows the size range obtainable from some of the more frequently used methods whose details will be discussed in following chapters.

Table 2.6 Microcapsule Size Range Produced by Various Production Procedures

Production process	Size range (μm)
Coacervation phase separation	1–2000
Interfacial polycondensation	2–2000
Pan coating	200–5000
Air suspension	50–1500
Spray drying	5–800

Table 2.7 Approximate Number of Microcapsules per Gram of Material at Various Diameters

Microcapsule diameter (μm)	Approximate number per gram (density assumed = 1)
5	15,279,000,000
25	122,230,000
50	15,279,000
100	1,910,000
200	239,000
300	71,000
400	30.000
500	15,000
600	9,000
700	5,500
800	3,500
900	2,500
1,000	2,000

Source: Ref. 96.

Table 2.7 shows the approximate number of microcapsules per gram of material at various diameters. The data are useful for obtaining a preliminary estimate of the likely number of microcapsules to be filled into an outer hard gelatin shell at a given dosage weight.

Different size fractions of microcapsules prepared by a particular method are normally separated by mechanical screening on a suitable nest of sieves. The efficiency of this size separation, particularly for fine and irregular-shaped microcapsules, is questionable, as Lazarus et al. [97] and Orr [98] have both reported that fine particles may cling to large particles and fail to be adequately separated by sieving. Excessive sieving may give rise to damage to delicate microcapsules.

2.6.3 Encapsulation and Tableting of Microcapsules

After preparation, microcapsules are commonly filled at appropriate dosage into outer hard gelatin capsules as described by Hostetler and Bellard [99]. This facilitates handling of different microcapsules, which may be blended in suitable proportions to achieve a controlled release of medicament upon ingestion. The outer gelatin shell starts to dissolve and splits at the shoulders within a few minutes to release its contents, as reported by Ludwig and Van Ooteghem [100], who studied the in vitro disintegration of hard gelatin capsules in 0.1 *N* HCl at 37°C.

The presence of certain colorants in pharmaceutical gelatins used to prepare hard capsules may retard their disintegration, as reported by Cooper et al. [101].

Recently, Hunter et al. [102] studied the dispersion of ^{99m}Tc-labeled ion-exchange resin particles in the human stomach from hard gelatin capsules using a gamma camera. In the presence of food the contents of a capsule generally appeared to be released rapidly and dispersed well in the stomach, capsule content being apparent in the duodenum from 20 min onward after ingestion. However, when capsules were administered with a small amount of fluid to fasting subjects, they did not disperse well in the stomach. In one subject, capsule lodgement in the esophagus was noted. Having the subject in the upright position, rather than supine, had an inhibitory effect on gastric emptying [103]. Alpsten et al. [104] labeled enteric-coated microcapsules with ^{51}Cr and monitored their in vivo release from hard gelatin capsules. The time for dispersion of the contents of the capsules in fasting human subjects was noted to be greater than that observed in official in vitro tests, having a mean disintegration time of 18 min with individual variation of 12 to 26 min.

Microcapsules may also be filled into outer flexible gelatin shells. D'onofrio et al. [105] used a coacervation-phase separation process whereby aspirin particles were coated in ethylcellulose and recovered in liquid paraffin, which allowed the slurry to be filled directly into soft gelatin capsules without the usual filtering, drying, and redispersion steps.

Microcapsules are rarely tableted, as their coatings tend to disrupt during the compression cycle involved, unless they contain very pliable materials. Where microcapsules have been tableted, they tend to form matrices that may be produced far more easily and cheaply by conventional tableting techniques involving mixing and granulation of the core and coating components. Walker et al. [106], for example, tableted potassium chloride microcapsules in an attempt to avoid having to use a very large and difficult-to-swallow gelatin shell to accommodate the intended 940 mg dose of microcapsules. Addition of 10% dicalcium phosphate (Emcompress) allowed satisfactory tablets to be produced, but retard properties were lost due to damage to the microcapsule coatings. Minimal loss of retard effect was obtainable by the use of polyethylene glycol and microcrystalline cellulose (Avicel). The polyethylene glycol probably acted as an internal lubricant, improving packing and reducing interparticle friction between microcapsules during compression.

Estevenel et al. [107], in a patent assigned to Choay SA, France, prepared triple-layered tablets whose central layer was composed of microcapsules. They claimed that the outer layers in adequate thickness protected the microcapsules during compression.

REFERENCES

1. J. W. McGinity, A. Martin, G. W. Cuff, and A. B. Combs, Influence of matrixes on nylon-encapsulated pharmaceuticals, *J. Pharm. Sci. 70*:372–375 (1981).
2. H. R. Marston, U.S. Patent 3,056,724 (October 2, 1962).
3. H. Bechgaard and O. Antonsen, Dispersal of pellets throughout the gastrointestinal tract. The influence on transit time exerted by the density or diameter of pellets, in *F.I.P. Abstracts, 37th International Congress of Pharmaceutical Sciences, The Hague, September*, 1977, p. 69.
4. H. Bechgaard and K. Ladefoged, Distribution of pellets in the gastrointestinal tract. The influence of transit time exerted by the density or diameter of pellets, *J. Pharm. Pharmacol. 30*:690–692 (1978).
5. H. Bechgaard and K. Ladefoged, Gastrointestinal transit time of single-unit tablets, *J. Pharm. Pharmacol. 33*:791–792 (1981).
6. H. Umezawa, U.S. Patent 4,101,650 (July 18, 1978).
7. Y. Raghunathan, L. Amsel, O. Hinsvark, and W. Bryant, Sustained-release drug delivery system 1. Coated ion-exchange resin system for phenylpropanolamine and other drugs, *J. Pharm. Sci. 70*:379–384 (1981).
8. R. C. Rowe, The expansion and contraction of tablets during film coating—a possible contributory factor in the creation of stresses within the film?, *J. Pharm. Pharmacol. 32*:851 (1980).
9. R. C. Rowe, The cracking of film coatings on film-coated tablets—a theoretical approach with practical implications, *J. Pharm. Pharmacol. 33*:423–426 (1981).
10. K. Lehmann, Programmed drug release from oral dosage forms, *Pharma. Int. 3*:34–41 (1971).
11. J. Spitael and R. Kinget, Factors affecting the dissolution rate of enteric coatings, in *F.I.P. Abstracts, 37th Internation Congress of Pharmaceutical Sciences, The Hague, September*, 1977, p. 44.
12. F. Alhaique, M. Marchetti, F. M. Riccieri, and E. Santucci, A polymeric film responding in diffusion properties to environmental pH stimuli: a model for a self-regulating drug delivery system, *J. Pharm. Pharmacol. 33*:413–418 (1981).
13. G. W. Johnston, R. I. Malani, and M. W. Scott, U.S. Patent 3,274,061 (September 20, 1966).
14. C. A. Signorino, U.S. Patent 3,738,952 (June 12, 1973).
15. C. A. Signorino, U.S. Patent 3,741,795 (June 26, 1973).
16. M. Donbrow and M. Friedman, Permeability of films of ethyl cellulose and PEG to caffeine, *J. Pharm. Pharmacol. 26*:148–150 (1974).
17. Y. Samuelov, M. Donbrow, and M. Friedman, Effect of pH on salicylic acid permeation through ethyl cellulose-PEG 4000 films, *J. Pharm. Pharmacol. 31*:120–121 (1979).

18. R. S. Okor, Effect of polymer cation content on certain film properties, *J. Pharm. Pharmacol.* *34*:83–86 (1982).
19. W. A. Hunke and L. E. Matheson, Mass transport properties of co(polyether) polyurethane membranes. II: Permeability and sorption characteristics, *J. Pharm. Sci.* *70*:1313–1318 (1981).
20. M. Donbrow and Y. Samuelov, Zero order drug delivery from double-layered porous films: release rate profiles from ethylcellulose, hydroxypropyl cellulose and polyethylene glycol mixtures, *J. Pharm. Pharmacol.* *32*:463–470 (1980).
21. S. Borodkin and F. E. Tucker, Linear drug release from laminated hydroxypropyl cellulose-polyvinyl acetate films, *J. Pharm. Sci.* *64*:1289–1294 (1975).
22. S. Borodkin and F. E. Tucker, Drug release from hydroxypropyl cellulose-polyvinyl acetate films, *J. Pharm. Sci.* *63*:1359–1364 (1974).
23. A. F. Kydonieus and A. R. Quisumbing, Multilayered laminated structures, in *Controlled Release Technologies: Methods, Theory and Applications, Vol. 1* (A. F. Kydonieus, ed.), CRC Press, Boca Raton, Fla., 1980, pp. 183–238.
24. M. S. Harris, Preparation and release characteristics of potassium chloride microcapsules, *J. Pharm. Sci.* *70*:391–394 (1981).
25. N. J. Morris and B. Warburton, Three-ply walled w/o/w microcapsules formed by a multiple emulsion technique, *J. Pharm. Pharmacol.* *34*:475–479 (1982).
26. R. E. Baier, Blood compatibility of synthetic polymers: perspective and problems, in *Polymers in Medicine and Surgery* (R. L. Kronenthal, Z. Oser, and E. Martin, eds.), Plenum, New York, 1975, pp. 139–159.
27. R. K. Kulkarni, Brief review of biochemical degradation of polymers, *Polym. Eng. Sci.* *5*:227–230 (1965).
28. J. Autian, Plastics in pharmaceutical practice and related fields, Part 1, *J. Pharm. Sci.* *52*:1–23 (1963).
29. Y. W. Chien, Methods to achieve sustained drug delivery. The physical approach: implants, in *Sustained and Controlled Release Drug Delivery Systems* (J. R. Robinson, ed.) Dekker, New York, 1978, pp. 211–349.
30. Martindale, *The Extra Pharmacopoeia*, 28th ed., Pharmaceutical Press, London, 1982.
31. R. C. Rowe, The molecular weight and molecular weight distribution of hydroxypropyl methylcellulose used in the film coating of tablets, *J. Pharm. Pharmacol.* *32*:116–119 (1980).
32. R. C. Rowe, Molecular weight studies on ethyl cellulose used in film coating, *Acta Pharm. Suec.* *19*:157–160 (1982).

33. J. R. Nixon, P. P. Georgakopoulos, and J. E. Carless, Diffusion from gelatin-glycerin-water gels, *J. Pharm. Pharmacol. 19*:246–252 (1967).
34. N. F. Cardarelli, Monolithic elastomeric materials, in *Controlled Release Technologies: Methods, Theory and Applications, Vol. 1* (A. F. Kydonieus, ed.), CRC Press, Boca Raton, Fla., 1980, pp. 73–127.
35. P. Bueche, *Physical Properties of Polymers*, Interscience, New York, 1962, pp. 85-111.
36. C. A. Entwistle and R. C. Rowe, Plasticization of cellulose ethers used in the film coating of tablets, *J. Pharm. Pharmacol. 31*:269–272 (1979).
37. W. H. Chu and T. L. Smith, Rodlike superstructures in and mechanical properties of biaxially orientated polyethylene terephthalate film, in *Structure and Properties of Polymeric Films* (R. W. Lenz and R. S. Stein, eds.), Plenum, New York, 1973, pp. 67–79.
38. R. M. Fung and E. L. Parrott, Measurement of film-coating adhesiveness, *J. Pharm. Sci. 69*:439–441 (1980).
39. R. C. Rowe, The measurement of the adhesion of film coatings to tablet surfaces: the effect of tablet porosity, surface roughness and film thickness, *J. Pharm. Pharmacol. 30*:343–346 (1978).
40. R. C. Rowe, Rate effects in the measurement of adhesion of film coatings to tablet surface, *J. Pharm. Pharmacol. 32*:214–215 (1980).
41. P. B. Deasy, M. R. Brophy, B. Ecanow, and M. M. Joy, Effect of ethylcellulose grade and sealant treatments on the production and in vitro release of microencapsulated sodium salicylate, *J. Pharm. Pharmacol. 32*:15–20 (1980).
42. G. S. Banker, Film coating theory and practice, *J. Pharm. Sci. 55*:81–89 (1966).
43. J. D. Crowley, G. S. Teague, and J. W. Lowe, A three-dimensional approach to solubility, *J. Paint Technol. 38*:269–280 (1966).
44. P. D. Nadkarni, D. O. Kildsig, P. A. Kramer, and G. S. Banker, Effect of surface roughness and coating solvent on film adhesion to tablets, *J. Pharm. Sci. 64*:1554–1557 (1975).
45. D. J. Kent and R. C. Rowe, Solubility studies on ethyl cellulose used in film coating, *J. Pharm. Pharmacol. 30*:808–810 (1978).
46. S. C. Porter and K. Ridgway, The permeability of enteric coatings and the dissolution rates of coated tablets, *J. Pharm. Pharmacol. 34*:5–8 (1982).
47. Methocel cellulose ethers—aqueous systems for tablet coating, Dow Chemical Form 192-622-77, 1977.

48. G. Banker, G. Peck, S. Jan, P. Pirakitikulr, and D. Taylor, Evaluation of hydroxypropyl cellulose and hydroxypropyl methyl cellulose as aqueous based film coating, *Drug Develop. Indust. Pharm.* 7:693–716 (1981).
49. J. P. Delporte, J. M. De Seille, and F. Jaminet, Equipments and aqueous coating procedures with low viscosity HPMC at laboratory scale, *J. Pharm.Belg.* *36*:337-347 (1981).
50. J. P. Delporte, Some physical properties of two low viscosity hydroxypropylmethylcelluloses used for aqueous coating of pharmaceutical forms, *J. Pharm. Belg.* *35*:417–426 (1980).
51. M. E. Aulton, M. H. Abdul-Razzak, and J. E. Hogan, The mechanical properties of hydroxypropylmethylcellulose films derived from aqueous systems, Part 1: The influence of plasticisers, *Drug Develop. Indust. Pharm.* 7:649–668 (1981).
52. K. Lehmann and D. Dreher, The use of aqueous synthetic-polymer dispersions for coating pharmaceutical dosage forms, *Drugs—Made in Germany* *16*:126–136 (1973).
53. K. Lehmann, Polymer coating of tablets—a versatile technique. *Manuf. Chem. Aerosol News* *45*:5, 48 and 50 (1974).
54. H. S. Hall, K. D. Lillie, and R. E. Pondell, Comparison—aqueous vs solvent based ethylcellulose films, Coating Place, Inc., Verona, Wisc.
55. J. W. Stafford, Enteric film coating using completely aqueous dissolved hydroxypropyl methyl cellulose phthalate spray solutions, *Drug Develop. Ind. Pharm.* *8*:513–530 (1982).
56. A. T. Florence and D. Attwood, *Physiocochemical Principles of Pharmacy*, Macmillan, London, 1981, pp. 125–175.
57. I. Mellan, *The Behavior of Plasticizers*, Pergamon, New York, 1961.
58. T. Vemba, J. Gillard, and M. Roland, Influence of solvents and plasticizers on the permeability and the rupture force of ethylcellulose film coatings, *Pharm. Acta Helv.* 55:65–71 (1980).
59. C. A. Entwistle and R. C. Rowe, Plasticization of cellulose ethers used in the film coating of tablets—the effect of plasticizer molecular weight, *J. Pharm. Pharmacol.* *30 (Suppl):* 27P (1978).
60. J. Autian, Plastics in pharmaceutical practice and related fields, Part II, *J. Pharm. Sci.* *52*:105–122 (1963).
61. R. S. Okor and W. Anderson, Variation of polymer film composition and solute permeability, *Abstracts, British Pharmaceutical Conference, Exeter, September*, 1979, p. 78.
62. R. Gurny, P. Buri, H. Sucker, P. Guitard, and H. Leuenberger, Development of drug forms with release controlled by methacrylic films, *Pharm. Acta Helv.* *52*:175–181 (1977).
63. W. H. Lawrence, Phthalate esters: the question of safety, *Clin. Toxicol.* *13*:89–139 (1978).

64. A. L. Fites, G. S. Banker, and V. F. Smolen, Controlled drug release through polymeric films, *J. Pharm. Sci. 59*:610–613 (1970).
65. J. H. Dopper, E. M. Middelbeek, H. Larsen, and K. Wiedhaup, The application of "contact angle versus time curves" for the determination of experimental conditions during a film-coating process, in *F.I.P. Abstracts, 37th International Congress of Pharmaceutical Sciences, The Hague, September*, 1977, p. 43.
66. T. Maierson, U.S. Patent 3,436,452 (April 1, 1969).
67. M. Donbrow and M. Friedman, Enhancement of permeability of ethyl cellulose films for drug penetration, *J. Pharm. Pharmacol. 27*:633–646 (1975).
68. R. Sjökvist, A. Wirbrant, A. Agren, and G. Aström, Penetration through nylon film containing a carrier, *Acta Pharm. Suec. 15*: 419–430 (1978).
69. S. Motycka and J. G. Nairn, Influence of wax coatings on release rate of anions from ion-exchange resin beads, *J. Pharm. Sci. 67*:500–503 (1978).
70. R. A. Cain and N. J. Federici, U.S. Patent 3,402,240 (September 17, 1968).
71. M. R. Brophy and P. B. Deasy, Influence of coating and core modifications on the in vitro release of methylene blue from ethylcellulose microcapsules produced by pan coating procedure, *J. Pharm. Pharmacol. 33*:495–499 (1981).
72. M. E. Aulton, M. H. Abdul-Razzak and J. E. Hogan, The influence of solid inclusions on the mechanical properties of hydroxypropylmethylcellulose films, in *F.I.P. Abstracts, 41st International Congress of Pharmaceutical Sciences, Vienna, September*, 1981, p. 150.
73. E. L. Rowe, U.S. Patent 3,336,155 (August 15, 1967).
74. J. Spitael and R. Kinget, Preparation and evaluation of free films: influence of method of preparation and solvent composition upon the permeability, *Pharm. Acta Helv. 52*:47–50 (1977).
75. J. L. Kanig and H. Goodman, Evaluative procedures for film-forming materials used in pharmaceutical applications, *J. Pharm. Sci. 51*:77–83 (1962).
76. B. J . Munden, H. G. Dekay, and G. S. Banker, Evaluation of polymeric materials 1. Screening of selected polymers as film coating agents, *J. Pharm. Sci. 53*:395–401 (1964).
77. I. W. Kellaway, C. Marriott, and J. A. J. Robinson, The mechanical properties of gelatin films I. The influence of water content and preparative conditions, *Can. J. Pharm. Sci. 13*:83–86 (1978).
78. I. W. Kellaway, C. Marriott, and J. A. J. Robinson, The mechanical properties of gelatin films II. The influence of titanium dioxide and dyes, *Can. J. Pharm. Sci. 13*:87–90 (1978).

79. J. P. Delporte, Influence of some additives on the mechanical properties of free low viscosity hydroxypropylmethylcellulose films, *J. Pharm. Belg.* *36*:27–37 (1981).
80. P. Stanley, R. C. Rowe, and J. M. Newton, Theoretical considerations of the influence of polymer film coatings on the mechanical strength of tablets, *J. Pharm. Pharmacol.* *33*:557–560 (1981).
81. J. T. Fell, R. C. Rowe, and J. M. Newton, The mechanical strength of film coated tablets, *J. Pharm. Pharmacol.* *31*:69–72 (1979).
82. H. B. Hopfenberg, Formulation and structure of synthetic membranes, films and microcapsules, in *Polymeric Delivery Systems* (R. J. Kostelnik, ed.), Gordon and Breach, New York, 1978, pp. 1–23.
83. G. S. Banker, A. Y. Gore, and J. Swarbrick, Water vapour transmission properties of free polymer films, *J. Pharm. Pharmacol.* *18*:457–466 (1966).
84. V. Stannett, H. B. Hopfenberg, and J. L. Williams, Gas, vapor, and water transport in polymer films, in *Structure and Properties of Polymeric Films* (R. W. Lenz and R. S. Stein, eds.), Plenum, New York, 1973, pp. 321–338.
85. H. B. Hopfenberg (ed.), *Permeability of Plastic Films and Coatings to Gases, Vapors and Liquids*, Plenum, New York, 1974.
86. J. R. Nixon and A. Nouh, The effect of microcapsule size on the oxidative decomposition of core material, *J. Pharm. Pharmacol.* *30*:533–537 (1978).
87. J. F. Pickard, J. E. Rees, and P. H. Elworthy, Water vapour permeability of poured and sprayed polymer films, *J. Pharm. Pharmacol.* *24(Suppl.)*:139P (1972).
88. J. Gillard, T. Vemba, and M. Roland, Role of humidity on formulation of an ethylcellulose film, *Pharm. Acta. Helv.* *55*:231–234 (1980).
89. Y. Matsuda, K. Kouzuki, M. Tanaka, Y. Tanaka, and J. Tanigaki, Photostability of gelatin capsules: effect of ultraviolet irradiation on the water vapor transmission properties and dissolution rates of indomethacin, *Yakugaku Zassi* *99*:907–913 (1979).
90. C. Thies, Physiocochemical aspects of microencapsulation, *Polym. Plast. Technol. Eng.* *5*:1–22 (1975).
91. R. C. Rowe and S. F. Forse, The effect of film thickness on the incidence of the defect bridging of intagliations on film coated tablets, *J. Pharm. Pharmacol.* *32*:647–648 (1980).
92. R. C. Rowe, and S. F. Forse, The effect of plasticizer type and concentration on the incidence of bridging of intagliations on film coated tablets, *J. Pharm. Pharmacol.* *33*:174–175 (1981).

93. G. R. B. Down, An alternative mechanism responsible for the bridging of intagliations on film coated tablets, *J. Pharm. Pharmacol. 34*:281–282 (1982).
94. R. C. Rowe, The effect of some formulation and process variables on the surface roughness of film-coated tablets, *J. Pharm. Pharmacol. 30*:669–672 (1978).
95. R. C. Rowe, The effect of the particle size of an inert additive on the surface roughness of a film-coated tablet, *J. Pharm. Pharmacol. 33*:1–4 (1981).
96. J. Bakan, Microencapsulation using coacervation/phase separation techniques, in *Controlled Release Technologies: Methods, Theory and Applications, Vol. II* (A. F. Kydonieus, ed.), CRC Press, Boca Raton, Fla., 1980, pp. 83–105.
97. J. Lazarus, M. Pagliery, and L. Lachman, Factors influencing the release of a drug from a prolonged-action matrix, *J. Pharm. Sci. 53*:798–802 (1964).
98. N. Orr, Quality control and pharmaceutics of content uniformity of medicines containing potent drugs with special reference to tablets, in *Progress in the Quality Control of Medicines* (P. B. Deasy and R. F. Timoney, eds.), Elsevier Biomedical, Amsterdam, 1981, pp. 193–256.
99. V. Hostetler and J. Q. Bellard, Capsules. Part 1. Hard capsules, in *The Theory and Practice of Industrial Pharmacy* (L. Lachman, H. A. Lieberman, and J. L. Kanig, eds.) Lea & Febiger, Philadelphia, 1976, pp. 389–404.
100. A. Ludwig and M. Van Ooteghem, Disintegration mechanism of hard gelatin capsules, in *F.I.P. Abstracts, 39th International Congress of Pharmaceutical Sciences, Brighton, September,* 1979, p. 28.
101. J. W. Cooper, H. C. Ansel, and D. E. Cadwallader, Liquid and solid solution interactions of primary certified colorants with pharmaceutical gelatins, *J. Pharm. Sci. 62*:1156–1164 (1973).
102. E. Hunter, J. T. Fell, R. T. Calvert, and H. Sharma, "In vivo" disintegration of hard gelatin capsules in fasting and non-fasting subjects, *Int. J. Pharm. 4*:175–183 (1980).
103. E. Hunter, J. T. Fell, and H. Sharma, The gastric emptying of pellets contained in hard gelatin capsules, *Drug Develop. Indust. Pharm. 8*:751–757 (1982).
104. M. Alpsten, C. Bogentoft, G. Ekenved, and L. Sölvell, On the disintegration of hard gelatin capsules in fasting volunteers using a profile scanning technique, *J. Pharm. Pharmacol.* 31: 480–481 (1979).
105. G. P. D'onofrio, R. C. Oppenheim, and N. E. Bateman, Encapsulated microcapsules, *Int. J. Pharm. 2*:91–99 (1979).

106. S. E. Walker, J. Ganley, and T. Eaves, Tableting properties of potassium chloride microcapsules, in *F.I.P. Abstracts, 37th International Congress of Pharmaceutical Sciences, The Hague, September*, 1977, p. 31.
107. Y. F. Estevenel, M. H. Thely, and W. A. Coulon, U.S. Patent 4,113,816 (September 12, 1978).

3

Coacervation–Phase Separation Procedures Using Aqueous Vehicles

3.1 INTRODUCTION

Many microencapsulation procedures have been developed for the coating of pharmaceuticals. These procedures often originated from the coating of nonpharmaceutical products and have been adapted for the encapsulation of drugs. In this and subsequent chapters the major microencapsulation procedures are discussed. A number of microencapsulation procedures are difficult to classify simply under any one heading and consequently have been arbitrarily assigned to the most suitable classification. Where a technique involves a combination of procedures, it will be discussed after the separate elements of the coating operation have been reviewed. The extensive literature cited from patent and nonpatent sources will deal mainly with microencapsulation procedures for pharmaceuticals. However, suitable reference will be made to other applications of microencapsulation for historical reasons and to provide a better understanding of the theory and development of a procedure. Many details of concentrations of reagents have been omitted so that the basis of the encapsulation procedure may be more readily appreciated without distracting particulars. Strengths of reagents and other conditions used should be determined by reference to the original literature or by direct experimentation.

Microencapsulation by coacervation-phase separation using aqueous manufacturing vehicles was first developed commercially by the National Cash Register Co. (NCR), Dayton, Ohio, in 1954. In a series of patents [1-3] ascribed to Green and Schleicher, gelatin and gelatin-acacia (gum arabic) coating systems were described for the encapsulation of dye material used in the manufacture of carbonless carbon paper. Because of the lack of toxicity of the coating materials employed and the comparative simplicity of the procedures, the technique was quickly developed by NCR in association with its subsidiary Eurand and other commercial and noncommercial groups for the coating of pharmaceuticals.

A number of reviews [4–9] on the use of phase separation or coacervation have recently been published. Figure 3.1 shows the typical steps involved in a coacervation process and Fig. 3.2 shows an encapsulated liquid and solid.

The term coacervate, from the Latin *acervus*, meaning a heap or aggregation, was first introduced into scientific literature by two Dutch scientists, Bungenberg de Jong and Kruyt [11], to describe the process of phase separation in liquids containing colloidal solute. In a subsequent publication [12] a number of coacervation processes were described in detail, though their application to microencapsulation was not developed until the later studies of Green and co-workers.

In order to understand fully the process of coacervation for microencapsulation it is necessary to review aspects of the stabilization of hydrophilic colloids such as gelatin and acacia. Gelatin is a macromolecular protein derived from collagen and is composed of 18 different types of amino acids linked by carboxylic and alpha-amino groups to form a branched chainlike structure with free amino-guanidine basic and carboxyl acidic groups. Gelatins prepared from different sources of collagen such as pig skin, ox skin, or ox bone contain varying amounts

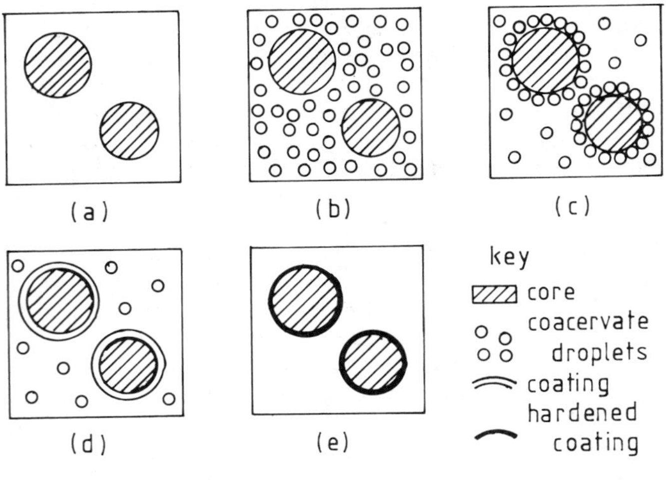

Figure 3.1 Typical steps in a coacervation method of microencapsulation. (a) Core particles dispersed in solution of polymer by agitation. (b) Coacervation visible as droplets of colloid-rich phase induced by one or more agents. (c) Deposition of coacervate droplets on surface of core particles. (d) Mergence of coacervate droplets to form the coating. (e) Shrinkage and crosslinking of the coating to rigidize it as necessary.

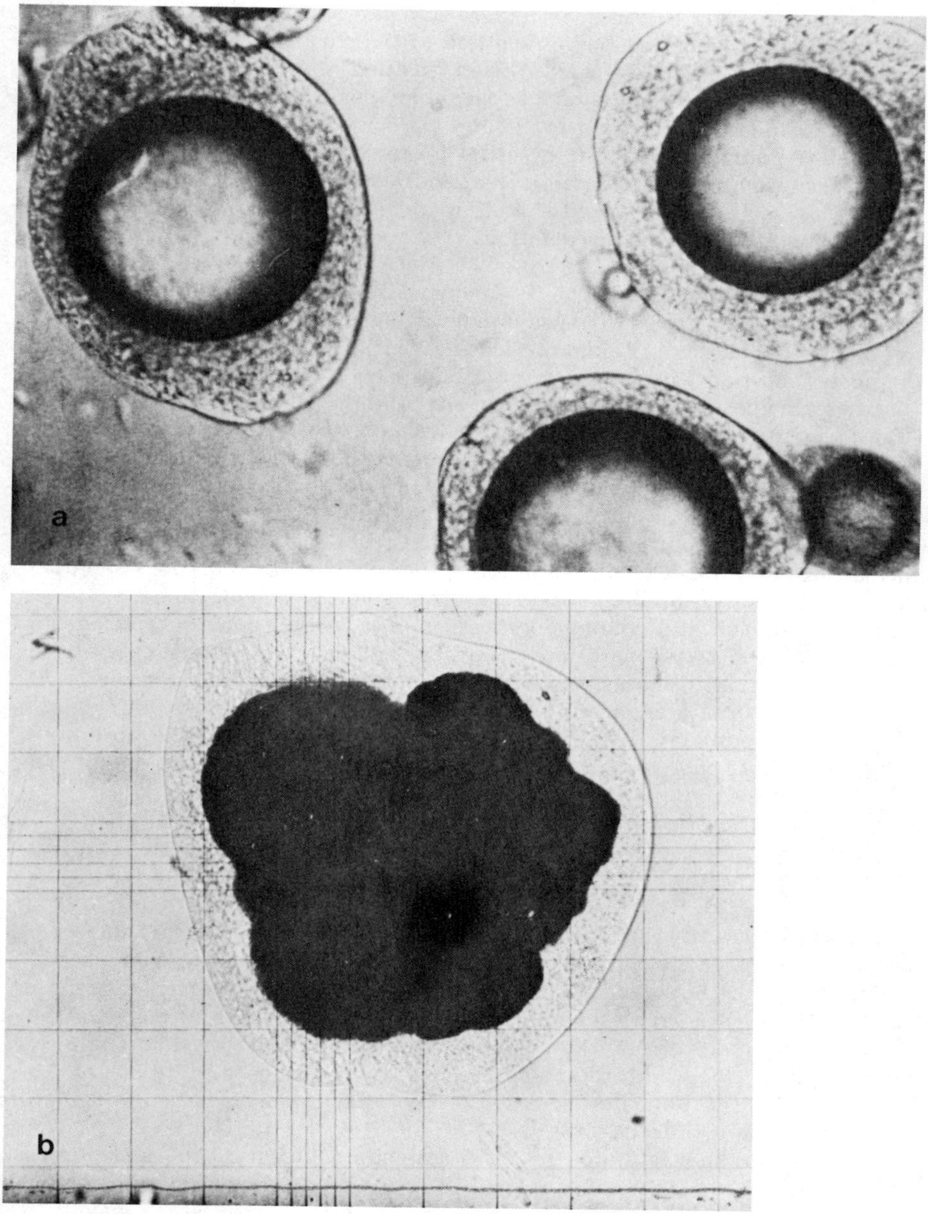

Figure 3.2 Photographs of encapsulated liquid (a) and solid (b). (Reproduced with permission from Ref. 10 and courtesy of Dr. J. Bakan, Eurand America Inc., Dayton, Ohio.)

of the amino acids in their molecules, and their average molecular weight and purity are affected by the method of preparation. Pretreatment of raw materials with alkali (e.g., limed gelatins) tends to hydrolyze the amide groups of glutamine and asparagine, resulting in a net increase in the number of free carboxyl groups and reduction in the isoelectric (isoionic) point (IEP) of the gelatin produced. Gelatins of greater purity and higher isoelectric point are generally prepared from pig skin using mild acid pretreatment. Eastoe and Leach [13] reviewed the chemical composition of various gelatins.

Acacia is composed chiefly of 30.3% L-arabinose, 11.4% L-rhamnose, 36.8% D-galactose, and 13.8% D-glucuronic acid. As it contains only free carboxylic groups, it always carries a negative charge except at very low pH values. Below the isoelectric point, amphoteric gelatin in aqueous solution will acquire a positive charge by dissociation of amino groups, which attract protons from the water and help bind water to the gelatin molecules. Hydrophilic colloids may also acquire charge from other sources, in particular by adsorption of ions from electrolyte present in the surrounding medium. Obviously any factor that influences hydration or charge, such as solvent addition, pH or temperature change, or electrolyte addition, will affect the amount of water bound, which in turn will affect the solubility or dispersibility of the colloid.

In coacervation a colloidal dispersion can be caused to separate into colloid-rich and colloid-poor regions by careful control of temperature, pH, electrolyte addition, or other factors. The coacervate or colloid-rich region forms as droplets, which make the system opaque and sediment to form a separate lower layer unless prevented by stirring. The phase boundary formed is unlike that between immiscible liquids in that the solvent is continuous on both sides of the interface, facilitating the free migration of solute between layers.

3.2 SIMPLE AND COMPLEX COACERVATION

There are two types of coacervation: simple and complex. Simple coacervation involves the use of only one colloid, e.g., gelatin in water, and involves removal of the associated water from around the dispersed colloid by agents with a greater affinity for water, such as various alcohols and salts. The dehydrated molecules of polymer tend to aggregate with surrounding molecules to form the coacervate. Figure 3.3 shows two three-phase diagrams for the coacervation of a typical gelatin by ethanol or sodium sulfate.

When ethanol is added to a 10% aqueous solution of gelatin, point x, Fig. 3.3a, above its gelation temperature, phase separation will begin to occur at point y. The coacervate forms as droplets, which make the system opaque when the ethanol concentration is between 50 and 60%. Further increase in ethanol concentration is undesirable, as it tends to cause flocculation of the gelatin as a precipitated mass. Methanol or

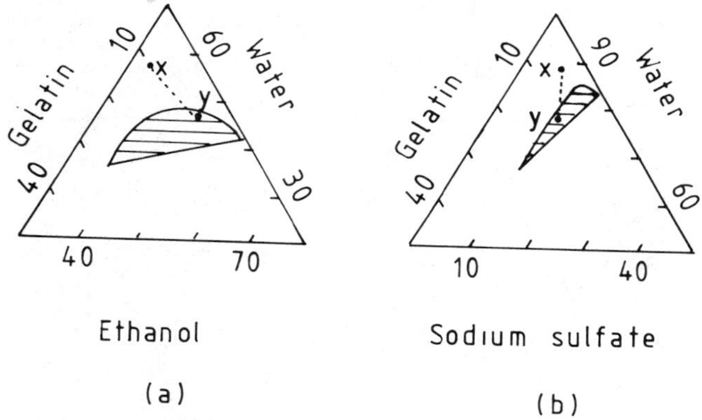

Figure 3.3 Three-phase diagrams for the simple coacervation of gelatin in water by (a) ethanol or (b) sodium sulfate. Shaded region denotes coacervation.

acetone can be used instead of ethanol to cause a similar effect. Any suitable organic liquid can be used, provided that it is miscible with water and is a nonsolvent for the polymer. Thus aqueous solutions of the following hydrophilic polymers can be caused to coacervate under suitable conditions by (1) acetone for agar or methylcellulose, (2) isopropanol for pectin, and (3) propanol for polyvinyl alcohol.

Coacervation of various hydrophilic colloids can also be induced by addition of large amounts of various salts. When 25% aqueous sodium sulfate is added to a 5% aqueous solution of gelatin, point x, Fig. 3.3b, phase separation will begin to occur at point y. Again, overaddition of coacervating agents is undesirable. Various other salts can be used, particularly effective being those containing cations with high hydration affinity, such as Na, K, or NH_4, and anions ranked roughly in accordance with the Hofmeister series of decreasing flocculating power, i.e., $SO_4 >$ citrate $>$ tartrate $>$ etc. Likewise, electrolyte and heating can be used to coacervate an aqueous solution of methylcellulose because the solubility of this polymer decreases with a rise in temperature.

Complex coacervation involves the use of more than one colloid. Gelatin and acacia in water are most frequently used, and the coacervation is accomplished mainly by charge neutralization of the colloids carrying opposite charges rather than by dehydration. From the previous discussion, it follows that if a pig-skin gelatin of high isoelectric point (8.0 to 8.5) is mixed with acacia at pH 4.0 to 4.5 in water, the marked net positive charge on the gelatin will tend to neutralize the negative charge on the acacia, forming a complex coacervate of the colloids. The charge neutralization is accompanied by loss of bound water by the polymers. Figure 3.4 shows a three-phase diagram for the coacervation of a

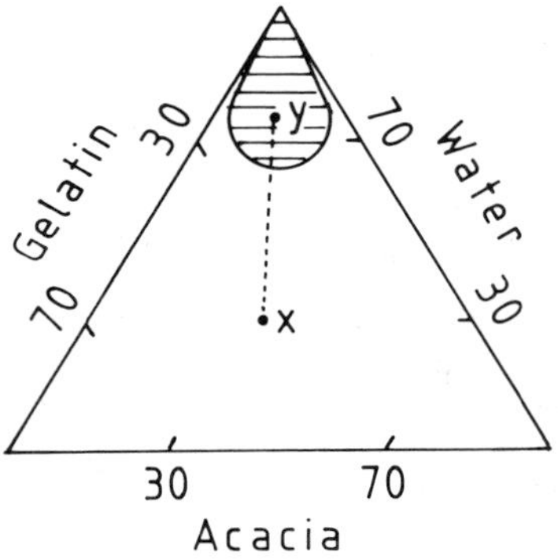

Figure 3.4 Three-phase diagram for the complex coacervation of a gelatin-acacia-water system. Shaded region denotes coacervation.

gelatin-acacia-water system at elevated temperature. Obviously, if warm water is added to the system having a point composition denoted by x, Fig. 3.4, dilution will cause coacervation to occur at point y. Coacervation is usually marked when the concentration of the polymers is less than 3%. The coacervate forms as droplets about 100 μm in diameter, easily visible under the microscope. The system develops a turbid appearance and the coacervate will settle as a bottom layer unless prevented by agitation.

Obviously, a key element in the development of a successful coacervation system is the construction of a suitable three-phase diagram for the selected materials. Phares and Sperandio [14] described a procedure for the preparation of such phase diagrams. They used refractive index and specific gravity measurements rather than chemical analysis to determine percent weight-in-weight concentrations of gelatin, water, and sodium sulfate for coacervation. Using similar procedures, Nixon et al. [15] constructed three-phase diagrams for the coacervation of gelatin-ethanol-water. They found that the amount of ethanol required decreased with increasing molecular weight of gelatin used and that the coacervate phase had a constant ethanol content independent of molecular weight of gelatin and overall concentration of alcohol in the system. Electrophoresis studies can also be used to predict coacervation conditions.

Core material for encapsulation in the coacervate may be liquid or solid and must obviously be water-insoluble or very poorly water-soluble or it will simply dissolve in the water used. Usually an oil phase is encapsulated by mechanically dispersing it in the form of droplets in the aqueous phase containing the colloid(s) as an oil-in-water (o/w) emulsion. Auxiliary adjuvants to stabilize the emulsion are rarely needed because of the emulsifying and thickening properties of the colloids. Likewise, dispersed solids rarely require the addition of suspending agents. At least one of the colloids must be gelable in that it undergoes sol gel transformation with change in temperature. Gelatin is most frequently used because aqueous solutions above 1% are viscous gels at room temperature and become mobile sols at a temperature of 25 to 30°C or higher.

Coacervation of the hydrophilic colloid(s) at elevated temperature in the sol state is then brought about as previously indicated. The coacervate deposits colloid around the oil droplets or solid particles as a coating. Two mechanisms have been proposed by Luzzi [4] for the formation of the microcapsules: (1) individual coacervate droplets may be attracted to and coalesce around core particles immiscible in the system, and/or (2) a single coacervate droplet may encompass one or more core particles. Coacervation may also produce compound microcapsules whereby groups of smaller, previously formed capsules are coated. The deposition of coating material is aided by a reduction in total interfacial energy of the system consequent upon the decrease in the surface area of the coating material as its droplets coalesce around the core material. The coating is then gelled by lowering the temperature and hardened by the addition of a crosslinking agent such as formaldehyde or glutaraldehyde.

Figure 3.5 shows a typical flow diagram for the encapsulation of an oil or solid by gelatin using simple coacervation.

The particle size of the microcapsules formed can be controlled to some degree by using core material of suitable size range. Also, the size of the pores in the coating may be controlled to some degree by the rate of the gelling process. Rapid cooling tends to form a finer pore size, which reduces escape of core material. However, it is difficult to control precisely the capsule size obtained, and microcapsules tend to adhere together to form agglomerates, particularly during the hardening and drying stages.

When complex coacervation occurs in a gelatin-acacia system at elevated temperature, the phase-separated colloid-rich droplets usually contain an approximately 1:1 ratio of the two colloids in about 20% concentration. In contrast, the colloid-poor aqueous equilibrium phase contains less than 0.5% of the polymers and consequently does not gel with decrease in temperature, unlike the coacervate. Figure 3.6 shows a typical flow diagram for the encapsulation of an oil by gelatin and acacia using complex coacervation. A similar procedure could be used to encapsulate solid particles by initially dispersing them in the gelatin solution.

Coacervation is caused to occur by pH adjustment and dilution if necessary. It is also promoted by decreasing the temperature. The coacervate droplets deposit around the core particles, which serve as nuclei, until the coacervate finally surrounds the core material completely. If only one particle is enclosed in coacervate, a mononuclear microcapsule is obtained; whereas if more than one particle is enclosed,

Disperse oil or solid
in aqueous solution
of gelatin at 50°C.

↓

Coacervate by addition
of ethanol or sodium
sulfate.

↓

Reduce temperature to
10°C to gel coating
while maintaining
agitation.

↓

Wash with water and
filter to remove
coacervating agent.

↓

Add aqueous solution
of formaldehyde or
glutaraldehyde as
hardening agent.

↓

Wash with water and
filter to remove residual
hardening agent.

↓

Dry product.

↓

Oil or solid containing
microcapsules

Figure 3.5 Typical steps of a simple coacervation process for microencapsulation.

then multinuclear microcapsules are formed. When the system is cooled, the coating gels. This may then be fixed by the addition of a suitable hardening agent such as formaldehyde, elevation of the pH to 9 by alkali, and gradual raising of the temperature to 50°C to complete the curing of the capsule walls. Finally, after washing off residual hardening agent, the microcapsules may be recovered by filtration, centrifugation, or decantation, and may be converted into a dry powder by removal of residual water using a solvent, spray or freeze drying, or a fluidized bed dryer.

Kondo [8] investigated the influence of coacervation pH over the range 4.1 to 4.55 on the gelatin and acacia content of microcapsule walls

Oil phase ↘ ↙ Water phase with solution of 10% pig-skin gelatin ⊖, IEP 8–8.5

↓

Emulsify above 35°C to form o/w emulsion of desired droplet size.

↓

Add 10% aqueous solution of acacia ⊖ at elevated temperature, mix.

↓

Adjust pH to 4–4.5 using acid, whereupon gelatin becomes ⊕, dilute with water if necessary, mix. Coacervate forms by charge neutralization and deposits on oil droplets.

↓

Cool to 5°C to gel coating.

↓

Add aqueous solution of formaldehyde or glutaraldehyde, raising the pH to 9 and the temperature to 50°C to aid hardening.

↓

Wash, filter, and dry.

↓

Oil containing microcapsules

Figure 3.6 Typical steps of a complex coacervation process for microencapsulation.

prepared under identical conditions when the initial ratio of the polymers was 1:1. At pH 4.1 the gelatin content of the walls was higher than the acacia content. As the pH rose, the total amount of wall material deposited decreased and the relative gelatin content decreased. Finally, at pH 4.55 the acacia content of the walls was greater than the gelatin content.

Much research into simple and complex coacervation has been carried out in the last 20 years to elucidate details of the procedures and to help overcome associated problems. The following three sections review the relevant literature, with particular emphasis on the encapsulation of drugs.

3.3 MICROENCAPSULATION OF DRUGS BY SIMPLE GELATIN COACERVATION

The references in this section are arranged in chronological order. In 1964 Phares and Sperandio [16] used a gelatin-water-sodium sulfate system to coat riboflavin, cod liver oil, procaine penicillin G, micro ion-exchange resin, aspirin, acetanilide, and castor oil by coacervation. They had found in prior studies that ethanol produced much larger coacervate droplets than sodium sulfate and also that if the temperature was raised above 50°C, the coacervate droplets became smaller. Recovery of the gelled-only capsules was difficult, as they tended to block conventional filter media, and was best accomplished by allowing the capsules to settle and then decanting the upper phase. If such capsules were air-dried, they tended to bind together to form a solid mass. Treatment with a combination of formaldehyde and alcohol lessened aggregation problems.

Khalil et al. [17] found that coacervation occurred only within a pH range close to the isoelectric point in a gelatin-water-ethanol system, indicating that charge effects have an important role here as in complex coacervation. In a gelatin-water-sodium sulfate system, coacervation occurred over a wide pH range, from 2.1 to 10.2. These results have important implications for simple coacervation, as the added core material may have an influence on the pH of the system.

In a subsequent paper, Nixon et al. [18] examined the preparation of sulfamerazine microcapsules using gelatin water and ethanol or sodium sulfate, and experienced difficulties in the recovery of the microcapsules as a fine powder. The spray-dried product was almost entirely unencapsulated. In contrast, isopropanol treatment of the centrifugal slurry with its mild dehydration effect and subsequent hardening with a formalin-isopropanol mixture resulted in the production of a fine, free-flowing powder with no rupturing of the microcapsules. It was noted that the release of drug was slower from ethanol-coacervated microcapsules than from the sodium sulfate-prepared ones and decreased with increase in formaldehyde treatment and wall thickness.

Paradissis and Parrott [19] microencapsulated a number of drugs, including aspirin and sulfadiazine, by dispersing them with mineral oil in an aqueous gelatin solution and inducing phase separation by addition of isopropanol and temperature reduction. Hardening was done using formaldehyde solution. In vivo studies based on urinary excretion data from two human subjects showed that the microencapsulated forms of both drugs exhibited a sustained-release effect for more than 24 hr in comparison to nonmicroencapsulated controls.

In 1969 Navari et al. [20] examined the rheological properties of a suspension containing gelatin-castor oil microcapsules, suspended in water using carboxymethylcellulose. The suspension had pseudoplastic flow behavior and other physical properties similar to mammalian whole blood at 37°C. The material was suggested as a possible agent for modeling blood flow.

Nixon and Walker [21] microencapsulated sulfadiazine using a gelatin-water-sodium sulfate system and showed that the temperature of coacervation and duration of the hardening process with formalin had little influence on the size of the product obtained. Individual capsules tended to be multinuclear, with particles of drug arranged toward the center and with the surface tending to take the shape of the dispersed crystals of drug. Hardening or drying caused the coating to shrink on the surface of the microcapsules. Upon subsequent immersion in water, capsules swelled almost immediately due to the rehydration of the coating. However, increasing the duration of the hardening treatment tended to reduce the magnitude of the swelling effect. In vitro dissolution studies were difficult to model mathematically.

Later Matthews and Nixon [22] presented scanning electron photomicrographs of gelatin-water-ethanol microcapsules containing a solid sulfonamide-type diuretic (see Fig. 3.7). The capsules were 100 to 500 μm in size with folded and invaginated surfaces with no apparent cracks or pores. Large microcapsules were conglomerates of smaller capsules. In another paper, the same authors [23] used scanning electron microscopy to examine both ethanol- and sodium sulfate-coacervated gelatin-coated microcapsules containing sulfadiazine or hydroflumethiazide. The ethanol-coacervated product showed a variable degree of surface folding depending on the drug encapsulated and associated with vacuole collapse, but no evidence of pores or cracking. The sodium sulfate-coacervated product showed the surface covered with a nonuniform distribution of saltlike crystals of sodium sulfate, though the walls appeared intact.

Madan [24] reported that microcapsules prepared containing the liquid hypocholesterolemic drug, clofibrate, using a gelatin-sodium sulfate coacervation procedure, exhibited zero-order release. There was a linear correlation between the hardening time and the in vitro $t_{50\%}$ release time. However, if the microcapsules were hardened for longer than 8 hr, they tended to rupture. In a subsequent paper, Madan [25] reported that as the wall thickness of the microcapsules was reduced, their release rate deviated from zero-order.

3.4 MICROENCAPSULATION OF DRUGS BY COMPLEX GELATIN-ACACIA COACERVATION

In this section a selection of references on the microencapsulation of drugs with gelatin and acacia are arranged in chronological order.

Luzzi and Gerraughty [26] used gelatin-acacia complex coacervation and hardening with formaldehyde to encapsulate a range of oils such as light liquid paraffin or coconut oil with added oleic acid, lauric acid, or benzoic acid and various surfactants such as polysorbate 20 or sodium oleate. The capsules obtained after filtration and drying were composed of irregular aggregates of spherical oil droplets embedded in a matrix of coating material. Saponification values of the oils

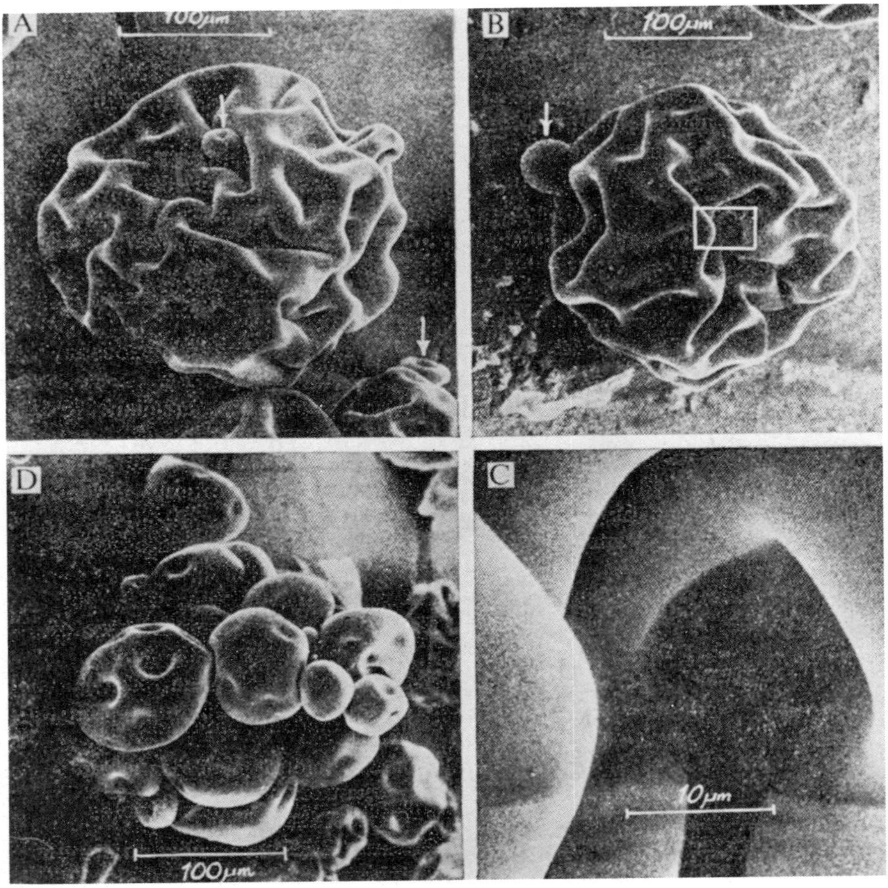

Figure 3.7 A and B. Scanning electron micrographs of gelatin-coacervated microcapsules. C. Enlargement of marked section of B. D. Cluster of microcapsules. (From Ref. 22.)

had little effect on encapsulation, whereas acid values and added surfactant decreased the degree of encapsulation as determined by core resistance to extraction by ethyl ether for periods up to 1 hr. In a subsequent paper, Luzzi and Gerraughty [27] used the same complex coacervation procedure to encapsulate sulfur, aspirin, phenacetin, barbituric acid, lycopodium spores, DDT, pentobarbituric acid, and carbon black. Dissolution studies in simulated gastric-intestinal fluids on pentobarbituric acid microcapsules indicated that drug release decreased at 37°C with decrease in core:coat ratio and adequate formaldehyde treatment. Excessive hardening treatment increased drug release, possibly by excessively denaturing and cracking the coating.

In another paper, the same authors [28] reported that addition of polysorbate 20 and a polyelectrolyte dispersing agent (Darvan 7) to the aqueous phase, prior to the addition of pentobarbituric acid particles and encapsulation, generally tended to increase the release of drug from the microcapsules obtained.

Morse [29], in a patent assigned to Merck and Co., Inc., microencapsulated indomethacin by complex coacervation using glutaraldehyde as the crosslinking agent to obtain microcapsules that had a sustained-release effect in simulated gastric fluid.

In an interesting result by Javidan et al. [30], complex coacervates containing gelatin and acacia were used as a model system for studying the effect of partitioning of the antimicrobial agent, phenylmercuric nitrate, between coacervates and their equilibrium colloid-poor aqueous medium on the growth of *Escherichia coli*. It was suggested that a similar procedure could be developed to model drug distribution in physiological systems.

Madan et al. [31] microencapsulated spherical particles of stearyl alcohol produced by vibrating reed or capillary methods using gelatin and acacia, provided that they were less than 250 μm in diameter. Encapsulation of large particles required a modified procedure and was probably due to direct interaction of gelatin and acacia on the surface of microcapsules rather than by the observed aggregation and coalescence of coacervated droplets around the smaller particles. In a later paper [32], they showed that only a small fraction of the colloids was used in the encapsulation process. Scanning electron microscopy studies showed that capsules had a scaly surface appearance. The results tended to confirm that the gelatin in the coacervate phase and the acacia adsorbed onto the stearyl alcohol particles interacted directly on the surface of the core to form the coating. In a subsequent paper, Madan et al. [33] separated pig-skin gelatin into four fractions using acetone to study the influence of gelatin fraction on complex coacervation of phenobarbituric acid microcapsules. There was some evidence obtained to suggest a relationship between decreasing molecular weight of gelatin fraction and decreasing release of drug from the various microcapsules. Fractions forming thicker walls tended to give slower release of the drug.

Newton et al. [34] prepared microcapsules containing sulfamerazine and various other drugs as well as unmedicated microglobules using complex coacervation. Multinuclear capsules were formed that were spherical when the diameter was 30 μm, but more irregular in shape as the size increased. The capsules readily dispersed when suspended in water or physiological solutions.

Palmieri [35] described how to produce gelatin-acacia films by the usual complex coacervation methods for use in diffusion studies. By suitable control of production conditions, films of thickness 95 to 100 μm could be produced that were apparently devoid of structural defects when observed using a stereo microscope.

Palmieri [36] reported on the dissolution of prednisolone from microcapsules prepared by complex coacervation of the drug as a powder or as a suspension in mineral oil. Drug encapsulated as a powder was released at a slower rate than drug encapsulated in oil, presumably because of the greater wall thickness of the encapsulated powder product. Both microencapsulated forms of the drug had slower release rates than pure prednisolone powder or its commercially available tablet.

The same author [37] microencapsulated undecenovanillylamide, a potential coyote deterrent for the protection of sheep, using complex coacervation with gelatin and acacia. In vitro dissolution studies indicated that increasing hardening time with 37% formaldehyde solution decreased the release of core material. However, all products readily released the drug, having a dissolution $t_{50\%}$ less than 30 min, indicating the general unsuitability of this encapsulation process to achieve the prolonged-release properties required for this drug formulation. It was interesting to note that addition of colloidal silica lessened aggregation of microcapsules and aided their recovery as free-flowing powder.

Nixon and Hassan [38] showed that increasing stirrer speed decreased the particle size of gelatin-acacia complex coacervated microcapsules of thiabendazole. Presumably, faster stirrer speeds would produce smaller coacervate droplets, which would entrap a drug particle without appreciable tendency to aggregate. At slower stirrer speeds, presumably, larger coacervate droplets would entrap multiple core particles and would tend to be in contact with one another for longer periods of time thus enhancing aggregation and the formation of larger microcapsules. Particularly at slower stirrer speeds, there was an obvious tendency for the particle size of microcapsules to increase as the concentration of the colloids increased. In another paper, Nixon and Hassan [39] investigated the effect of additives on the release of thiabendazole from compacts containing the microcapsules. Increasing compression force, stearic acid content, and sodium stearate content decreased tablet disintegration and drug release, whereas increasing starch content tended to produce the opposite effect.

Nixon and Agyilirah [40] investigated the effect of varying the ratio of gelatin and acacia used for the encapsulation of phenobarbitone. Equal ratio up to 3% of each were found to be most suitable for producing free-flowing microcapsules.

Takenaka et al. [41] microencapsulated micronized sulfamethoxazole using gelatin-acacia complex coacervation and hardening with formaldehyde treatment. The effects of coacervation pH and amount of crosslinking agent on micromeritic properties of the microcapsules were investigated. The particle size decreased as the pH increased and was larger for the formalized microcapsules. The highest core content occurred at pH 3.5. Formalized microcapsules also had a rough, netlike surface, whereas nonformalized ones were smooth. Appreciable residual formaldehyde levels were detected in formalized microcapsules, which could have important toxicological considerations. Conventionally dried microcapsules showed sulfamethoxazole in the core in its original crystalline form, whereas the use of spray drying caused the formation of folding and invaginations on the surface of the microcapsules and an amorphous appearance of the drug.

Electrophoretic properties of such microcapsules were reported in a subsequent paper [42] to elucidate the mechanism of encapsulation. Acacia appeared to be adsorbed onto the surface of core particles and formaldehyde reacted with gelatin, permitting release of acacia.

Complex gelatin-acacia coacervation was used by Nixon [43] to microencapsulate chlorothiazide, a diuretic. Tablets compressed from the microcapsules did not disintegrate during in vitro dissolution tests and were shown to be less effective, compared to free microcapsules, at enhancing urinary output in humans.

Indomethacin suspended in soya bean oil was microcapsulated in gelatin-acacia by Takeda et al. [44] using formaldehyde to harden the product. In vitro studies showed the microcapsules to have only a slight sustained-release effect compared to an unencapsulated control. When tested in dogs, the time for peak serum concentration of drug in both preparations was identical, but the microencapsulated product had a higher bioavailability, possibly because of enhanced drug absorption promoted by its contained oil phase.

3.5 OTHER ASPECTS OF MICROENCAPSULATION BY SIMPLE OR COMPLEX COACERVATION INVOLVING GELATIN AND ACACIA

A large number of patents assigned particularly to NCR for pressure-sensitive copying paper and to Fuji Photo Co. for photographic applications have been filed in the last 15 years. Many of these contain approaches that might be adaptable to the encapsulation of pharmaceuticals if the additives used do not represent an unacceptable toxicological hazard. A number of these patents, together with a small number of supportive papers published in journals, are briefly reviewed in this section under a number of subheadings.

3.5.1 Alteration of Colloid Charge

Taylor [45], in a patent assigned to Polaroid Corporation, described the preparation of gelatin derivatives containing excess carboxylic acid groups that permitted simple coacervation to be carried out by reduction in pH to approximately 3.6, followed by temperature decrease and hardening with glutaraldehyde.

Complex coacervation generally depends largely on charge neutralization at intermediary pH where both gelatin and acacia have appreciable opposite charge. Accordingly, a gelatin of high isoelectric point (over 8.0) is required so that at the pH of 4.0 to 4.5 of coacervation it is markedly positively charged and interacts strongly with the negatively charged acacia. To enhance this interaction Matsukawa and Saeki [46], in a patent assigned to Fuji Photo Co., used a sodium salt of naphthalenesulfonic acid/formalin condensate to increase the negative charge associated with the acacia, thereby aiding reaction with positive gelatin. The advantage of the process was that higher concentrations of colloids interacted to give thicker walls with reduced porosity. Coacervation also was induced at a lower temperature. As an example, lemon oil was encapsulated using the process. Likewise Taylor [47], in a patent assigned to Polaroid Corporation, replaced the positive colloid gelatin of high isoelectric point with a negative colloid that was a dicarboxylic acid anhydride of gelatin and still got a system in which a coacervate formed. The water-insoluble core material was encapsulated in a mixture of low-isoelectric-point gelatin and the gelatin derivative, both of which had negative charges though of different magnitude. The carboxylated gelatin derivative reacted with a portion of the amino groups of the gelatin to partially neutralize the positive charges associated with the latter. Capsule size could be controlled for the encapsulation of a nonpolar liquid by varying the droplet size using suitable agitation conditions.

3.5.2 Shock Prevention

Shock is the term used for the phenomenon that occurs prior to the hardening treatment of the coacervate-coated core material when the apparent viscosity of the system rises rapidly as the pH is adjusted near the isoelectric point of the gelatin used. The pH of the system is raised through the isoelectric point of the gelatin to make it alkaline to accelerate the subsequent reaction of the gelatin and aldehyde hardening agent. The effect becomes more evident as the temperature is dropped to promote gelation of the coating. Gelatins have higher apparent viscosities close to their isoelectric points because their linear molecules can interact to form a metastable gel more easily in the absence of repulsive charge. The rapid rise in apparent viscosity causes undesirable cohesion and aggregation of the microcapsules and may be lessened by using a shock-preventing agent. These agents are usually polyelectrolytes having anionic functional groups that when added in low concentration interact with the gelatin to lessen rise in viscosity by imparting charge repulsion to neighboring gelatin molecules. The material is usually added at a temperature lower than the gelation point of the gelatin used. Its usage is similar to that of employing a peptizing agent to control rise in apparent viscosity and degree of thixotropy in linear hydrophilic colloids used in the formulation of pharmaceutical gels and suspensions.

Katayama et al. [48], in a patent assigned to Fuji Photo Co., used sodium carboxymethylcellulose, sodium carboxymethylstarch, pectic acid and other materials as shock-preventing agents to obtain mononuclear capsules using gelatin-acacia complex coacervation. In a subsequent patent from the same company, Matsukawa et al. [49] used a phthalic anhydride derivative of gelatin as a shock-preventing agent and obtained capsules with enhanced heat resistance. In another patent assigned to Fugi by Matsukawa et al. [50], the use of acrylic acid-methacrylic acid copolymers as shock-preventing agents was described. Use of resorcinol or phenolic resin in association with sodium carboxymethylcellulose was shown by Matsukawa and Saeki [51, 52] in two other Fuji patents to further improve shock prevention, allowing microencapsulation to be conducted using higher concentrations of gelatin and acacia. The thicker and less permeable walls produced on the microcapsules gave the product superior heat resistance. In further patents assigned to the same company, the use of gelatin derivatives [53] and anionic surfactants [54] as shock-preventing agents was described.

3.5.3 Further Methods of Reducing Aggregation by Use of Cationic Surfactants or Inert Filters

Aggregation or clustering of microcapsules during preparation is a troublesome problem commonly encountered in simple and complex coacervation. Apart from the use of shock-preventing agents, a number of other approaches have been used to help overcome the problem. Maierson [55], in a patent assigned to NCR, used cationic surfactants to lessen aggregation during the isolation and drying stages of complex coacervation. The surfactant is strongly adsorbed at the interface between the coating and its surrounding aqueous medium, reducing the interfacial tension and leading to reduced tendency of the particles to aggregate. Also, by imparting a positive charge to the microcapsules, electrical repulsion between neighboring capsules should lessen aggregation. Surfactants used included primary fatty acid amine derivatives of tallow and various vegetables oils and fats.

Another approach to the reduction of clustering involved the use of an inert filter such as silica dioxide, as suggested by Maierson and Crainich [56] in a patent assigned to NCR. The filler separates the microcapsules and prevents cohesion, particularly during the separation and drying stages.

3.5.4 Wall-Hardening Agents

Many different agents are used to contract, harden, and render less water-soluble gelatin-containing walls of microcapsules formed by simple or complex coacervation. Crosslinking by agents that react with gelatin is synonymous with tanning, and the concentration of gelatin is very important because the distance between neighboring molecules will determine whether linkage will take place between molecules (intermolecular) or within individual molecules (intramolecular). Intermolecular linkage is desired and is favored in the coacervated wall material because of the high concentration of polymer present. It may be further aided by using a dehydrating agent such as ethanol or isopropanol to withdraw water from the coacervate and reduce intermolecular distance between protein molecules. However, the tighter structure produced in the gel makes it difficult for hardening (crosslinking) agents of large molecular weight to penetrate effectively. The most frequently used hardening agents for crosslinking gelatin-coated microcapsules are formaldehyde and glutaraldehyde, both of which have small molecular weights, which aids their penetration.

The low concentration of gelatin in the equilibrium colloid-poor layer is not appreciably crosslinked by the added hardening agent and consequently there is very little rise in viscosity of the manufacturing vehicle.

As early as 1945, Fox and Opferman [57] patented a process involving formalin treatment of hard gelatin capsule shells for filling with powders. The treated capsules had enteric properties, releasing their content in the small intestine and not in the stomach.

Tanaka et al. [58] investigated the effect of 10% formalin in isopropanol on the properties of gelatin micropellets containing sulfanilamide and riboflavin. Micropellets were hardened at 2 to 5°C for up to 3 days and showed prolonged release properties in simulated gastric juice containing protease. In vivo studies in dogs and humans confirmed these findings and indicated that by extending the duration of hardening treatment, a proportional increase in prolonged action effect of riboflavin in humans was obtained. This is surprising in view of the fact that the drug is reported to be absorbed mainly in the upper small intestine and is not a good candidate for a sustained-release product (see Chap. 1).

Glassman [59] reported that exposure of gelatin to formaldehyde gas to make it acid-resistant has often resulted in the formation of a crosslinked gelatin that is completely insoluble in gastric juice. Residual amounts of formaldehyde can continue to react with the gelatin, rendering a coating composed of it completely insoluble throughout the gastrointestinal tract, with failure to release the core material. In the patent described, the depth of hardening was controlled by immersion in dilute aqueous or alcoholic formaldehyde solution, interspaced by immersion in a dehydrating agent such as absolute alcohol for short periods of time before being dried in warm air. Water aids the penetration of the formaldehyde into the coating, and immersion in the dehydrating agent extracts free water and formaldehyde from the wall to suppress further hardening. Subsequent immersion in aqueous-alcoholic formaldehyde solution promotes further controlled hardening, which again may be terminated using the dehydrating agent.

Clarke and Courts [60] reviewed the mechanism of reaction of gelatin with formaldehyde. Initially, formaldehyde reacts with the amino groups of the gelatin as follows:

$$\mathrm{HCHO} + \text{gelatin}\text{—}\mathrm{N^{+}H_3}\text{—} \rightleftarrows \text{gelatin}\text{—}\mathrm{NH}\text{—}\mathrm{CH_2OH} + \mathrm{H^{+}}$$

In adequately high concentration (proximity), two molecules of the product can react to crosslink by means of a dimethylene ether bridge or a methylene bridge as follows:

$$\text{gelatin}\text{—}\mathrm{NH}\text{—}\mathrm{CH_2OH} \;+\; \mathrm{HOCH_2}\text{—}\mathrm{NH}\text{—}\text{gelatin}$$

$$\searrow \quad \text{gelatin}\text{—}\mathrm{NH}\text{—}\mathrm{CH_2}\text{—}\mathrm{NH}\text{—}\text{gelatin}$$

$$\swarrow \quad \text{gelatin}\text{—}\mathrm{NH}\text{—}\mathrm{CH_2}\text{—}\mathrm{O}\text{—}\mathrm{CH_2}\text{—}\mathrm{NH}\text{—}\text{gelatin}$$

Shimosaka and Suzuki [61] described the use of sodium thiocyanate solution to accelerate the hardening produced by formaldehyde in a gelatin-acacia system. Obviously, because of the toxicity of thiocyanate, the process would have limited use in pharmaceuticals, though its suitability for perfumes, deodorants, and other products was discussed. Saeki and Matsukawa [62], in a patent assigned to Fuji, used a combination of a dialdehyde, such as glutaraldehyde or glyoxal, and formaldehyde to improve the hardening of microcapsules prepared by complex coacervation. Hardening could be hastened by heating at 40 to 60°C for a short period, rather than by treating at room temperature for a long period. The hardened capsules tended not to yellow upon storage, not to aggregate, and to have enhanced heat resistance. Also Egawa et al. [63], in a patent assigned to NCR, used a combination of a slow-acting hardening agent such as formaldehyde or glyoxal and a quick-acting one such as glutaraldehyde or 2-methylglutaraldehyde to form improved mononuclear gelatin-coated microcapsules.

Glutaraldehyde is considered a better crosslinking agent for protein than formaldehyde. Aqueous solutions of glutaraldehyde are largely polymeric and form unsaturated aldehydes, which, as reported by Richards and Knowles [64], probably crosslink protein rapidly as follows:

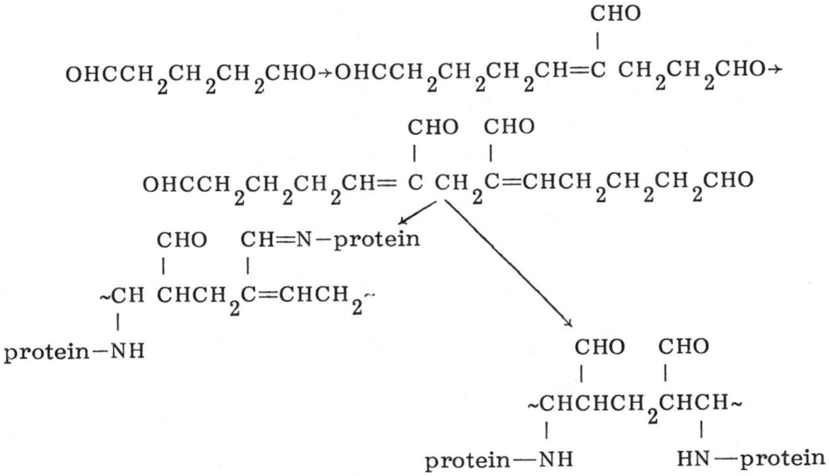

Recently Takenaka et al. [65] investigated the effect of amount of formaldehyde as hardening agent on the in vitro release of sulfamethoxazole from microcapsules prepared by gelatin-acacia complex coacervation. They showed that increasing formalin treatment and the pH of coacervation from 2.5 to 4.0 both tended to decrease the dissolution rate of the drug from the microcapsules in water, pH 1.2 solution, or pH 7.5 solution. Reduction in the pH of coacervation was associated with reduction in wall thickness, though the effect of variable formaldehyde treatment was more pronounced. The possibility of formaldehyde forming a soluble complex with the drug, which enhanced drug release when mild formalin treatment was used, was discussed. Increasing tortuosity was found to occur when an increasing amount of hardening agent was used. There are many other hardening or cross-linking agents used for gelatin, and Clarke and Courts [60] review a wide range of these materials.

3.5.5 Enahncement of the Speed of Encapsulation

Schnoring and Schon [66], in a patent assigned to Bayer AG, described how the process of complex coacervation could be speeded up by passing solutions of the polymers for purification through a bed of mixed ion-exchange resins at elevated temperature. Cations and anions contaminating the polymers were exchanged for H^+ or OH^- ions, and coacervation occurred more rapidly when the resultant gelatin and acacia solutions were mixed.

3.5.6 Colored, Pearlized, or Dual-Walled Microcapsules

Sirine [67], in a patent assigned to NCR, described how capsule walls could be colored by incorporating carbon black so as to protect an encapsulated oil from decomposition by sunlight. Overincorporation of carbon black into the walls led to poor retention of the encapsulated material. Likewise Marinelli [68], in another patent assigned to NCR, described how to incorporate mica particles coated with titanium dioxide into the walls of gelatin-acacia microcapsules to give them a pearlized appearance. The capsules were considered useful for the formulation of shampoos and other toiletries to enhance esthetic appearance.

Baxter [69], in a patent assigned to Moore Business Forms, described the preparation of dual-walled capsules composed of a thin inner gelatin wall surrounded by a thicker gelatin-acacia wall. An acid chloride was initially dispersed in the oil phase to be encapsulated and reacted with the inner coating of deposited gelatin. The acid liberated lowered the pH, whereby complex gelatin-acacia coacervation took place to cause formation of the outer layer of coating.

3.5.7 Core Exchange

As it is not possible to encapsulate polar liquids or solids in gelatin wall containing microcapsules prepared by simple or complex coacervation, a number of techniques have been developed for exchanging the nonpolar core material for a polar one after the capsule walls have been formed but prior to the hardening step. Brynko and Olderman [70] and Striley and Williams [71], in two patents assigned to NCR, described how to replace nonpolar liquid encapsulated material by polar liquid material. The replacement was facilitated by using a mutual solvent such as ethanol or dioxane of dielectric constant preferably less than 20. The solvent aided diffusion of the exchanging materials. Repeated replenishment of the outer polar phase ensured that the final equilibrium resulted in exchange of almost all the nonpolar core material. After the exchange process had been completed, the capsule walls could be dehydrated to facilitate sealing by immersion in excess anhydrous ethanol or anhydrous dioxane. Alternatively, they could be treated with concentrated aqueous solutions of mono- or disaccharide sugars, which allowed only partial hydration of the wall. This prevented dilution of the core material by inward diffusion of water, as the buildup of the polyhydric sugar tended to seal the walls. A list of suitable solvents with low dielectric constants and other details of the process were given by Kondo [8]. Recently Jalsenjak and Kondo [72] exchanged an olive oil core in gelatin-acacia hardened microcapsules for acetone, which in turn was replaced by water.

3.6 OTHER WALL-FORMING POLYMERS

Though simple and complex coacervation generally involve the use of gelatin and gelatin-acacia, respectively, various other polymers may be used for microencapsulation by phase separation from an aqueous system. The original patent of Green and Schleicher [2] also recommended albumin, agar, alginates, carboxymethylcellulose, casein, pectins, and starch. Any ionizable hydrophilic colloid can be used provided that it coacervates upon dilution, pH change, temperature change, solvent addition, or electrolyte addition.

3.6.1 Sodium Silicate or Sodium Polymetaphosphate

Hörger [73], in a patent assigned to NCR, described the use of an inorganic polysilicate such as sodium silicate or water-glass as the negative component and gelatin, polyvinylpyrrolidone, or albumin as the positive component to cause complex coacervation. The microcapsules obtained had hard, glassy walls that were relatively insensitive to water. In a subsequent patent assigned to the same company, Fogle and Hörger [74] outlined the use of sodium polymetaphosphate as the negative component in conjunction with the above-mentioned positively charged polymers. Unlike the previous patent, chemical hardening with glutaraldehyde was unnecessary.

3.6.2 Methylcellulose

Methylcellulose is slowly soluble in cold water but insoluble in hot water. Therefore this hydrophilic coating material has the property of decreasing in solubility as the temperature is increased. Bakan [75], in a patent assigned to NCR, used a dilute aqueous solution of this polymer to encapsulate lemon oil, menthol, aspirin, and various other drugs. He induced phase separation of the polymer so that the coacervate droplets formed the coating on the dispersed core material by two procedures. First he added a complementary, more hydrophilic polymer such as dextran, polyvinyl alcohol, or polyvinylpyrrolidone to help dehydrate the methylcellulose; second, the temperature of the agitated system was raised to about 60°C. In order to prevent dissolution of the coating into the manufacturing vehicle as the temperature was decreased again, a melamine formaldehyde resin was added at elevated temperature to react with and render the methylcellulose coating water-insoluble.

In a subsequent patent by Himmel [76], also assigned to NCR, a low concentration of a second hydrophilic polymer such as carboxymethylcellulose, sodium carboxymethylcellulose, or acacia was added to the dilute aqueous solution of the methylcellulose at room temperature together with a dispersion of the intended core material maintained by constant agitation. Phase separation to cause formation of droplets of a complex coating coacervate of the two polymers was induced by raising the temperature to 60°C and adding a concentrated electrolyte solution of sodium chloride/sodium sulfate or ammonium chloride/ammonium sulfate to further aid dehydration. The microcapsules were recovered by decanting the supernatant or by filtration at elevated temperature followed by hot air drying to remove residual moisture. A modification of this approach was used by Salib et al. [77] to prepare methylcellulose micropellets containing sulfadiazine.

3.6.3 Cellulose Acetate Phthalate

Cellulose acetate phthalate (CAP) is probably the most widely used enteric polymer for the coating of pharmaceuticals. The polymer is insoluble in water and acidic solutions but dissolves in buffer solutions of pH 5.6 to 6.0 and above. Merkle and Speiser [78] dissolved CAP in an equivalent concentration of an aqueous disodium hydrogen phosphate solution. Sodium sulfate together with temperature reduction was used to induce coacervation-phase separation of the CAP as a coating on the dispersed water-insoluble drug, phenacetin. A three-phase diagram was constructed to develop the optimum encapsulation conditions. Increasing the core:coat ratio increased drug release into 0.1 *M* hydrochloric acid, which indicated that a polymer diffusion-controlled mechanism of release operated. When the coating was plasticized by inbibing glycerol, the release rate appeared to be controlled by drug dissolution in the microcapsules.

3.6.4 Poly(Methyl Vinyl Ether/Maleic Anhydride)

Poly(methyl vinyl ether/maleic anhydride) (PVM/MA) may be used in part or in whole as a substitute for acacia as the negative polymer for use in conjunction with gelatin (Bakan [79] and Yurkowitz [80], both patents assigned to NCR). PVM/MA is reported to be relatively nontoxic [81] and has been suggested for use as a granulating agent and film former for compressed tablets.

The *n*-butyl half-ester of PVM/MA was used by Mortada [82] to encapsulate phenacetin using a simple coacervation procedure whereby the polymer was dissolved in sodium acetate solution and phase-separated onto suspended core particles by addition of sodium sulfate. A three-phase diagram showing optimum coacervation conditions was presented. A certain amount of aggregation of the microcapsules formed was noted, and the product showed sustained-release properties when tested in vitro.

3.6.5 Acrylates

Schon et al. [83], in a patent assigned to Bayer, described an aqueous-phase separation procedure that initially involved the copolymerization of the monomers acrylic acid and acrylamide (approximately 50:50 by weight) using a redox catalyst and by mixing in water under nitrogen at an elevated temperature of 40°C for about 2 hr. Water-insoluble core material such as liquid paraffin was then dispersed as droplets in the copolymer solution formed at the elevated temperature using a high-speed stirrer. Provided that the pH of the copolymer solution was adjusted if necessary to below 3.5, phase separation of the copolymer occurred as the temperature of the system was decreased below 30°C to form a coating around the core droplets. In order to rigidize the coating formed, methanol, ethanol, or acetone can be added to dehydrate the coacervate shell followed by crosslinking of the coating using polyvalent metals ions (as their salts, e.g., aluminium sulfate), aldehydes, or other agents.

Matsukawa and Saeki [84], in a patent assigned to Fuji, described the use of sodium polyacrylate as a negative polymer additive to a gelatin-acacia system to cause coacervation. Microcapsules with thicker and less porous walls were obtained.

Sternberg et al. [85], in a patent assigned to Amicon Corp., described an encapsulation procedure whereby a core material was dissolved or dispersed in a solution of a polymer in an organic solvent. To effect phase separation of the polymer as a coating, the system was passed through a syringe needle to form droplets that entered a vehicle that was a cosolvent for the organic solvent but a nonsolvent for the polymer. A large number of polymer solvent systems were listed in the patent as being suitable for the process, including various acrylic copolymers. The exterior of the particles formed was covered by a relatively dense, semipermeable layer of polymer 0.1 to 5 μm thick overlaying a much more porous polymer interior. An example of the process involved encapsulation of the enzyme catalase by first dissolving it in water to which was then added a large excess of dimethylformamide. Acrylonitrile-vinyl chloride copolymer was then dissolved in this and the resultant system was added dropwise into an excess of agitated water containing a surfactant. Spherical particles 2 to 3 mm in diameter were formed in which the enzyme was active but immobilized.

3.6.6 Carboxyvinyl Polymer

Elgindy and Elegakey [86] used a carboxyvinyl polymer (Carbopol) as the negative hydrophilic colloid in association with gelatin to produce complex coacervation. Carbopol 941 and a gelatin of high IEP 8.4 in concentrations of 0.1 and 1.0%, respectively, were found to produce optimum coacervation. Almost uniform spherical coacervate droplets of mean diameter 59 μm were obtained using a stirrer speed of 300 to 350 rpm. The encapsulation of no drug was described in this preliminary report.

3.7 SOLVENT EVAPORATION PROCESS

A range of different microspheres have been prepared by dissolving or dispersing drug in a solution of a polymer in a single or mixed organic solvent having a low boiling point. The phase is then emulsified into a continuous aqueous phase containing a low concentration of hydrophilic colloid or surfactant to stabilize the oil-in-water (o/w) emulsion formed. Reduced pressure and/or heat is then often applied to the emulsion while stirring to evaporate off the organic solvent, and the microspheres formed are collected by filtration or centrifugation. The process is the subject of a patent by Morishita et al. [87] assigned to Toyo Jozo Co. Ltd., Japan. Mortada [88] dispersed sulfathiazole in a solution of ethylcellulose in trichloromethane. This was then emulsified into 0.04% sodium lauryl sulfate solution and the organic solvent evaporated by stirring at room temperature for 5 hr to form drug-containing ethylcellulose microspheres. A similar approach was also used by Wakiyama et al. [89, 90] to encapsulate butamben, tetracaine, or dibucaine. The drug and polylactic acid were dissolved in methylene chloride, methyl acetate, or ethyl acetate and emulsified into 1% gelatin or 1% sodium alginate solution in water. The organic solvent was evaporated and the drug-polymer microspheres formed by phase separation were recovered.

In order to produce microcapsules with distinct walls, a multiple water-in-oil-in-water (w/o/w) emulsion may be formed whereby the inner water phase contains the drug and the oil phase contains the polymer in a volatile solvent that is evaporated during the manufacturing process. Such an approach has been used to produce polystyrene microcapsules containing placental alkaline phosphatase by Takenaka et al. [91] and α-amylase or sodium salicylate by Nozawa and Higashida [92]. Polymethyl methacrylate microcapsules of methylglyoxal and derivatives were likewise prepared by Nozawa and Fox [93]. Using a solution of ethylcellulose in dichloromethane, Tateno et al. [94] similarly prepared ethylcellulose microcapsules containing aqueous solutions of polyelectrolytes.

In an interesting series of patents [95–97] assigned to Gaevert Photo Producten NV, aqueous solutions or dispersions of drugs were emulsified into hydrophobic polymer material dissolved in a water-immiscible solvent having a boiling point below 100°C, e.g., polystyrene in methylene chloride. An initial w/o emulsion was formed and this emulsion was in turn emulsified into an aqueous solution of a hydrophilic colloid (e.g., gelatin or a cellulose derivative) to form a w/o/w emulsion. By raising the temperature the organic solvent evaporated, causing phase separation of the polymer as a coating around the inner aqueous droplets. Alternatively, the organic solvent could be removed by adding a liquid that is miscible with both the solvent and the water but that is a nonsolvent for the polymer and core material. Various other polymers, such as acrylates, cellulose derivatives, and polyamides, could also be used in association with suitable organic solvents. Loss of water-soluble core substances tended to occur during encapsulation by their migration from the inner water droplets across the organic solvent layer and into the outer continuous water phase before the polymer coating was properly deposited. To prevent this problem occurring, a saturated concentration of the intended core material could be dissolved in the outer aqueous phase, provided that it does not precipitate the dissolved hydrophilic colloid present.

A similar approach was used by Morris and Warburton [98–100] to form three-walled microcapsules. An aqueous acacia and drug solution was emulsified into a solution of polychloroprene (a rubber) in xylene to form a w/o emulsion. This emulsion was then reemulsified into further aqueous acacia solution to form a w/o/w emulsion. The xylene was then evaporated by bubbling air through the multiple emulsion to give an acacia-polychloroprene-acacia three-ply coating around aqueous cores. To get a more pharmaceutically acceptable intermediary ply, a solution of ethylcellulose in ethyl acetate could be used as the oil phase. Since ethyl acetate is soluble in water, it could be removed by dialysis of the microcapsules. The microcapsules could be recovered as a free-flowing powder by evaporation of water from their concentration suspension.

3.8 GELATIN NANOPARTICLES

Considerable interest has been shown in recent years in the development of solid colloidal drug delivery systems whose particle size is measurable in the nanometer range (nm = mμm). These so-called nanoparticles and nanocapsules are so small that their suspensions in water appear clear or slightly opalescent and they do not sediment because of Brownian motion. Such suspensions are readily injected without producing obstruction through the 21-gauge hypodermic needle normally used. Also the small particle size is desirable in parenteral products to minimize adverse tissue reactions at the site of injection [101, 102]. These carriers have been investigated for the inclusion of various materials such as enzymes, gamma globulins, and cytotoxic agents. A number of reviews on the subject of nanoparticles and nanocapsules have been published [103–106]. In this chapter only the use of gelatin as the encapsulating polymer will be considered; parts of other chapters, notably Chaps. 9 and 10, will review the use of other polymers for this purpose.

The production of biodegradable gelatin nanoparticles involves a modification of the simple coacervation procedure previously discussed in this chapter, usually using either sodium sulfate or ethanol as coacervating agent. To a dilute aqueous solution of gelatin (1 to 3% w/v) containing a suitable concentration of surfactant such as polysorbate

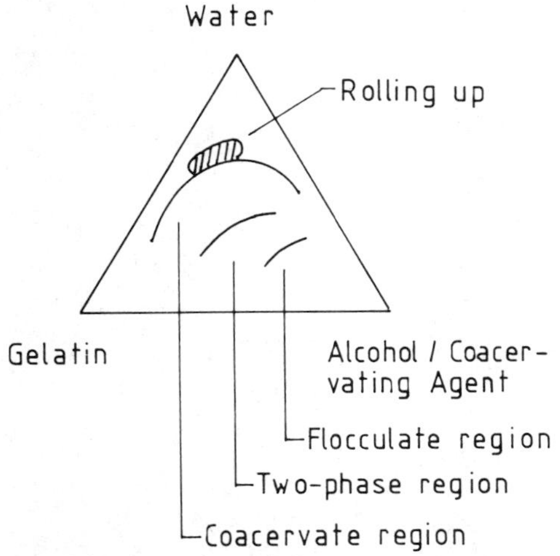

Figure 3.8 Three-phase diagram for water, gelatin, and coacervating agent. (From Ref. 104.)

20 is gradually added the coacervating agent with constant stirring at about 35°C and with appropriate pH control, until turbidity indicative of coacervation is observed. The three-phase diagram is shown diagramatically in Fig. 3.8, indicating that continued addition of coacervating agent will produce a two-phase region and finally flocculation. However, if the coacervate is partly resolvated again by addition of, for example, isopropanol until the turbidity just disappears, but corresponding to a predetermined intensity of light scatter as measured in a suitable instrument, the ternary system produced corresponds with the rolling-up region shown in Fig. 3.8.

It may not be necessary to employ this overshoot procedure involving addition of resolvating agent if optimum conditions have been established previously. In dilute aqueous solution gelatin molecules are in an expanded linear form either singly or in small aggregates as shown diagramatically in Fig. 3.9a. When partially solvated the molecules decrease their solubility by rolling up as shown in Fig. 3.9b. To prevent the rolled-up gelatin moelcules from aggregating together, a laboratory homogenizer is introduced into the system to provide vigorous agitation.

The rolled-up molecules have a size of about 100 to 200 nm and can entrap drug molecules or other substances introduced into the system. A hardening agent, preferably glutaraldehyde, is added and sodium sulfite or sodium metasulfite is then added after a suitable interval to quench the hardening reaction while continuing to maintain vigorous agitation to prevent aggregation.

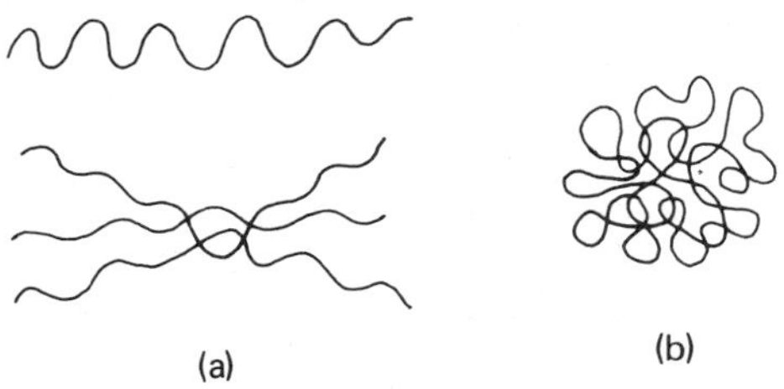

Figure 3.9 Gelatin molecules in (a) aqueous solution, single or small aggregates; (b) rolling up region.

A gel chromatography clean-up procedure can be used to separate low-molecular-weight impurities, unreacted drug, and surfactant from the product. This may then be freeze-dried, diluting with water if there is a high concentration of ethanol present so as to prevent melting during the initial stages of drying. The active material may be incorporated either before or after addition of coacervating agent to form a true or solubilized system. Active materials of low aqueous solubility, including those solubilized and those that bind strongly with gelatin, are better retained in the nanoparticles as they are less readily removed by washing during the clean-up step. Scanning electron micrographs show the product to be composed of roughly spherical-shaped particles having no distinct core or coating and having a size in the range 50 to 500 nm. Hence the term nanoparticle rather then nanocapsule is appropriate to describe the product. The nanoparticles can be reconstituted from the freeze-dried form by addition of water and have been sterilized by autoclaving at 121°C for 15 min without any gross change being observed. Further details of the production and possible applications of these nanoparticles are contained in a paper by Marty et al. [107], who indicated that other macromolecules such as albumin, zein, and chemically modified celluloses could be similarly used. The process is also the subject of a patent [108].

However, there are two major problems associated with gelatin nanoparticles. The first is the risk of antigenic reactions following injection, though none have been observed after single or repeated injection into experimental animals. The second is rapid clearance after intravenous (IV) injection by the reticuloendothelial system, mainly into the liver and to a lesser extent the spleen of experimental animals, which which is undesirable unless therapy is specifically required in these organs. Oppenheim et al. [109] reported this finding using ^{99m}Tc-labeled gelatin nanoparticles injected intravenously into rodents. However, as some tumor cells are also actively phagocytic, these nanoparticles containing cytotoxic drugs could be useful for the treatment of certain cancers that often show resistance to the uptake of free drug.

The feasibility of this approach was shown by Oppenheim and Stewart [110], who labeled gelatin nanoparticles with fluorescein isothiocyanate bound to their surface. The nanoparticles, unlike the free dye, were phagocytized by certain experimental tumor lines, though labeled albumin nanoparticles, whose preparation was described, were not (see Chap. 10 for discussion of albumin nanoparticles).

Spherical gelatin nanocapsules prepared by a congealable emulsification procedure have also been produced as described in Chap. 12.

REFERENCES

1. B. K. Green and L. Schleicher, U.S. Patent 2,730,456 (January 10, 1956).
2. B. K. Green and L. Schleicher, U.S. Patent 2,730,457 (January 10, 1956).
3. B. K. Green and L. Schleicher, U.S. Patent 2,800,457 (July 23, 1957).
4. L. A. Luzzi, Microencapsulation, *J. Pharm. Sci. 59*:1367–1376 (1970).
5. J. E. Vandegaer, Encapsulation by coacervation, in *Microencapsulation—Processes and Applications* (J. E. Vandegaer, ed.), Plenum, New York, 1974, pp. 21–37.
6. J. A. Bakan and J. L. Anderson, Microencapsulation, in *The Theory and Practice of Industrial Pharmacy*, 2nd ed. (L. Lachman, H. A. Lieberman, and J. L. Kanig, eds.), Lea & Febiger, Philadelphia, 1976, pp. 420–438.
7. P. L. Madan, Microencapsulation 1. Phase separation or coacervation, *Drug Develop. Indust. Pharm. 4*:95–116 (1978).
8. A. Kondo, *Microcapsule Processing and Technology*, (J. Wade Van Valkenburg, ed.), Dekker, New York, 1979, pp. 70-94.
9. J. A. Bakan, Microencapsulation using coacervation/phase separation techniques, in *Controlled Release Technologies: Methods, Theory and Applications*, Vol. 1 (A. F. Kydonieus, ed.), CRC Press, Boca Raton, Fla., 1980, pp. 83–105.
10. J. A. Bakan, Microcapsule drug delivery systems, in *Polymers in Medicine and Surgery* (R. L. Kronenthal, Z. Oser, and E. Martin, eds.), Plenum, New York, 1975, pp. 213–235.
11. H. G. Bungenberg de Jong and H. R. Kruyt, Chemistry—coacervation (partial miscibility in colloid systems) (preliminary communication), *Proc. Kon. Ned. Akad, Wetensch. 32*:849–856 (1929).
12. H. G. Bungenberg de Jong, in *Colloid Science*, Vol. 2 (H. R. Kruyt, ed.), Elsevier, Amsterdam, 1949, pp. 248–487.
13. J. E. Eastoe and A. A. Leach, Chemical constitution of gelatin, in *The Science and Technology of Gelatin* (A. G. Ward and A. Courts, eds.), Academic, London, 1977, pp. 73–107.

14. R. E. Phares and G. J. Sperandio, Preparation of a phase diagram for coacervation, *J. Pharm. Sci.* *53*:518–521 (1964).
15. J. R. Nixon, S. A. H. Khalil, and J. E. Carless, Phase relationships in the simple coacervation system isoelectric gelatin:ethanol:water, *J. Pharm. Pharmacol.* *18*:409–416 (1966).
16. R. E. Phares and G. J. Sperandio, Coating pharmaceuticals by coacervation, *J. Pharm. Sci.* *53*:515–518 (1964).
17. S. A. H. Khalil, J. R. Nixon, and J. E. Carless, Role of pH in the coacervation of the systems: gelatin-water-ethanol and gelatin-water-sodium sulphate, *J. Pharm. Pharmacol.* *20*:215–225 (1968).
18. J. R. Nixon, S. A. Khalil, and J. R. Carless, Gelatin coacervate microcapsules containing sulphamerazine: their preparation and the in vitro release of the drug, *J. Pharm. Pharmacol.* *20*:528–538 (1968).
19. G. N. Paradissis and E. L. Parrott, Gelatin encapsulation of pharmaceuticals, *J. Clin. Pharmacol.* *8*:54–59 (1968).
20. R. M. Navari, J. L. Gainer, and O. L. Updike, Blood flow modeling with microcapsular suspensions, *Ind. Eng. Chem. Fund.* *8*:615–620 (1969).
21. J. R. Nixon and S. E. Walker, The in vitro evaluation of gelatin coacervate microcapsules, *J. Pharm. Pharmacol.* *23*:147S–155S (1971).
22. B. R. Matthews and J. R. Nixon, Surface characteristics of gelatin microcapsules by scanning electron microscopy, *J. Pharm. Pharmacol.* *26*:383–384 (1974).
23. J. R. Nixon and B. R. Matthews, The surface characteristics of gelatin coacervate microcapsules by scanning electron microscopy, in *Microencapsulation* (J. R. Nixon, ed.), Dekker, New York, 1976, pp. 173–183.
24. P. L. Madan, Clofibrate microcapsules, in *Microencapsulation: New Techniques and Applications* (T. Kondo, ed.), Techno, Inc., Tokyo, 1979, pp. 11–34.
25. P. L. Madan, Clofibrate microcapsules II: Effect of wall thickness on release characteristics, *J. Pharm. Sci.* *70*:430–433 (1981).
26. L. A. Luzzi and R. J. Gerraughty, Effects of selected variables on the extractability of oils from coacervate capsules, *J. Pharm. Sci.* 53:429–431 (1964).
27. L. A. Luzzi and R. J. Gerraughty, Effects of selected variables on the microencapsulation of solids, *J. Pharm. Sci.* *56*:634–638 (1967).
28. L. A. Luzzi and R. J. Gerraughty, Effect of additives and formulation techniques on controlled release of drugs from microcapsules, *J. Pharm. Sci.* *56*:1174–1177 (1967).
29. L. D. Morse, U.S. Patent 3,557,279 (January 19, 1971).

30. S. Javidan, R. Haque, and R. G. Mrteck, Microbiological determination of drug partitioning 1: gelatin-acacia complex coacervate system, *J. Pharm. Sci. 60*:1825–1829 (1971).
31. P. L. Madan, L. A. Luzzi, and J. C. Price, Factors influencing microencapsulation of a waxy solid by complex coacervation, *J. Pharm. Sci. 61*:1586–1588 (1972).
32. P. L. Madan, L. A. Luzzi, and J. C. Price, Microencapsulation of a waxy solid: wall thickness and surface appearance studies, *J. Pharm. Sci. 63*:280–284 (1974).
33. P. L. Madan, J. C. Price, and L. A. Luzzi, Encapsulation of spherical particles as a characteristic of gelatin fractions, in *Microencapsulation—Processes and Applications* (J. E. Vandegaer, ed.), Plenum, New York, 1974, pp. 39–56.
34. D. W. Newton, J. N. McMullen, and C. H. Becker, Characteristics of medicated and unmedicated microglobules recovered from complex coacervates of gelatin-acacia, *J. Pharm. Sci. 66*:1327–1330 (1977).
35. A. Palmieri, Production of a coacervate film for microcapsule diffusion studies, *Drug Develop. Indust. Pharm. 3*:309–314 (1977).
36. A. Palmieri, Dissolution of prednisone microcapsules in conditions simulating the pH of the gastrointestinal tract, *Can. J. Pharm. Sci. 12*:88–89 (1977).
37. A. Palmieri, Microencapsulation and dissolution parameters of undecenvanillylamide: a potential coyote deterrent, *J. Pharm. Sci. 68*:1561–1562 (1979).
38. J. R. Nixon and M. Hassan, The effect of preparative technique on the particle size of thiobendazole microcapsules, *J. Pharm. Pharmacol. 32*:856–857 (1980).
39. J. R. Nixon and M. Hassan, The effect of tableting on the dissolution behaviour of the thiobendazole microcapsules, *J. Pharm. Pharmacol. 32*:857–859 (1980).
40. J. R. Nixon and G. A. Agyilirah, The influence of colloidal proportions on the release of phenobarbitone from microcapsules, *Int. J. Pharm. 6*:277–283 (1980).
41. H. Takenaka, Y. Kamashima, and S. Y. Lin, Micromeritic properties of sulfamethoxazole microcapsules prepared by gelatin-acacia coacervation, *J. Pharm. Sci. 69*:513–516 (1980).
42. H. Takenaka, Y. Kawashima, and S. Y. Lin, Electrophoretic properties of sulfamethoxazole microcapsules and gelatin-acacia coacervates, *J. Pharm. Sci. 70*:302–305 (1981).
43. J. R. Nixon, In vitro and in vivo release of microencapsulated chlorothiazide, *J. Pharm. Sci. 70*:376–378 (1981).
44. Y. Takeda, N. Nambu, and T. Nagai, Microencapsulation and bioavailability in beagle dogs of indomethacin, *Chem. Pharm. Bull. 29*:264–267 (1981).

45. L. D. Taylor, U.S. Patent 3,369,900 (February 20, 1968).
46. H. Matsukawa and K. Saeki, U.S. Patent 3,869,406 (March 4, 1975).
47. L. D. Taylor, U.S. Patent 3,702,829 (November 14, 1972).
48. S. Katayama, H. Matsukawa, M. Yamamoto, and J. Matsuyama, U.S. Patent 3,687,865 (August 29, 1972).
49. H. Matsukawa, K. Saeki, and T. Shimada, U.S. Patent 3,769,231 (October 30, 1973).
50. H. Matsukawa, K. Saeki, and S. Katayama, U.S. Patent 3,789,015 (January 29, 1974).
51. H. Matsukawa and K. Saeki, U.S. Patent 3,803,045 (April 9, 1974).
52. H. Matsukawa and K. Saeki, U.S. Patent 3,965,033 (June 22, 1976).
53. H. Matsukawa, K. Saeki, and T. Shimada, U.S. Patent 3,944,502 (March 16, 1976).
54. H. Matsukawa, S. Katayama, and M. Kiritani, U.S. Patent 3,970,585 (July 20, 1976).
55. T. Maierson, U.S. Patent 3,436,452 (April 1, 1969).
56. T. Maierson and V. A. Crainich, U.S. Patent 3,494,872 (February 10, 1970).
57. S. H. Fox and L. P. Opferman, U.S. Patent 2,390,088 (December 4, 1945).
58. N. Tanaka, S. Takino, and I. Utsumi, A new oval gelatinized sustained-release dosage form, *J. Pharm. Sci.* *52*:664-667 (1963).
59. J. A. Glassman, U.S. Patent 3,275,519 (September 27, 1966).
60. R. C. Clarke and A. Courts, The chemical reactivity of gelatin, in *The Science and Technology of Gelatin* (A. G. Ward and A. Courts, eds.), Academic, London, 1977, pp. 209–247.
61. Y. Shimosaka and H. Suzuki, U.S. Patent 3,753,922 (August 21, 1973).
62. K. Saeki and H. Matsukawa, U.S. Patent 3,956,172 (May 11, 1976).
63. S. Egawa, M. Sakamoto, and T. Matsushita, U.S. Patent 4,082,688 (April 4, 1978).
64. F. M. Richards and J. R. Knowles, Glutaraldehyde as a protein cross-linking agent, *J. Mol. Biol.* *37*:231–233 (1968).
65. H. Takenaka, Y. Kawashima, and S. Y. Lin, The effects of wall thickness and amount of hardening agent on the release characteristics of sulfamethoxazole microcapsules prepared by gelatin-acacia complex coacervation, *Chem. Pharm. Bull.* *27*:3054–3060 (1979).
66. H. Schnoring and N. Schon, U.S. Patent 3,743,604 (July 3, 1973).
67. G. F. Sirine, U.S. Patent 3,510,435 (May 5, 1970).
68. N. Marinelli, U.S. Patent 4,115,315 (September 19, 1978).

69. G. Baxter, U.S. Patent 3,578,605 (May 11, 1971).
70. C. Brynko and G. M. Olderman, U.S. Patent 3,516,943 (June 23, 1970).
71. D. J. Striley and J. E. Williams, U.S. Patent 3,520,821 (July 21, 1970).
72. I. Jalsenjak and T. Kondo, Effect of capsule size on permeability of gelatin-acacia microcapsules towards sodium chloride, *J. Pharm. Sci. 70:* 456–457 (1981).
73. G. Hörger, U.S. Patent 3,692,690 (September 19, 1972).
74. M. V. Fogle and G. Hörger, U.S. Patent 3,697,437 (October 10, 1972).
75. J. A. Bakan, U.S. Patent 3,567,650 (March 2, 1971).
76. R. K. Himmel, U.S. Patent 3,594,326 (July 20, 1971).
77. N. N. Salib, M. E. El-Menshawy, and A. A. Ismail, Preparation and evaluation of the release characteristics of methyl cellulose micropellets, *Pharm. Ind. 38:* 577–580 (1976).
78. H. P. Merkle and P. Speiser, Preparation and in vitro evaluation of cellulose acetate phthalate coacervate microcapsules, *J. Pharm. Sci. 62:* 1444–1448 (1973).
79. J. A. Bakan, U.S. Patent 3,436,355 (April 1, 1969).
80. I. L. Yurkowitz, U.S. Patent 3,533,958 (October 13, 1970).
81. GAF Technical Bulletin 9670-006.
82. S. A. Mortada, Preparation of microcapsules using the *n*-butyl half ester of PVM/MA coacervate system, *Pharmazie 36:* 420–423 (1981).
83. N. Schon, H. Schnoring, J. Witte, and G. Pampus, U.S. Patent 3,712,867 (January 23, 1973).
84. H. Matsukawa and K. Saeki, U.S. Patent 3,840,467 (October 8, 1974).
85. S. Sternberg, H. J. Bixler, and A. S. Michaels, U. S. Patent 3,639,306 (February 1, 1972).
86. N. A. Elgindy and M. A. Elegakey, Carbopol-gelatin coacervation: influence of some variables, *Drug Develop. Indust. Pharm. 7:* 587–603 (1981).
87. M. Morishita, Y. Inaba, M. Fukushima, Y. Hattori, S. Kobari, and T. Matsuda, U.S. Patent 3,960,757 (June 1, 1976).
88. S. M. Mortada, Preparation of ethyl cellulose microcapsules using the complex emulsification method, *Pharmazie 37:* 427–429 (1982).
89. N. Wakiyama, K. Juni, and M. Nakano, Preparation and evaluation in vitro of polylactic acid microspheres containing local anesthetics, *Chem. Pharm. Bull. 29:* 3363–3368 (1981).
90. N. Wakiyama, K. Juni, and M. Nakano, Influence of physiocochemical properties of polylactic acid on the characteristics and in vitro release patterns of polylactic acid microspheres containing local anesthetics, *Chem. Pharm. Bull. 30:* 2621–2628 (1982).

91. H. Takenaka, Y. Kawashima, Y. Chikamatsu, and Y. Ando, Reactivity and stability of microencapsulated placental alkaline phosphatase, *Chem. Pharm. Bull.* *30*:695–700 (1982).
92. Y. Nozawa and F. Higashide, Drug release from and properties of poly(styrene) microcapsules, in *Polymeric Delivery Systems* (R. J. Kostelnik, ed.), Gordon & Breach, New York, 1978, pp. 101–110.
93. Y. Nozawa and S. W. Fox, Microencapsulation of methylglyoxal and two derivatives, *J. Pharm. Sci.* *70*:385–386 (1981).
94. A. Tateno, M. Shiba, and T. Kondo, Electrophoretic behavior of ethyl cellulose and polystyrene microcapsules containing aqueous solutions of polyelectrolytes, in *Emulsions, Latices and Dispersions* (P. Becker and M. N. Yudenfreund, eds.), Dekker, New York, 1978, pp. 279–288.
95. M. N. Vrancken and D. A. Claeys, U.S. Patent 3,523,906 (August 11, 1970).
96. M. N. Vrancken and D. A. Claeys, U.S. Patent 3,523,907 (August 11, 1970).
97. J. F. Van Besauw and D. A. Claeys, U.S. Patent 3,645,911 (February 19, 1972).
98. N. J. Morris and B. Warburton, Three-walled microcapsules, *J. Pharm. Pharmacol.* *32(Suppl.)*:24P (1980).
99. N. J. Morris and B. Warburton, Factors affecting the physical properties of three-ply walled microcapsules formed by a multiple emulsion method, *Abstracts, 4th International Conference on Surface and Colloid Science, Jerusalem, July*, 1981, p. 216.
100. N. J. Morris and B. Warburton, Three-ply walled w/o/w microcapsules formed by a multiple emulsion technique, *J. Pharm. Pharmacol.* *34*:475–479 (1982).
101. H. Nothdurft, Uber die nichtexistenz von metalkrebs im falle der edelmetalle, *Naturwissenschaften* *45*:549–550 (1958).
102. N. E. Stinson, Tissue reaction induced in guinea-pigs by particulate polymethyl-methacrylate, polythene and nylon of the same size range, *Brit. J. Exp. Path.* *46*:135–146 (1965).
103. J. J. Marty and R. C. Oppenheim, Colloidal systems for drug delivery. *Austr. J. Pharm. Sci.* *6*:65–76 (1977).
104. J. Kreuter, Nanoparticles and nanocapsules—new dosage forms in the nanometer size range, *Pharm. Acta Helv.* *53*:33–39 (1978).
105. R. C. Oppenheim, Nanoparticles, in *Drug Delivery Systems: Characteristics and Biomedical Applications* (R. L. Juliano, ed.), Oxford University Press, New York, 1980, pp. 177–188.
106. R. C. Oppenheim, Solid colloidal drug delivery systems: nanoparticles, *Int. J. Pharm.* *8*:217–234 (1981).

107. J. J. Marty, R. C. Oppenheim, and P. Speiser, Nanoparticles—a new colloidal drug delivery system, *Pharm. Acta Helv. 53*:17–23 (1978).

108. R. C. Oppenheim, J. J. Marty, and P. Speiser, U.S. Patent 4,107,288 (August 15, 1978).

109. R. C. Oppenheim, J. J. Marty, and N. F. Stewart, The labelling of gelatin nanoparticles with ^{99m}technetium and their in vivo distribution after intravenous injection, *Austr. J. Pharm. Sci. 7*:113–117 (1978).

110. R. C. Oppenheim and N. F. Stewart, The manufacture and tumour cell uptake of nanoparticles labelled with fluorescein isothiocyanate, *Drug Develop. Indust. Pharm.* 5:563–571 (1979).

4

Coacervation–Phase Separation Procedures Using Nonaqueous Vehicles

Many drugs are moderate to very water-soluble and would be unsuitable for encapsulation by the procedures described in the previous chapter. Accordingly various techniques have been developed for coating such drugs employing organic liquids in which the drug is insoluble but the coating polymer is soluble under certain conditions. Phase separation of the polymer may be induced by different methods such as temperature change, addition of incompatible polymer, or solvent alteration. The coacervated polymer encloses the core material to form the microcapsule wall. Usually low polymer concentrations are required for encapsulation by a coacervation effect involving separation into polymer-rich and polymer-poor regions. The phase separation must be gradual, enabling the concentrated polymer solution to deposit and flow uniformly over the surface of the core material to form a satisfactory coating. Higher polymer concentrations tend to give a rapid demixing effect upon phase separation, which is unsuitable for microencapsulation. Suitable polymers must be water-insoluble so that drug release from such microcapsules in aqueous environments is controlled mainly by diffusion of drug through the coating rather than by dissolution or erosion of the coating.

Many different coacervation-phase separation procedures involving nonaqueous manufacturing vehicles have been developed, some of which will be considered in this chapter because of their special applicability to the encapsulation of pharmaceuticals. Again, precise details of the procedures have been omitted for the sake of simplifying comprehension of the process and its applicability. The reader interested in obtaining greater detail should consult the original references.

4.1 ETHYLCELLULOSE

Most of the patents relating to microencapsulation with ethylcellulose by coacervation-phase separation procedures were assigned to NCR, IBM,

Merck, Fuji, and other companies in the 1960s and early 1970s. Like most patents, many details of the procedures involved were not adequately disclosed. However, particularly since 1975, a number of papers have been published by groups working in universities and similar institutions that give fuller details of the processes and assess their usefulness for the encapsulation of a wide range of drugs.

4.1.1 Coacervation Induced by Temperature Change in a Cyclohexane System—Waxy Sealant Treatment

In 1970 Fanger et al. [1], in a patent assigned to NCR, outlined a very simple process for the encapsulation of heat-stable drugs with ethylcellulose. The procedure has subsequently been investigated in detail by a number of research groups, including that of the author of this book. The process involves dissolving ethylcellulose in the nonideal solvent cyclohexane at 80 to 81°C and gradually cooling the solution so that the polymer separates as a liquid coacervate and encloses particles of core material that are dispersed by vigorous agitation in the system. The deposited wall material may be hardened by continuing to lower the temperature while maintaining vigorous agitation to prevent coalescence of microcapsules. When the system reaches 20 to 25°C, the microcapsules can be filtered off from the cyclohexane and dried. Fanger et al. reported that the process tended to give an aggregated product so that many microcapsules formed were multinuclear.

In another patent assigned to NCR, Powell [2] described how to treat preformed ethylcellulose microcapsules containing drugs such as paracetamol with solutions of waxy materials in cyclohexane. Upon drying, the capsule walls retained an amount of the waxy material that tended to alter the physicochemical properties of the coating by making it more hydrophobic. Paraffin wax was particularly recommended.

Salib [3] encapsulated phenobarbitone in ethylcellulose by coacervation from cyclohexane upon temperature change. In comparison to phase separation procedures involving gelatin and various flocculation techniques, the ethylcellulose encapsulation method was the most efficient at coating the core material, only 2.5% of drug being lost during preparation. In a subsequent paper, Salib et al. [4] microencapsulated chloramphenicol in a similar manner and showed that the product had sustained-release properties during dissolution studies in simulated gastric and intestinal media that were more pronounced than those of an ethylcellulose-flocculated complex of the drug.

A detailed appraisal of the phase separation procedure of ethylcellulose from cyclohexane by temperature change was carried out by Jalsenjak et al. [5] for the microencapsulation of phenobarbitone sodium. Though the process appears simple, slight changes in procedure can affect the quality of the product. Faster stirrer speeds tended to give a finer product, as less aggregation of microcapsules occurred. The size distribution of the product was somewhat affected by the geometry of the reaction vessel. Also, a slow rate of cooling was desirable in that it promoted a more even deposition of the coating material on the cores. Large microcapsules were observed to be aggregates of smaller microcapsules rather than single capsules with thicker walls, and microcapsules did not disintegrate during dissolution runs in water. Increase in coating content or particle size reduced drug release into water. Drug release tended to be rapid and diffusion-controlled, with more than 75% released within 1 hr, confirming the poor capacity of this coating polymer to retard release of highly water-soluble drugs over prolonged periods. In a subsequent paper by Nixon et al. [6], the phenobarbitone sodium microcapsules were tableted and shown to have a much slower in vitro rate of release. The tablets did not disintegrate during the dissolution experiments.

Deasy et al. [7] also encapsulated a highly water-soluble drug, sodium salicylate, using the same procedure. Figure 4.1 shows drug release curves into water using two different apparent viscosity grades of the polymer as coating material. At equivalent particle size range the 100 cP grade exhibited the slowest release, though both grades permitted over 80% drug release within 1 hr of the treatment.

Scanning electron micrographs (see Fig. 4.2) confirmed that larger microcapsules were aggregates of smaller ones, the surfaces of which had both rough and smooth areas with pores that might extend through the coating to the core. The microcapsules tended to swell and rupture during dissolution studies. Treatment of the microcapsules with paraffin wax or other waxy sealants markedly reduced drug release from the microcapsules.

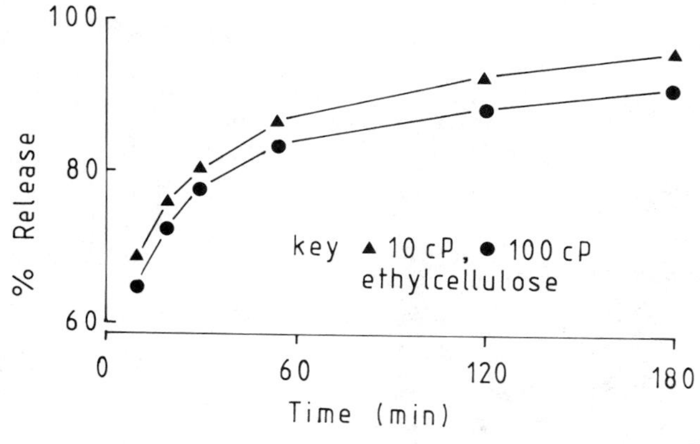

Figure 4.1 Release of sodium salicylate from 500–297 μm microcapsules coated with ethylcellulose standard. (From Ref. 7.)

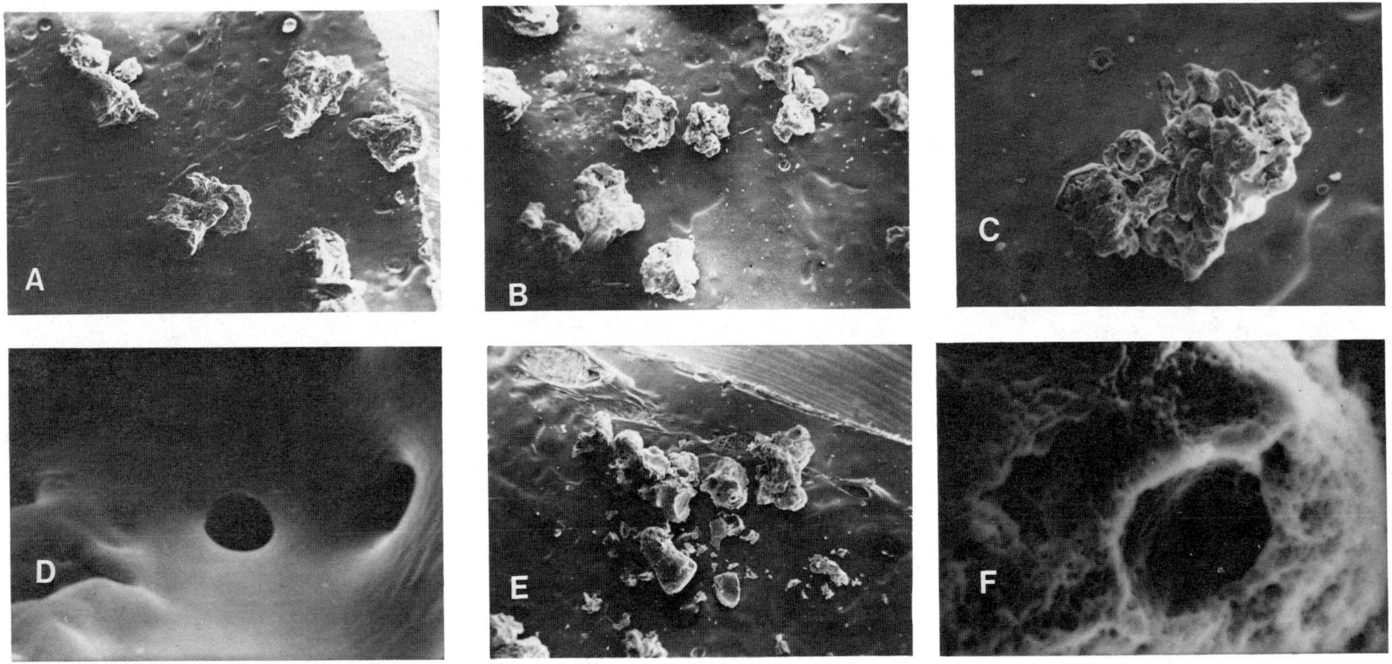

Figure 4.2 Scanning electron micrographs of 500–297 μm microcapsules containing sodium salicylate coated with ethylcellulose standard 10 cP (A) or 100 cP (B, C, D, E, F) before (A, B, C, D) or after (E, F) dissolution studies. (From Ref. 7.)

Mesiha and Sidhom [8] used the procedure to microencapsulate oxytetracycline hydrochloride and reported results that were qualitatively similar to those obtained by Jalsenjak et al. [6] and Deasy et al. [7]. However, they also reported binding between the drug and the polymer, which hindered the complete release of the core material from the microcapsules.

Jalsenjak et al. [9] also microencapsulated isoniazid by the same procedure. For this moderately water-soluble drug the time for 50% release was less than 1 hr, though tablets prepared from the microcapsules had longer $t_{50\%}$s. There was no drug left for release from the microcapsules after 5 hr dissolution treatment or after 24 hr for the tablets. In a subsequent paper by Senijkovic and Jalsenjak [10], the decreasing permeability of smaller-sized isoniazid microcapsules was attributed to the higher density of their ethylcellulose membranes.

Alpar and Walters [11] used the same procedure to microencapsulate phenethicillin potassium, another very water-soluble drug, for sustained-release and taste-masking purposes. To significantly retard drug release it was necessary to tablet the microcapsules.

4.1.2 Coacervation Induced by Addition of Polyethylene or Other Incompatible Polymers and Temperature Change in a Cyclohexane System—Modifications

Holliday et al. [12, 13], in patents assigned to Sterling Drug, Inc., described the microencapsulation of aspirin particles to produce a sustained-release dosage form. As indicated in the first chapter, the choice of aspirin for such a product is questionable and might better be regarded as a model compound useful for the illustration and development of the process.

A low concentration of ethylcellulose having an ethoxyl content of about 49% was dissolved in cyclohexane at 80 to 81°C in the presence of the core particles of aspirin and a low-molecular-weight polyethylene (7,500–10,000), as an incompatible polymer, using vigorous agitation. Acetic anhydride was added to prevent hydrolysis of the aspirin during the encapsulation at elevated temperature. The system was then gradually cooled and the ethylcellulose separated out encapsulating the core material. Cooling and agitation were continued until the system reached room temperature. After allowing the microcapsules to settle, the majority of the cyclohexane, containing suspended polyethylene particles, was decanted. The microcapsules obtained were dried to free them of residual cyclohexane. The microcapsules were then tableted. The precise function of the polyethylene was not disclosed. Presumably it aided the formation of the ethylcellulose coacervate and also acted as an inert filler to prevent neighboring microcapsules adhering during the initial coating stages when the polymer walls were still in the fluid state at elevated temperature.

In 1971 Morse [14], in a patent assigned to Merck & Co., Inc., described the microencapsulation of indomethacin, using the same procedure, to prolong its duration of action and to reduce its tendency to cause gastrointestinal irritation. Particle size of the drug was regulated by micronization if necessary, and different core:coat ratios were employed. The ethylcellulose encapsulation was estimated to be complete when the temperature was decreased to 45°C. The polyethylene was dispersed as minute particles in the cyclohexane and was decanted. The microcapsules were repeatedly washed with fresh cyclohexane until free of polyethylene and dried. The microcapsules released their content by diffusion and, depending on size and amount of drug contained, were blended to provide dosage units having different release rates. In another patent assigned to Merck, Morse and Hammes [15] used the same process to microencapsulate various B vitamins for taste-masking purposes.

Bitolterol is a bronchodilator that was microencapsulated to improve its duration of action and reduce its side effects by John et al. [16]. The drug was mixed with lactose, microcrystalline cellulose, and starch paste and formed into spherical core particles in the 500–1000 μm range by spheronization in a marumerizer. These cores were then microencapsulated with ethylcellulose using cyclohexane, causing it to form the coating film at a temperature where its viscosity was low enough to allow each core particle to be uniformly coated. Encapsulation of the cores with film comprising 2.6% of the microcapsule resulted in a $t_{50\%}$ of 40 min for release into 0.1 *N* hydrochloric acid and that could be further retarded by increasing the coating percentage. The microcapsules showed controlled-release properties when tested in dogs.

In a patent assigned to NCR, Anderson and Powell [17] used the same procedure to microencapsulate dried granules prepared from paracetamol as drug, cellulose acetate phthalate as an enteric polymer filler, and acacia solution as binder. The microcapsules obtained did not disintegrate because of their ethylcellulose coating and had a relatively slow release of drug in acidic media and a rapid but sustained release in alkaline media.

In another patent of Powell and Anderson [18] assigned to NCR, an acetylated monoglyceride was used to induce a liquid-phase separation at a temperature of about 70°C in an ethylcellulose-polyethylene system. The core material was a dispersion of amylobarbitone suspended in an aqueous solution of sucrose and acacia. This drug-containing aqueous phase was emulsified at high temperature as droplets of a water-in-oil (w/o) emulsion having a solution of the ethylcellulose in the cyclohexane together with the two coacervating agents as the continuous phase. Upon cooling, microencapsulation of the aqueous droplets occurred, which were subsequently dessicated by adding anhydrous silica gel or molecular sieves to the cold system. The dessicated microcapsules and dessicant were filtered off the vehicle, washed with fresh cyclohexane, dried, and separated from each other by sieving.

Bakan and Powell [19] reported that paracetamol, aspirin, potassium chloride, and theophylline microcapusles, prepared using an ethylcellulose-polyethylene cyclohexane system showed excellent stability upon storage for up to 5 years.

Donbrow and Benita [20] studied the effect of polyisobutylene as another polymer on the coacervation of ethylcellulose and the formation of microcapsules. Polyisobutylene is highly soluble and solvated in cyclohexane, and in increasing concentration it causes a progressive increase in the volume of ethylcellulose coacervate droplets formed in the cyclohexane by temperature change. It appears to act as a protective colloid, preventing the formation of large agglomerates of ethylcellulose by reducing the coacervate droplet size. Addition of polyisobutylene caused salicylamide microcapsules to release their core material more readily because of their thinner ethylcellulose walls. Further aspects of the release of theophylline [21], sodium salicylate or potassium dichromate [22], and salicylamide [23] from similarly prepared microcapsules were reported recently by the same authors.

Samejima et al. [24] have studied the role of coacervation-inducing agents such as butyl rubber, polyethylene, and polyisobutylene on the microencapsulation of vitamin C by ethylcellulose by phase separation from cyclohexane upon change in temperature. Polyisobutylene was found to be most effective at promoting the formation of smooth and thick-walled microcapsules, which showed little tendency to aggregate.

Powell et al. [25], in a patent assigned to NCR, used a solution of ethylcellulose having an ethoxyl content of about 47.5% in toluene: ethanol 8:2. Phase separation was induced by adding a second liquid polymer, such as polybutadiene (MW 8—10,000) or polydimethylsiloxane (500 cSt), which have little or no affinity for the core material but cause liquid droplets of coacervated ethylcellulose to deposit around dispersed liquid or solid core material. The coating formed contained mainly ethanol and a little toluene before being dried.

4.1.3 Coacervation Induced by Solvent Alteration

Reyes [26], in a patent assigned to IBM, described a process that could be modified for pharmaceuticals whereby gelable polymers and polar solvents were formed into a solution containing a dissolved drug. This water phase was emulsified in a solution of ethylcellulose in xylene to form a w/o emulsion. Good microcapsules were obtained when the gelable hydrophilic polymer was polyvinyl alcohol, alginic acid salts, or gelatin. Gelation of the disperse phase was obtained by cooling or by the addition of a gelling agent such as an inorganic salt. Addition of petroleum ether as an incompatible solvent and containing a microcrystalline wax caused precipitation of the ethylcellulose by solvent alteration as a coating around the droplets containing the drug. The microcapsules formed were isolated by filtration, washed with petroleum ether, and air-dried.

Kondo [27] listed a chloroform (solvent)-petroleum ether (incompatible solvent) and benzene (solvent)-corn oil (incompatible solvent) as suitable for ethylcellulose encapsulation without the need to use high temperatures, which are undesirable for heat-sensitive materials.

Likewise Kitajima et al. [28], in a patent assigned to Fuji and Toyo Jozo Co., Ltd., encapsulated aspirin by dispersing it with the aid of starch in a solution of ethylcellulose in benzene. This was emulsified as the disperse phase in an aqueous solution of disodium hydrogen phosphate, adjusted to pH 4 with phosphoric acid as continuous phase. When agitated under reduced pressure at a temperature of 30°C, the benzene was lost by evaporation and ethylcellulose-coated microcapsules of aspirin were recovered by filtration from the aqueous phase. In a similar approach Yoshida [29], in a patent assigned to NCR, dissolved ethylcellulose in acetone. Dispersol 81515, a distillate containing about 50% naphthenes and 50% paraffins, was added to the polymer solution together with suitable core material such as sodium chloride particles with constant agitation. The temperature of the system was raised to 50°C, causing the acetone to evaporate and the ethylcellulose to phase-separate and encapsulate the core material as a result of solvent change. When the acetone was completely evaporated, the microcapsules were separated off from the Dispersal 81515 and washed with a low-boiling-point petroleum distillate prior to drying.

In another modification of the ethylcellulose encapsulation process Kitajima et al. [30], in a patent assigned to Fuji, dissolved the polymer in methyl acetate to which was added a dispersion of aspirin particles as core material. This system was then emulsified as the disperse phase of an o/w emulsion into a concentrated aqueous salt solution of ammonium chloride, ammonium sulfate, or disodium phosphate, which prevented the aspirin from dissolving appreciably in the water phase because of a salting effect. While maintaining stirring, the organic solvent was evaporated off by raising the temperature to about 30°C. The encapsulated aspirin formed was filtered off, washed rapidly with water, and dried. The product had a prolonged release of aspirin in dissolution studies compared to an unencapsulated control. In a subsequent patent, Kitajima et al. [31] dissolved ethylcellulose in various solvents such as ethyl acetate and added aspirin particles as a dispersion to the solution. This system was then added to water saturated with aspirin and stirred to form an o/w emulsion. By continuing the agitation the ethyl acetate was evaporated off at room temperature to form an ethylcellulose-encapsulated aspirin.

Fukushima et al. [32], in a patent assigned to Toyo Jozo Co., Ltd., and Fuji Photo Film Co., Ltd., dissolved both the core and coating material in a solvent having a dielectric constant of about 10 that was poorly miscible with a polyhydric alcohol vehicle, so forming an o/w emulsion upon stirring. For example, aspirin and ethylcellulose could be dissolved in a mixed solvent of ethyl acetate:chloroform 1:1. When added to ethylene glycol and stirred, an o/w emulsion resulted and the disperse-phase solvent evaporated if stirring was continued for several hours at room temperature. The microcapsules formed were separated from the ethylene glycol by centrifugation, rinsed with water, and dried.

Two recent papers describing ethylcellulose microencapsulation involving polymer deposition by polymer nonsolvent addition have been published. In the first paper, by Kasai and Koishi [33], an ethylcellulose-dichloromethane (solvent)-*n*-hexane (incompatible solvent) system was used to encapsulate magnesium aluminium hydroxide hydrate granules. A three-phase diagram of the ethylcellulose-dichloromethane-*n*-hexane system showing the optimum coacervation conditions was presented. The deposition of ethylcellulose coacervate droplets on the hydrophobic surface of the model core particles was recorded on photomicrographs. Scanning electron micrographs of the coating showed an open network structure that appeared more uneven and porous than the coatings formed by temperature change from cyclohexane mentioned previously. A number of possible models for the deposition of coacervate droplets on the core material were presented. In the second paper, by Itoh and Nakano [34], sulfamethizole or 5-fluorouracil were coprecipitated with cellulose acetate using acetone and the resultant dried comminuted particles were microencapsulated using an ethylcellulose-diethyl ether (solvent)-*n*-heptane (incompatible solvent) system. Again scanning electron microscopy studies (see Fig. 4.3) showed that the surface of the microcapsules obtained was not very smooth in comparison to the phase separation procedure induced by temperature change. In vitro dissolution studies showed that the gel matrix containing cores themselves had significantly delayed release for each drug, and this effect was further enhanced upon microencapsulation, having a release $t_{50\%}$ of about 5 hr for the sulfamethizole and about 6 hr for the 5-fluorouracil microcapsules. The results of this study suggest that the use of a core with retarded drug-release properties that are further increased by application of a suitable coating appears to be an excellent way of obtaining microcapsules having very prolonged release properties.

D'Onofrio et al. [35] microencapsulated aspirin with ethylcellulose by a phase separation process from a solution of ethyl acetate induced by addition of light liquid paraffin. The process was very suitable for in-line production of soft gelatin capsules, as the slurry of the microcapsules obtained in the liquid paraffin after decanting excess vehicle could be filled directly into soft gelatin capsules without the usual isolation and formulation steps needed for filling microcapsules into hard gelatin capsules. The amount of residual ethyl acetate in the liquid paraffin vehicle for the microcapsules ranged from 0 to 0.8%, as most of it evaporated off, aiding hardening of the coat, during the agitation procedure. This level of ethyl acetate was regarded as having insignificant toxicity.

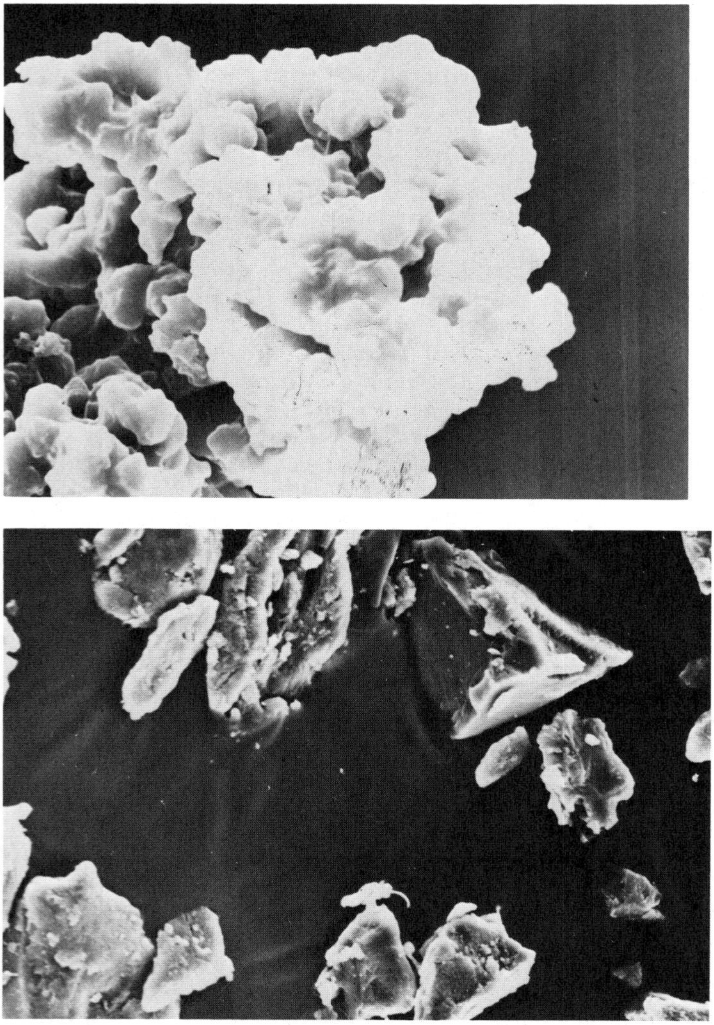

Figure 4.3 Scanning electron micrographs of a microcapsule (top) and the matrix milled (bottom). (From Ref. 34; courtesy of Dr. M. Itoh.)

4.1.4 Reduction of Aggregation

A recent patent of Morse et al. [36], assigned to Merck, was concerned with the reduction of aggregation that occurs with ethylcellulose microcapsules prepared by temperature change in a cyclohexane system. Residual solvated cyclohexane in the ethylcellulose walls was displaced by reslurrying with pentane, hexane, heptane, octane, or mixtures thereof, such as petroleum ether, prior to the drying stage. The product obtained had less tendency to clump, giving small discrete microcapsules.

4.1.5 "Blank" Microcapsules

In another patent by Morse et al. [37] assigned to Merck, "blank" ethylcellulose microcapsules were prepared by phase separation from cyclohexane upon temperature change. When the system had cooled to 50°C the ethylcellulose masses formed were still plastic enough to allow most of the riboflavin added at this stage as core material to enter the coating material and be enveloped, this being aided by raising the temperature to 57°C. The system was then cooled to 10°C, filtered off, washed with hexane to reduce aggregation, and dried in a fluidized bed dryer.

4.2 CELLULOSE ACETATE PHTHALATE

Cellulose acetate phthalate (CAP) is a favored film former for the production of dosage forms having enteric properties. It is practically insoluble in water and acidic solutions but will dissolve in buffer solutions with a pH above 5.5 to 6.0. Kitajima et al. [28] described various encapsulation procedures for heat-sensitive materials, one of which involved dispersing a powdered antiphlogistic enzyme in a CAP solution in acetone. This system formed the disperse phase of an o/w emulsion having a 35% ammonium sulfate solution in water as the continuous phase. By heating the emulsion at 30°C while stirring, the acetone was evaporated off to promote phase separation of CAP-coated microcapsules that had enteric properties.

In a slight modification of the procedure Kitajima et al. [38], in a patent also assigned to Fuji and Toyo Jozo, emulsified a CAP solution in acetone containing a dispersed powered enzyme into liquid paraffin using a suitable emulsifier. Evaporation of acetone was accomplished by raising the temperature to about 25°C while maintaining stirring, and the filtered microcapsules obtained were freed of residual liquid paraffin by washing with benzene. The resultant microcapsules showed very little loss of enzymic activity of the core material and had enteric properties.

CAP-coated microcapsules could also be prepared as described by Morishita et al. [39] in a patent assigned to Toyo Jozo Co., Ltd., Japan. The polymer was dissolved in dimethylsulfoxide and pancreatin powder was dispersed in the system, which was emulsified into liquid paraffin. Water:acetone 4:1 was then added as a nonsolvent to the emulsion to cause the formation of microcapsules, which were filtered off, washed with *n*-hexane, and dried.

4.3 CELLULOSE ACETATE BUTYRATE

Cellulose acetate butyrate (CAB) is another polymer with enteric properties. Kondo [27] described how CAB encapsulation of suitable core material could be accomplished using a solvent alteration procedure. The core could be dispersed in a solution of CAB in methyl ethyl ketone. Addition of isopropyl ether as an incompatible solvent caused alteration of the nature of the solvent with phase separation of the CAB as a coating on the core material.

Gardner et al. [40] described a nonaqueous phase separation procedure (no details given) for the encapsulation of a number of different steroids with CAB. It was proposed to use the microcapsules as an intrauterine contraceptive or for other medical purposes because of the reported transcervical migration of small particles. The in vitro release of progesterone from the microcapsules was very slow and, apart from an initial rapid release rate, zero-order release rate was obtainable because of the maintenance of a saturated solution of the steroid in equilibrium with undissolved particles of steroid within the microcapsules. In vivo release of drug in the rabbit uterus was found to be much more rapid (see Fig. 4.4).

4.4 HYDROXYPROPYLMETHYLCELLULOSE PHTHALATE

Hydroxypropylmethylcellulose phthalate (HPMCP) is another cellulose derivative that has enteric properties. In the patent of Fukushima

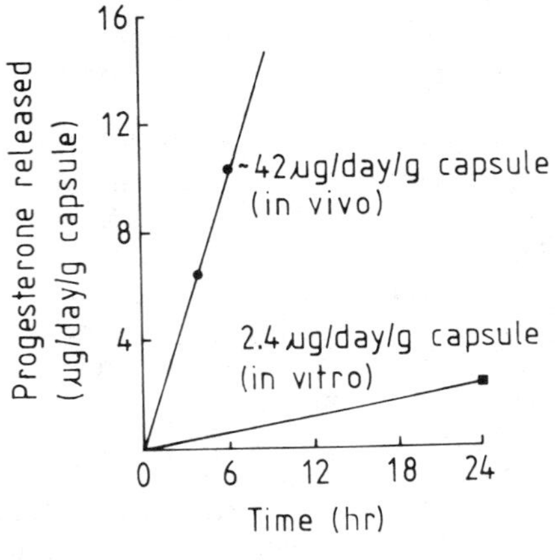

Figure 4.4 Comparison of in vitro and in vivo progesterone release rates from 420–500 μm diameter microcapsules. (Reprinted with permission from Ref. 40. Copyright 1976 American Chemical Society.)

et al. [41] assigned to Fuji and Toyo Jozo, the polymer was dissolved in methylene chloride into which was dispersed particles of the enzyme preparation, pancreatin. This system was then emulsified into ethylene glycol to form the disperse phase of an o/w emulsion. Stirring was maintained at room temperature until the methylene chloride had evaporated, producing the coating by phase separation of the polymer. After removal from the ethylene glycol and washing with water, the microcapsules obtained were found to retain almost 100% of the proteolytic activity of the core material, confirming the nondestructive nature of the encapsulation process for this material.

4.5 CARBOXYMETHYLETHYLCELLULOSE AND POLYLACTIC ACID

Nakano et al. [42] prepared enteric-coated microcapsules whose release rate should depend on the approximate first-order emptying rate from the stomach. Carboxymethylethylcellulose (CMEC) is a polymer suitable for enteric coating, and unlike CAP it is not susceptible to hydrolysis by atmospheric moisture. CMEC and polylactic acid (PLA) were dissolved in ethyl acetate and powdered sulfamethizole was dispersed in the solution. To induce phase separation of the two polymers as coating, ethyl ether was added as an incompatible solvent to the system. The decanted microcapsules formed were washed with diethyl ether before drying. The addition of the polylactic acid appeared to act as a plasticizer and resulted in the formation of microcapsules with improved drug release characteristics. In simulated gastric juice the microcapsules had a sustained drug release extending up to 10 hr. This finding indicates that enteric-coated microcapsules do not exhibit complete absence of drug release in the stomach. In simulated intestinal juice, drug release was complete in 20 min. Analysis of pharmacokinetic data based on urinary excretion data obtained in four human volunteers indicated a sustained-release effect with only slightly reduced bioavailability of the microcapsules compared to the same dose of unencapsulated drug in tablet form.

4.6 CELLULOSE NITRATE AND POLYSTYRENE

Collodion, U.S.P., contains pyroxylin, a nitrated cellulose, dissolved in 3 parts of ether and 1 part of alcohol. Chang et al. [43] described how this material could be used to produce "artificial cells" by an interfacial coacervation method. An aqueous erythrocyte hemolysate, buffered to neutralize acidic impurities in the cellulose nitrate, was emulsified into ether using Span 85 as emulgent at low temperature to form a w/o emulsion. Collodion with its alcohol content replaced by ether to lessen protein denaturation was then stirred into the emulsion. During a period of 45 min the cellulose nitrate phase separated at the interface of each droplet to form an ill-defined coating. Using Span 85 to prevent aggregation, the partially coated microcapsules were isolated by centrifugation and transferred to *n*-butyl benzoate. Residual ether was allowed to evaporate off to allow the microcapsule coating to set. The microcapsules were then recovered from the organic liquid phase by centrifugation and resuspended in water using Tween 20 to facilitate their transfer from the organic to the aqueous phase for storage. A similar procedure using polystyrene dissolved in benzene, with benzene replacing ether, could also be used to form polystyrene-coated microcapsules by interfacial coacervation.

Enzymes such as catalase were encapsulated in cellulose nitrate-coated "artificial cells" [44]. Such microcapsules could be made nonthrombogenic by incorporating heparin into the membrane as described by Chang [45] in a patent assigned to Research Corp. A number of applications of these microcapsules, which could be albumin-coated to make them blood-compatible, such as to encapsulate charcoal for use in an artificial kidney, have been discussed by Chang and coworkers [46–48].

4.7 ACRYLATE

El-Sayed et al. [49] microencapsulated riboflavin in an acrylate polymer (Eudragit RLPM) by suspending the drug in a solution of the polymer in benzene or isopropyl alcohol. Phase separation was induced by addition of light petroleum ether and the microcapsules were separated by filtration and air dried. In vitro dissolution studies on tablets prepared from the microcapsules showed a sustained-release effect.

4.8 POLY(ETHYLENE-VINYL ACETATE) AND CHLORINATED RUBBER

A system was described by Bayless et al. [50] in a patent assigned to NCR for microencapsulation with partly hydrolyzed poly(ethylene-vinyl acetate) [poly (EVA)]. Polar or nonpolar drugs either alone or as a solution or suspension in a polar liquid were dispersed with agitation in a solution of the poly(EVA) in toluene at elevated temperature. Phase separation was induced by adding an incompatible polymer such as polybutadiene or polydimethylsiloxane and lowering the temperature. If desired, microcapsules formed could be hardened by adding a suitable crosslinking agent before final recovery and drying. A number of modifications to the procedure were described in a further patent of Bayless [51] assigned to Capsular Systems, Inc.

Purcell [52], in a patent assigned to NCR, described the use of poly(EVA) as an incompatible polymer and temperature increase to cause phase separation of chlorinated rubber from solution in cyclohexane and toluene as a coating containing both polymers on dispersed aqueous droplets or hydrophilic particles of core material. The resultant microcapsules could be treated with a waxy sealant dissolved in cyclohexane to reduce loss of core material by diffusion through the coating. Toxicological considerations would obviously limit the applicability of this process for pharmaceutical use.

4.9 HARDENED OILS AND FATS

Various hardened derivatives of castor oil, whale oil, beef tallow oil, etc., with melting points between 40 and 80°C are suitable for the encapsulation of water-soluble drugs such as amino acids. Maekawa et al. [53], in a patent assigned to Fuji, described a process that initially involved dispersing a powered amino acid in a melted hardened oil, with or without a dispersion of fine inert powder such as kaolin to alter the density of the microcapsules produced. This system was then emulsified above the melting point of the hardened oil into a continuous phase composed of a mixture of methanol and ethylene glycol or propylene glycol and glycerin. When the desired droplet size had been obtained, the temperature of the system was reduced to solidify the hardened oil droplets containing dispersed core material.

In a more recent patent assigned to Kondo and Nakano [54] of Takeda Chemical Industries, Ltd., a solvent was used that was capable of dissolving an oil or fat derivative, normally solid at room temperature, when hot but not capable of significantly dissolving the dispersed core material. Table 4.1 lists some of the oleaginous materials, with melting

Table 4.1 Suitable Solvents for Hardened Oils and Fats at High Temperature

Oleaginous material	Solvent (at high temperature)
Hydrogenated beef tallow	Ethanol
Hydrogenated castor oil	Methanol Ethanol Isopropanol Ethyl acetate Acetone *n*-Hexane Diethyl ether
Glyceryl monostearate	Ethanol
Theobroma oil	Ethanol

points above 30°C and preferably above 50°C, and suitable solvents that will dissolve at least 5% of the oleaginous material at high temperature.

A second polymer, such as a cellulose derivative, may also be dissolved in the solvent to help disperse the core material at high temperature in the oil phase, and to modify release characteristics of the product. When the temperature of the system was decreased while stirring, at least 50% of the oleaginous material previously in solution coacervated as a coating onto the surface of the core material to form microcapsules. Completion of coacervation at room temperature in such systems can take several hours while maintaining agitation. An inert filler such as lactose, starch, talc, or silica could be added after coacervation to prevent aggregation of the microcapsules, particularly if a second polymer had been used. The microcapsules formed were separated from the manufacturing vehicle and dried. The process was recommended for the encapsulation of antibiotics, enzymes, vitamins, and various other drugs using particularly the hydrogenated castor oil/ethanol system.

4.10 MISCELLANEOUS POLYMERS

In the patent of Kitajima et al. [38], a wide range of polymer-organic solvent systems suitable for the encapsulation of dispersed core material by a phase separation procedure were described. Table 4.2 lists some of the polymers and their solvents. As these solvents have dielectric constants between 10 and 40, they tend to be immiscible with liquid silicone oil. The dispersion of the core material in the polymer solvent systems was emulsified into either of the two oily liquids to produce droplets of the required particle size range. The disperse-phase solvent could then be evaporated by stirring at ambient or elevated

Table 4.2 Polymer-Solvent System for Use with Liquid Paraffin or Silicone Oil

Polymer	Solvent
Acrylic acid-acrylic acid ester copolymer	Acetone:ethanol (1:1 by weight)
Cellulose acetate-N,N-di-*n*-butylhydroxypropyl ether	Acetone:ethanol (1:1 by weight)
Polyvinylpyrrolidone	Acetone Ethanol
Polyvinyl alcohol	Ethanol

Table 4.3 Combinations of Polymers, Solvents, and Incompatible Solvents

Polymer	Solvent	Incompatible solvent
Benzylcellulose	Trichloroethylene	Propanol
Caprolactam	Ethylene glycol (5)	Ethylene glycol (3)
Cellulose nitrate	Acetone	Water
Polyethylene	Xylene	Amyl chloride
Polystyrene	Xylene	Petroleum ether
Polyvinyl chloride	Tetrahydrofuran	Water

temperature with or without reduced pressure to obtain a suspension of the microcapsules in the liquid paraffin or silicone oil, which might subsequently be separated.

Kondo [27] listed a large number of combinations of solvents and incompatible solvents suitable for encapsulation by various polymers and drawn from a review of the patent literature. Table 4.3 lists some of the combinations mentioned.

The core material is dispersed in a solution of the polymer in the solvent. Addition of the incompatible solvent, which is soluble with the solvent but immiscible with the polymer, causes phase separation of a coating of the polymer under suitable conditions as a coating on the dispersed core material. The toxicological problems associated with some of these polymers and their residual solvents may limit the applicability of these processes for the microencapsulation of drugs.

Scheu et al. [55] used a nonaqueous coacervation procedure to apply crosslinked polyethyleneimine to water-soluble core particles of gold sodium thiosulfate, amaranth, or sodium chloride. The cores were dispersed in a glycerin-dioxane system containing the polymer, polyethyleneimine. The pH was controlled and ethylcellulose was added as an anticlumping agent. Glutaraldehyde solution was used to crosslink and render insoluble the polymer coating. Microcapsules formed were centrifuged, washed in dioxane, and dried at elevated temperature. The gold-197 and/or sodium-23 could be activated by exposure to a neutron flux. The products were investigated as a timed-release parenteral dosage form and as a radiopharmaceutical imaging agent.

Biodegradable microcapsules containing ampicillin or gentamicin in a polyglycolic acid-polylactic acid copolymer coating for topical application have been recently reported by Tice et al. [56]. However, no details of the phase separation procedure used were given.

A disadvantage of most phase separation procedures involving ethylcellulose or other polymers is that they lack flexibility in that it is not possible to incorporate other materials in the coating such as co-film formers, plasticizers, etc. These additives might improve the quality of the deposited film and modify its release characteristics.

REFERENCES

1. G. O. Fanger, R. E. Miller, and R. G. McNiff, U.S. Patent 3,531,418 (September 29, 1970).
2. T. C. Powell, U.S. Patent 3,623,997 (November 30, 1971).
3. N. N. Salib, A study of microencapsulation and flocculation techniques in pharmaceutical formulation 1. The evaluation of some microencapsulation and flocculation techniques, *Pharm. Ind. 34*:671–674 (1972).
4. N. N. Salib, M. E. El-Menshawy, and A. A. Ismail, Ethyl cellulose as a potential sustained release coating for oral pharmaceuticals, *Pharmazie 31*:721–723 (1975).
5. I. Jalsenjak, C. F. Nicolaidou, and J. R. Nixon, The in vitro dissolution of phenobarbitone sodium from ethylcellulose microcapsules, *J. Pharm. Pharmacol. 28*:912–914 (1976).
6. J. R. Nixon, I. Jalsenjak, C. F. Nicolaidou, and M. Harris, Release of drugs from suspended and tabletted microcapsules, *Drug Develop. Indust. Pharm. 4*:117–129 (1978).
7. P. B. Deasy, M. R. Brophy, B. Ecanow, and M. Joy, Effect of ethylcellulose grade and sealant treatments on the production and in vitro release of microencapsulated sodium salicylate, *J. Pharm. Pharmacol. 32*:15–20 (1980).
8. M. S. Mesiha and M. B. Sidhom, The interaction of oxytetracycline hydrochloride with ethyl cellulose in microcapsules, *J. Pharm. Pharmacol. 32(Suppl.)*:26P (1980).
9. I. Jalsenjak, J. R. Nixon, R. Senjkovic, and I. Stivic, Sustained-release dosage forms of microencapsulated isoniazid, *J. Pharm. Pharmacol. 32*:678–680 (1980).
10. R. Senjkovic and I. Jalsenjak, Effect of capsule size and membrane density on permeability of ethyl cellulose microcapsules, *Pharm. Acta Helv. 57*:16–19 (1982).
11. H. O. Alpar and V. Walters, The prolongation of the in vitro dissolution of a soluble drug (phenethicillin potassium) by microencapsulation with ethyl cellulose, *J. Pharm. Pharmacol.* 33:419–422 (1981).
12. W. M. Holliday, M. Berdick, S. A. Bell, and G. C. Kiritsis, U.S. Patent 3,488,418 (January 6, 1960).
13. W. M. Holliday, M. Berdick, S. A. Bell, and G. C. Kiritsis, U.S. Patent 3,524,910 (August 18, 1970).
14. L. D. Morse, U.S. Patent 3,557,279 (January 19, 1971).
15. L. D. Morse and P. A. Hammes, U.S. Patent 3,860,733 (January 14, 1975).

16. P. M. John, H. Minatoya, and F. J. Rosenberg, Microencapsulation of bitolterol for controlled release and its effect on bronchodilator and heart rate activities in dogs, *J. Pharm. Sci. 68*:475–481 (1979).
17. J. L. Anderson and T. C. Powell, U.S. Patent 3,909,444 (September 30, 1975).
18. T. C. Powell and J. L. Anderson, U.S. Patent 3,576,759 (April 27, 1971).
19. J. A. Bakan and T. C. Powell, Long term stability of controlled release pharmaceutical microcapsules prepared by phase separation techniques, *Abstracts, 8th International Symposium on Controlled Release of Bioactive Materials, Ft. Lauderdale, July,* 1981, pp. 158–161.
20. M. Donbrow and S. Benita, The effect of polyisobutylene on the coacervation of ethyl cellulose and the formation of microcapsules, *J. Pharm. Pharmacol. 29(Suppl.)*:4P (1977).
21. S. Benita and M. Donbrow, Release kinetics of sparingly soluble drugs from ethyl cellulose-walled microcapsules: theophylline microcapsules, *J. Pharm. Pharmacol. 34*:77–82 (1982).
22. S. Benita and M. Donbrow, Dissolution rate control of the release kinetics of water-soluble compounds from ethyl cellulose film-type microcapsules, *Int. J. Pharm. 12*:251–264 (1982).
23. M. Donbrow and S. Benita, Release kinetics of sparingly soluble drugs from ethylcellulose-walled microcapsules: salicylamide microcapsules, *J. Pharm. Pharmacol. 34*:547–551 (1982).
24. M. Samejima, G. Hirata, and Y. Koida, Studies on microcapsules. 1. Role and effect of coacervation-inducing agents in the microencapsulation of ascorbic acid by a phase separation method, *Chem. Pharm. Bull. 30*:2894–2899 (1982).
25. T. C. Powell, M. E. Steinle, and R. A. Yoncoskie, U.S. Patent 3,415,758 (December 10, 1968).
26. Z. Reyes, U.S. Patent 3,405,070 (October 8, 1968).
27. A. Kondo, *Microcapsule Processing and Technology* (J. Wade Van Valkenburg, ed.), Dekker, New York, 1979, pp. 95–105.
28. M. Kitajima, T. Yamaguchi, A. Kondo, and N. Muroya, U.S. Patent 3,691,090 (September 12, 1972).
29. N. H. Yoshida, U.S. Patent 3,657,144 (April 18, 1972).
30. M. Kitajima, Y. Tsuneoka, and A. Kondo, U.S. Patent 3,703,576 (November 21, 1972).
31. M. Kitajima, A. Kondo, and F. Arai, U.S. Patent 3,951,851 (April 20, 1976).
32. M. Fukushima, Y. Inaba, S. Kobari, and M. Morishita, U.S. Patent 3,891,570 (June 24, 1975).
33. S. Kasai and M. Koishi, Studies on the preparation of ethylcellulose microcapsules containing magnesium aluminium hydroxide hydrate, *Chem. Pharm. Bull. 25*:314–320 (1977).

34. M. Itoh and M. Nakano, Sustained release of drugs from ethylcellulose microcapsules containing drugs dispersed in matrices, *Chem. Pharm. Bull. 28*:2816–2819 (1980).
35. G. P. D'Onofrio, R. C. Oppenheim, and N. E. Bateman, Encapsulated microcapsules, *Int. J. Pharm. 2*:91–99 (1979).
36. L. D. Morse, M. J. Boroshok, and R. W. Grabner, U.S. Patent 4,107,072 (August 15, 1978).
37. L. D. Morse, W. G. Walker, and P. A. Hammes, U.S. Patent 4,123,382 (October 31, 1978).
38. M. Kitajima, A. Kondo, M. Morishita and J. Abe, U.S. Patent 3,714,065 (January 30, 1973).
39. M. Morishita, Y. Inaba, M. Fukushima, S. Kobari, A. Nagata, and J. Abe, U.S. Patent 3,943,063 (March 9, 1976).
40. D. L. Gardner, D. J. Fink, A. J. Patanus, W. C. Baytos, and C. R. Hassler, Steroid release via cellulose acetate butyrate microcapsules, in *Controlled Release Polymeric Formulations* (D. R. Paul and F. W. Harris, eds.), American Chemical Society, Washington, D.C., 1976, pp. 171–181.
41. M. Fukushima, Y. Inaba, S. Kobari, and M. Morishita, U.S. Patent 3,891,570 (June 24, 1975).
42. M. Nakano, M. Itoh, K. Juni, H. Sekikawa, and T. Arita, Sustained urinary excretion of sulfamethizole following oral administration of enteric coated microcapsules in humans, *Int. J. Pharm. 4*:291–298 (1980).
43. T. M. S. Chang, F. C. MacIntosh, and S. G. Mason, Semipermeable aqueous microcapsules 1. Preparation and properties, *Can. J. Physiol. Pharmacol. 44*:115–128 (1966).
44. T. M. S. Chang and M. J. Poznansky, Semipermeable microcapsules containing catalase for enzyme replacement in acatalasaemic mice, *Nature 218*:243–245 (1968).
45. T. M. S. Chang, U.S. Patent 3,522,346 (July 28, 1970).
46. T. M. S. Chang and N. Malave, The development and first clinical use of semipermeable microcapsules (artifical cells) as a compact artifical kidney, *Trans. Amer. Soc. Artif. Int. Organs 16*:141–148 (1970).
47. T. M. S. Chang, J. F. Coffey, C. Lister, E. Taroy, and A. Stark, Methaqualone, methyprylon and glutethimide clearance by the ACAC microcapsule artificial kidney: in vitro and in patients with acute intoxication, *Trans. Amer. Soc. Artif. Int. Organs 19*:87–91 (1973).
48. T. M. S. Chang and M. Migchelsen, Characterization of possible "toxic" metabolites in uremia and hepatic coma based on the clearance spectrum for larger molecules by the ACAC microcapsule artifical kidney, *Trans. Amer. Soc. Artif. Int. Organs 19*: 314–319 (1973).

49. A. A. El-Sayed, A. A. Badawi, and A. M. Fouli, Evaluation of solvent used in the preparation of solid dispersions and microcapsules on the dissolution of drugs, *Pharm. Acta Helv.* 57:61–64 (1982).
50. R. G. Bayless, C. P. Shank, R. A. Botham, and D. W. Werkmeister, U.S. Patent, 3,674,704 (July 4, 1972).
51. R. G. Bayless, U.S. Patent 4,107,071 (August 15, 1978).
52. A. M. Purcell. U.S. Patent 3,640,892 (February 8, 1972).
53. Y. Maekawa, S. Miyano, and A. Kondo, U.S. Patent 3,726,806 (April 10, 1973).
54. S. Kondo and H. Nakano, U.S. Patent 4,102,806 (July 25, 1978).
55. J. D. Scheu, G. J. Sperandio, S. M. Shaw, R. R. Landolt, and G. E. Peck, Use of microcapsules as timed-release parenteral dosage form: application as radiopharmaceutical imaging agent, *J. Pharm. Sci. 66*:172–177 (1977).
56. T. R. Tice, W. E. Meyers, D. H. Lewis, and D. R. Cowsar, Controlled release of ampicillin and gentamicin from biodegradable microcapsules, *Abstracts, 8th International Symposium on Controlled Release of Bioactive Materials, Ft. Lauderdale, July,* 1981, pp. 108–111.

5

Interfacial Polycondensation

Interfacial polycondensation involves the reaction of various monomers at the interface between two immiscible liquid phases to form a film of polymer that encapsulates the disperse phase. Usually two reactive monomers are employed, one dissolved in the aqueous disperse phase containing a solution or dispersion of the core material, and the other dissolved after the emulsification step in the nonaqueous continuous phase. The water-in-oil (w/o) emulsion formed requires the addition of a suitable emulgent as stabilizer. Figure 5.1 shows a diagramatic representation of the process, which is often referred to as interfacial polymerization. The monomers diffuse together and rapidly polymerize at the interface between the phases to form a thin coating, and the by-product of the reaction is neutralized by added material such as an alkaline buffer. The degree of polymerization can be controlled by the reactivity of the monomers chosen, their concentration, the composition of either phase vehicle, and by the temperature of the system. Variation in particle size of the disperse phase controls the particle size of the product. The reaction between the monomers is quenched by depletion of monomer, which is frequently accomplished by adding excess continuous-phase vehicle to the emulsion.

Interest in the technology has grown enormously since 1959, when Du Pont, at a national American Chemical Society meeting, demonstrated how polyamide fiber could be prepared by interfacial polymerization. In the last 25 years many different monomer combinations have been investigated for the microencapsulation of pharmaceuticals. Table 5.1 lists polymers formed from various monomer combinations that will be discussed in greater detail in later sections of this chapter.

Of particular importance in the development of the published pharmaceutical literature in this field have been the many research papers of Chang and co-workers in Montreal and Kondo and co-workers in Tokyo. As with most microencapsulation processes, many of the interfacial polycondensation techniques are covered by patents, a number of which will be referred to in this chapter. A number of reviews

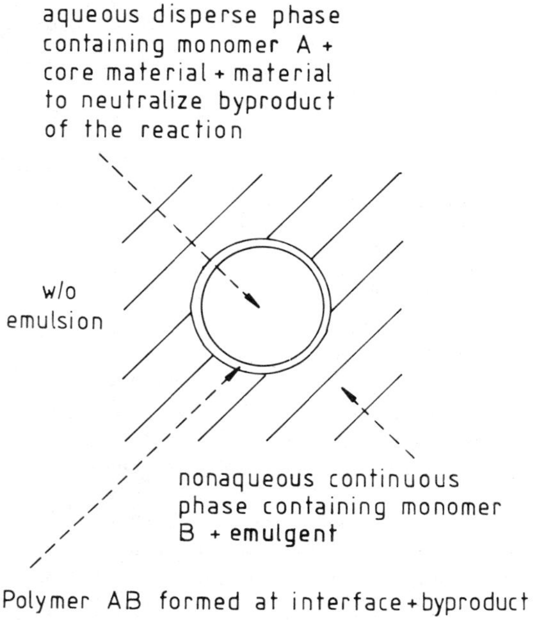

Figure 5.1 Schematic representation of microencapsulation of a droplet by interfacial polycondensation.

on interfacial polycondensation have also been published [1-5] that refer to many uses apart from those relating to pharmaceuticals.

However, despite the considerable interest in medical applications of the process, very few have been commercially exploited. This frequently arises because of (1) toxicity problems associated with unreacted monomer, the polymer, or other constitutents of the system; (2) excessive drug degradation caused by reaction with monomer; (3) the high permeability of the coating formed to low-molecular-weight species, including most drugs; (4) the fragility of the microcapsules formed; and (5) the lack of biodegradability of the product. Polyamide encapsulation is the most extensively investigated of the procedures involving interfacial polymerization for the coating of high-molecular-weight material such as enzymes. Accordingly this procedure will be reviewed in depth so as to explain details of the production and its limitations for the microencapsulation of pharmaceuticals.

Table 5.1 Principal Monomer Combinations Investigated for the Microencapsulation of Pharmaceuticals by Interfacial Polycondensation

Aqueous phase monomer A	Nonaqueous phase monomer B	Polymer AB wall material formed
1. Polyamine	Polybasic acid halide	Polyamide
e.g., 1,6-hexamethylene diamine	sebacoyl chloride	nylon 6-10
piperazine	terephthaloyl chloride	polyterephthal-amide
L-lysine	terephthaloyl chloride	poly(terephtha-loyl L-lysine)
2. Polyphenol	Polybasic acid halide	Polyester
e.g., 2,2-bis(4-hydroxy-phenyl)propane	sebacoyl chloride	polyphenyl ester
3. Polyamine e.g., 1,6-hexamethylene-diamine	Bischloroformate 2,2-dichlorodiethyl ether	Polyurethane polyurethane

5.1 POLYAMIDE

5.1.1 Nylon 6-10 Microcapsules

Wittbecker and Morgan [6] were the first to publish details of an interfacial polycondensation process involving a Schotten-Baumann type of reaction between an acid dichloride and a compound containing reactive hydrogen atoms (—NH, —OH, —SH). As an example of this type of nucleophilic reaction, a poly(hexamethylene sebacamide) polymer was formed at the interface between a solution of 1,6-hexamethylene diamine in water and sebacoyl chloride in a water-immiscible solvent as follows:

$$H_2N(CH_2)_6NH_2 + ClC(=O)(CH_2)_8C(=O)Cl \longrightarrow \left[\sim HN(CH_2)_6NHC(=O)(CH_2)_8C(=O)\sim\right]_n + HCl$$

1,6-Hexamethylene diamine — Sebacoyl chloride — Nylon 6-10

An inorganic base such as sodium bicarbonate or sodium hydroxide in the aqueous phase was used to neutralize the hydrogen chloride formed in the condensation. The polyamide polymer formed is called nylon 6-10, the first and second numbers representing the number of carbon atoms in the diamine and acid dihalide, respectively.

Figure 5.2 shows a flow diagram for the production of nylon 6-10 coated microcapsules based on the above polymerization procedure and on the publications of Chang et al. [7-9]. Suitable core material should form a suspension or a solution of macromolecules in the water

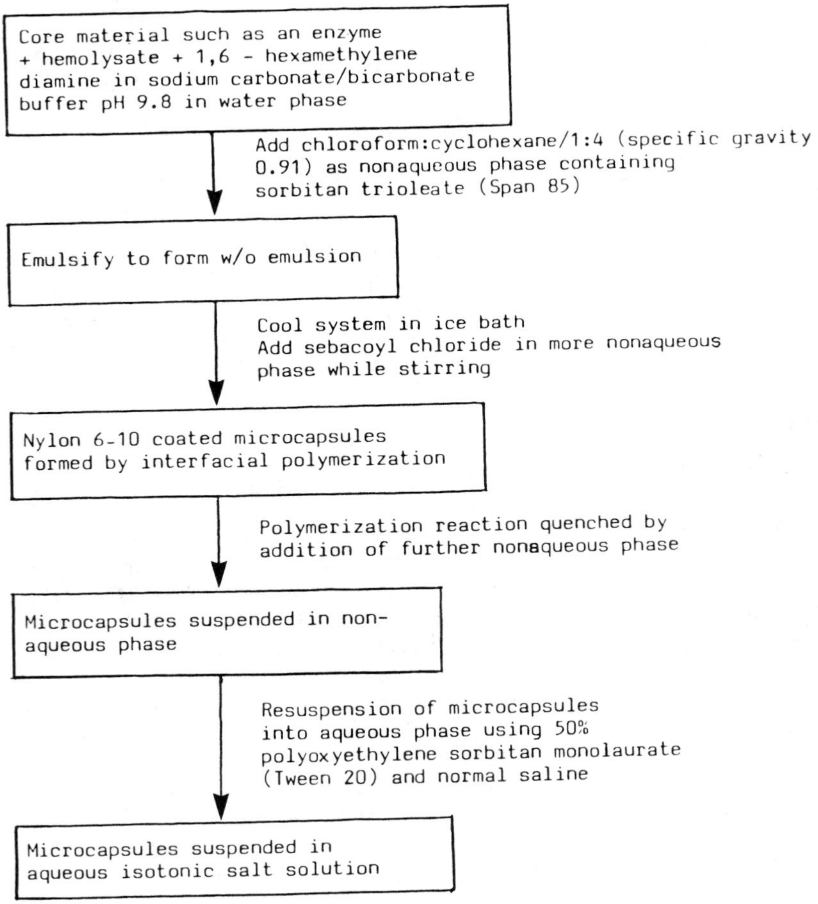

Figure 5.2 Typical steps of a nylon 6-10 microencapsulation process.

phase, as Chang and Poznansky [10] have reported that low-molecular-weight ions such as urea, creatinine, glucose, and salicylic acid readily permeate the nylon wall formed. Accordingly, enzymes are frequently chosen as core material, but these high-molecular-weight substances tend to undergo chemical reaction between their active hydrogen groups and the acid dihalide. In order to lessen the resultant inactivation of the core material during encapsulation, a hydrophilic colloid with a competitive source of reactive groups such as a hemolysate of red blood corpuscles may be added with the core material. Despite this protection, often more than 50% of the enzyme activity is lost upon encapsulation. The bicarbonate buffer is included to neutralize the hydrogen chloride liberated in the polymerization reaction, as it might otherwise damage the encapsulated material.

Considerable care needs to be exercised in the choice of the nonaqueous-phase composition. Organic liquids such as chloroform, cyclohexane, and carbon tetrachloride do not readily denature proteins and in suitable combination can be matched to the density of the microcapsules so as to reduce their rate of sedimentation. Excessive downward sedimentation of microcapsules during production is undesirable because it leads to aggregation problems, whereas excessive upward sedimentation can also aid aggregation and can cause difficulties in the separation of the microcapsules during subsequent centrifugation procedures. The partition coefficient of the diamine between the aqueous and nonaqueous phase is very important and is influenced very much by the composition of the latter. Sebacoyl chloride is almost insoluble in water, unlike the diamine, which has appreciable solubility in both phases, and tends to diffuse into the nonaqueous phase to replace that consumed in the polymerization reaction. The polymer therefore tends to form in the organic phase, but because the rate of monomer reaction exceeds the rate of diffusion of the diamine into the nonaqueous phase, the nylon tends to be formed close to the interface but in the organic, nonpolar phase. Also, Morgan and Kwolek [11] reported that a nonaqueous solvent of chloroform alone gave thick coarse films, whereas cyclohexane gave thin, smooth films. In suitable combination as indicated in Fig. 5.2, both solvents gave thin, strong, and relatively nonporous films.

Emulsification of the aqueous and nonaqueous phases is aided by inclusion of a suitable nonionic emulgent such as 1% sorbitan trioleate, H.L.B. 1.8 (Span 85), which helps the transfer of both diamine and acid dihalide across the interface. Unlike ionic emulgents, Span 85 appears to cause little damage to the protein. Emulsification is accomplished using an electrically driven stirrer, higher speeds giving a smaller and narrower size range of droplets for encapsulation. Droplets smaller than 20 μm in diameter are usually prepared using up to 15% Span 85 to lessen aggregation and also by using a suitable homogenizer.

The emulsion formed is cooled by immersion in an ice bath prior to the addition of the acid dihalide in more nonpolar solvent, so as to absorb the heat liberated during the polymerization reaction and to slow the rate of reaction. If not absorbed, the liberated heat might tend to denature the core material. If the rate of polymerization is too fast, an excessively porous membrane tends to form; whereas if the reaction is overly prolonged, a coarse membrane containing excessive low-molecular-weight polymer of lower equivalent pore radius tends to form. The sebacoyl chloride must obviously react with the protein to denature it and form part of the membrane. However, since protein will not diffuse readily into the nonpolar phase, where polymerization mainly occurs, it is unlikely that crosslinked protein is a significant structural element of the wall.

After quenching the polymerization reaction by addition of further nonaqueous solvent, the newly formed microcapsules must be quickly removed from the nonpolar vehicle to lessen degradation of core material and further slight polymerization. This is accomplished by gently centrifuging the microcapsules, decanting the supernatant, and dispersing them in a 50% v/v solution of polysorbate 20 (Tween 20) in water. Lower concentrations of Tween do not effectively separate the microcapsules from the nonaqueous vehicle. The resultant system is then gently centrifuged to separate the microcapsules, which are rapidly resuspended in normal saline and repeatedly washed if necessary to free them from residual Tween 20. Microcapsules suspended in normal saline are normally stored at 2 to 4°C prior to use.

Microscopic examination of the microcapsules shows that they have a spherical shape with thin flexible walls about 0.2 μm thick. Some microcapsules enclose smaller microcapsules. In hypertonic solution the microcapsules tend to collapse, acquiring bizarre shapes. Hence the need to use high concentrations of protein or other macromolecules in the core so as to maintain the turgor of the microcapsules while being agitated in the hypertonic Tween solution. The microcapsules retain appreciable enzymic activity, the encapsulated enzyme being unable to escape from the microcapsules, but the wall being freely permeable to low-molecular-weight substrate, having an equivalent pore radius of 16 to 18Å. These nylon 6-10 microcapsules have no surface charge and are nonantigenic.

Because of their semipermeable and other properties, Chang and co-workers extensively investigated the use of these microcapsules as "artificial cells" and for other purposes. Chang [12] described an extracorporeal shunt system for perfusing blood over nylon-coated microcapsules, 90 μm in mean diameter, containing the enzyme urease, before being returned to the circulation in dogs. Blood urea was significantly lowered by the device, which, however, necessitated use of an anticoagulant and the cannulation of an artery and vein. Blockage problems also tended to occur by migration of microcapsules in the shunt in the direction of blood flow. The ion-exchange resin Dowex-50W-X12 was also nylon-encapsulated and was successfully used to remove blood ammonia when loaded into a shunt. Most of the carbonic anhydrase activity of the hemolysate survived encapsulation.

Chang [13] found that injection of semipermeable nylon microcapsules containing asparaginase to suppress the growth of an implanted asparaginase-dependent tumor in mice were much more effective than the free enzyme control because of their extended duration of action. In another paper Chang [14] reported that enzymes such as asparaginase, catalase, and urease could be stabilized by microencapsulation in the presence of a high concentration of protein solution. The stability of enzymes after microencapsulation at body temperature of 37°C could be increased by treatment with the crosslinking agent glutaraldehyde. The improved stability and extended duration of action of asparaginase and catalase following nylon microencapsulation were subsequently confirmed by Sui Chang and Chang [15] and Poznansky and Chang [16], respectively.

In 1972 Chang [17] published a book entitled *Artificial Cells* in which details of the preparation of these and other microcapsules and their biophysical properties were reviewed. He also discussed a large number of possible medical applications for the microcapsules. Subsequently, Chang [18, 19] again reviewed physicochemical and biological properties of these microcapsules and their clinical applications.

A number of other research workers have applied the Chang procedure with modifications for the nylon 6-10 microencapsulation of various pharmaceuticals. Thus Luzzi et al. [20] microencapsulated an aqueous solution of sodium pentobarbital and after decanting the nonaqueous supernatant of chloroform and cyclohexane, the microcapsules were recovered from the slurry and freed of residual moisture and organic solvent by spray or vacuum drying. Photomicrographs (see Fig. 5.3) showed that the spray-dried product was composed of irregularly shaped particles with several of them clumped together in most cases, whereas the microcapsules were spherical and almost unclumped before drying.

Release of drug into water and phosphate buffer from spray dried microcapsules was rapid, as might be expected. About 90% release of encapsulated drug occurred within 45 min of the dissolution treatment. The vacuum dried microcapsules initially released drug more rapidly but were otherwise similar. Tableting the spray dried microcapsules produced a slower drug release rate which could be decreased by increasing tablet hardness.

Mori et al. [21] also microencapsulated asparaginase with nylon 6-10 by a modification of the Chang procedure using casein, gelatin, hemoglobin, or albumin as protective proteins, and ethanol instead of Tween 20 solution for removal of the microcapsules from the nonaqueous vehicle. The particle size of the microcapsules could be controlled by the concentration of Span 85 and stirrer speed used. Under optimal

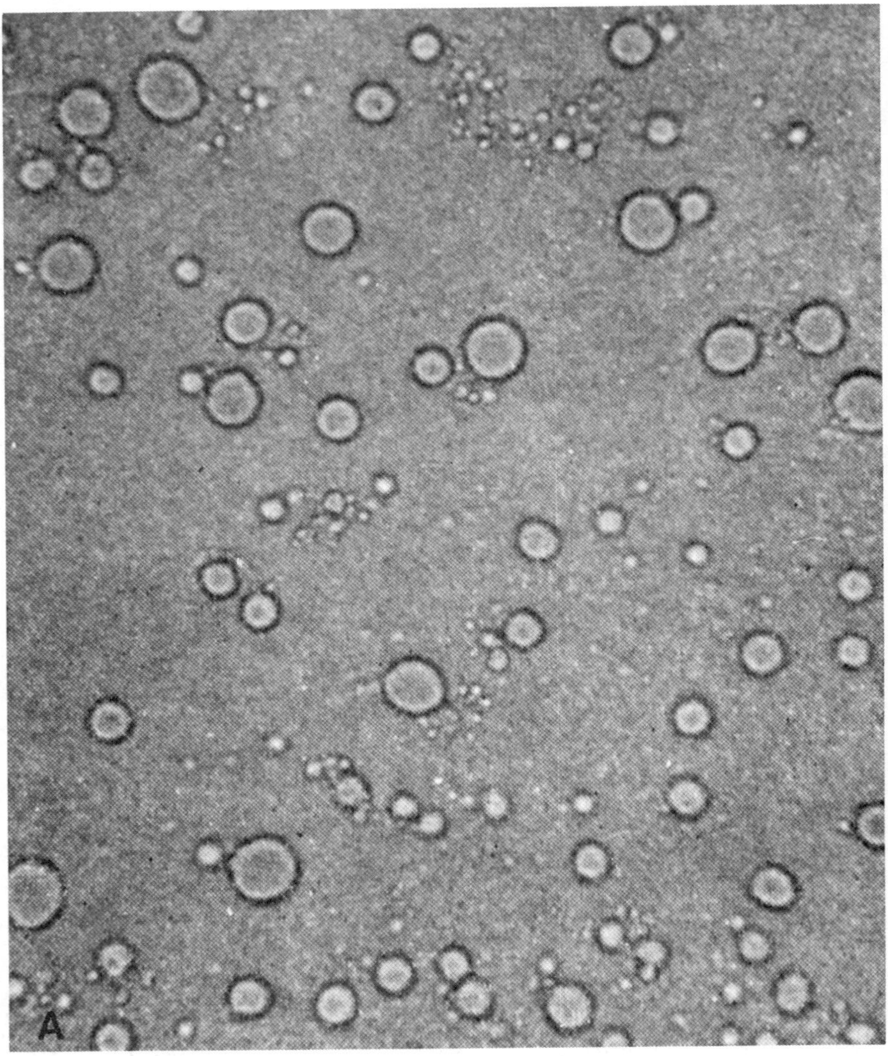

Figure 5.3 Photographs of nylon microcapsules in (A) liquid medium and (B) after drying. (Reproduced with permission of the copyright owner, American Pharmaceutical Association; from Ref. 20.)

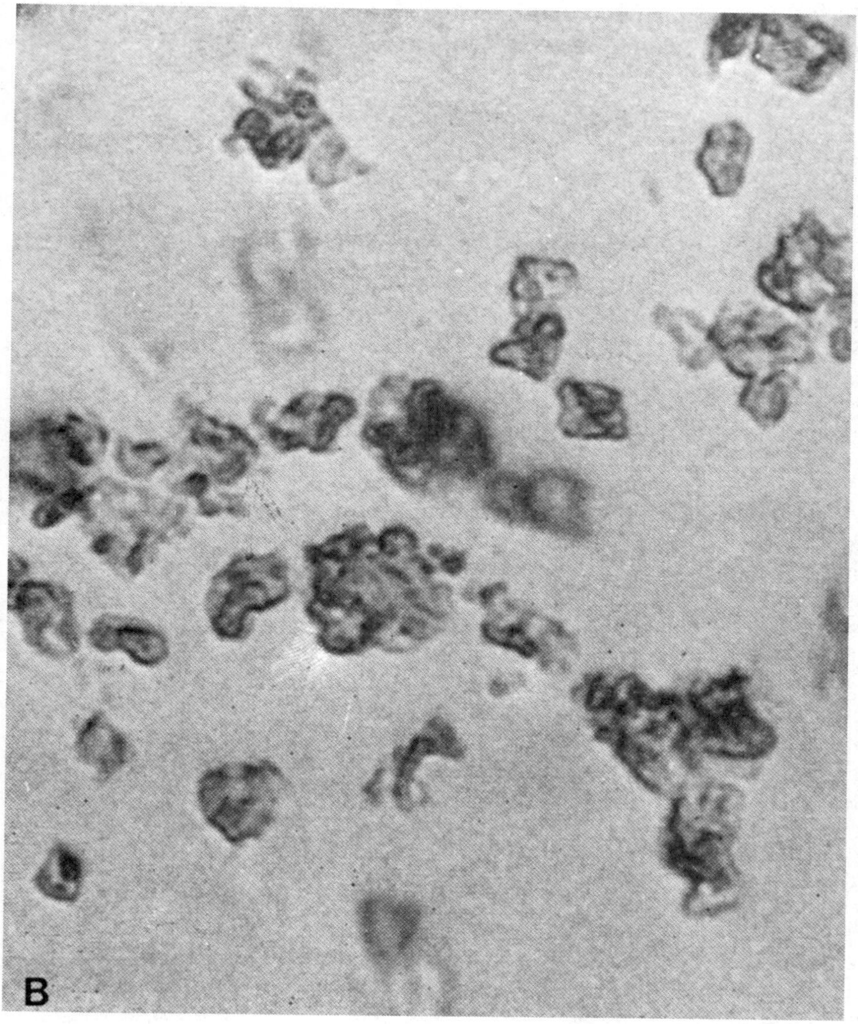

Figure 5.3 (Continued)

conditions of preparation only about 37% of the enzyme activity of the unencapsulated material was retained. However, no leakage of enzyme from the microcapsules or loss of activity after being used five times was reported.

Jenkins and Florence [22] examined drug containing nylon 6-10 coated microcapsules by scanning electron microscopy. The microcapsules were much smaller (2 to 5 μm) than those prepared by Chang and co-workers (20 to 200 μm), and were spherical with rough, porous surfaces. A number of microcapsules had splits that extended deep into the core that may have resulted from preparative procedures for microscopy. A number of other microcapsules were joined together with distorted faces at the points of contact. No significant change in diameter of the microcapsules was recorded after in vitro dissolution studies, though the pores were noticed to increase in size. Nylon films formed on planar interfaces under nonturbulent conditions were also porous in appearance. In a subsequent paper by the same authors [23], the phenothiazine tranquilizers trifluoperazine embonate and pericyazine embonate were dissolved in polyoxyethylene glycol 400 (PEG 400) prior to nylon 6-10 encapsulation. The drugs as their hydrochlorides were unsuitable because of reaction with the hexamethylene diamine. Microcapsules prepared by the method of Chang with a simple aqueous core had a liquid interior and discrete walls, unlike those prepared with the drug and diamine in solution in PEG, which were apparently solid throughout. In vitro drug release could be decreased by doubling the quantity of nylon monomers and decreasing the pH of the dissolution medium. The microcapsules showed sustained-release properties when injected into dogs. However, the authors considered them unlikely to be suitable for human use because of lack of evidence of biodegradability.

In 1975 McGinity et al. [24] published an improved method for the microencapsulation of soluble drugs. Nylon 6-10 was used to coat the aqueous droplet phase containing a solution of the drug sulfathiazole sodium and gelatin by the Chang procedure and the resulting capsules were hardened by formalin treatment. The microcapsules obtained did not adhere, flowed well, and could be varied in diameter by controlling the stirrer speed. Nylon-coated microcapsules containing unhardened gelatin, proteins, various cellulose derivatives, or alginates were generally difficult to separate. Dissolution studies indicated that the hardened gelatin matrix in the core significantly delayed drug release compared to a nylon-encapsulated core containing no gelatin. In a subsequent paper by McGinity et al. [25], other drugs, such as sodium salicylate, caffeine, theobromine, methantheline bromide, benzalkonium chloride, phenytoin, diphenhydramine hydrochloride, diazepam, theophylline, and morphine sulfate were encapsulated using the same procedure. In addition to formalized gelatin, calcium sulfate hemihydrate, and calcium alginate were also used as core materials. The capsules formed were dense and free-flowing. However, quarternary compounds were difficult to encapsulate, as they interfered appreciably with the formation of the nylon coating. Also diazepam, caffeine, theophylline, phenytoin, and morphine sulfate were not fully encapsulated, as they tended to partition from the microcapsules into the nonaqueous phase of chloroform and cyclohexane during manufacture. Scanning electron microscopy (see Fig. 5.4) showed that both formalized gelatin and calcium alginate matrix-containing microcapsules had a rough, porous, and platelike surface. In contrast, the microcapsules containing the calcium sulfate matrix were very porous and fibrous, probably because of the increase in volume associated with the hydration of the calcium sulfate in the core.

DeGennaro et al. [26] studied the effect of incorporating the crosslinking agent diethylenetriamine or triethylenetetramine progressively in place of 1,6-hexamethylenediamine upon the release of sodium pentobarbital from nylon 6-10 microcapsules. As nylon 6-10 is a linear polymer, it was hoped that the nonlinear polymer formed by addition of such crosslinking agents would have a marked effect on the release properties of the membrane. A modification of the Chang procedure was employed using glycerol in the aqueous phase, carbon tetrachloride as the nonaqueous phase, and recovery by filtration, oven drying, and vacuum desiccation. When diethylenetriamine was used to replace the diamine, the percentage of sodium pentobarbital released at various sampling intervals tended to show an irregular increase with increasing amount of replacement. A similar effect was found when triethylenetetramine was used to replace diamine, confirming that addition of either crosslinking agent is not a suitable means of prolonging drug release from nylon-coated microcapsules. Had a higher-molecular-weight core material been checked, it might have revealed differences in permeability of the coatings.

5.1.2 Polyphthalamide Microcapsules

Koishi et al. [27] described how polyphthalamide microcapsules could be prepared by an interfacial polymerization reaction between a water-soluble monomer 1,6-hexamethylenediamine or piperazine (diethylenediamine) and a monomer soluble in an immiscible organic solvent, such as phthaloyl chloride, isophthaloyl chloride, or terephthaloyl chloride. The reaction was based on the studies of Katz [28] and Shashoua and Eareckson [29] on interfacial polycondensation between a diamine and a phthaloyl dichloride. The aqueous solution of the water-soluble monomer containing sodium carbonate to neutralize the liberated hydrogen chloride was emulsified in chloroform:cyclohexane/1:4 or 1:3 to form a w/o emulsion using Span 85 as a nonionic emulgent and a high-speed stirrer. The phthaloyl dichloride in the mixed solvent was then added to the emulsion, whose container was immersed in ice to absorb the heat liberated during the polymerization reaction. Further mixed solvent was added to quench the reaction, and the newly formed

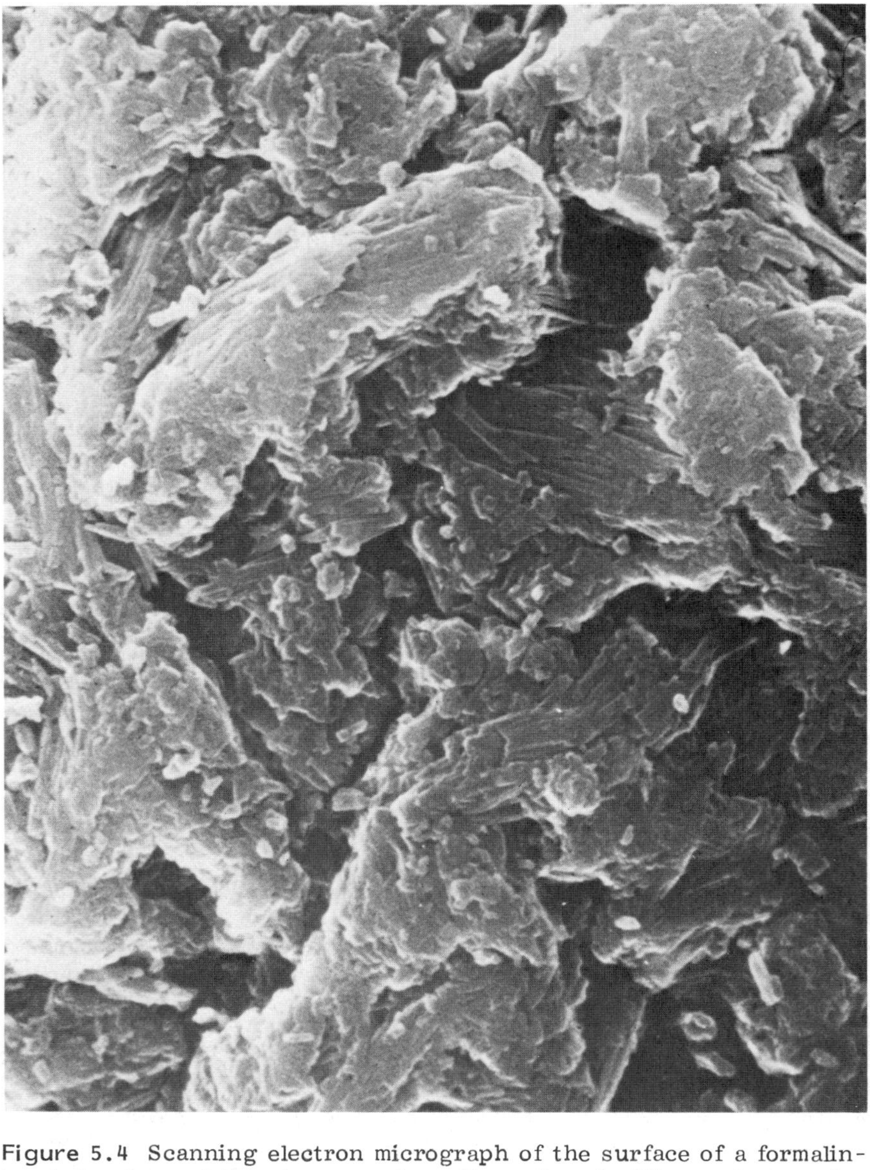

Figure 5.4 Scanning electron micrograph of the surface of a formalin-treated nylon gelatin microcapsule. (Reproduced with permission of the copyright owner, American Pharmaceutical Association; from Ref. 25 and courtesy of Dr. J. McGinity.)

microcapsules were separated by centrifugation, dispersed in 50% aqueous Tween 20 solution, washed by dilution with further water, and finally dispersed in dilute aqueous Tween 20 solution. Some typical polymerization reactions involved are shown below.

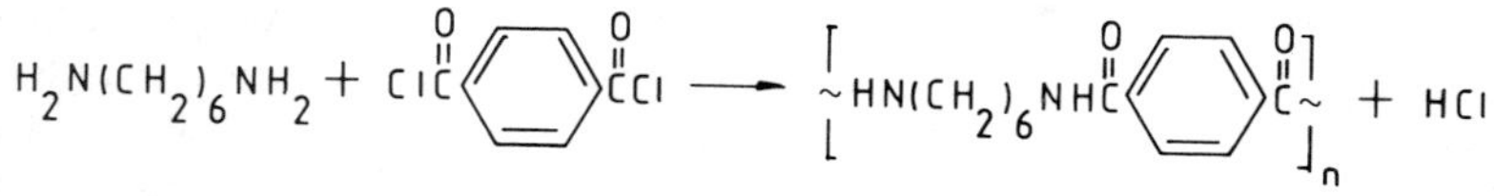

1,6-Hexamethylene diamine   Terephthaloyl chloride   Polyterephthalamide

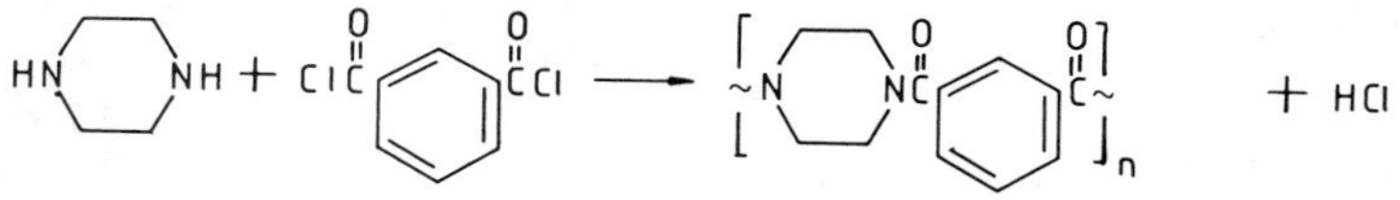

Piperazine   Isophthaloyl chloride   Poly(piperazineisophthalamide)

Spherical microcapsules that varied in size from <1 to 10μm were obtained. Koishi et al. [27] studied the partitioning of the diamines between aqueous and organic phases and the effect of Span 85 on them in an attempt to optimize polymerization at the interface. The piperazine yielded microcapsules more easily with the 1:3 mixture of chloroform and cyclohexane than with the 1:4 mixture, presumably due to its being more soluble in the former mixture. Increasing concentration of Span 85 up to 20% aided the formation of microcapsules, probably by increasing the solubility of the diamines in the organic phase due to solubilization. Also, the microcapsules prepared using piperazine were more resistant to centrifugation than those prepared using 1,6-hexamethylenediamine. Increasing concentration of Span 85 up to 5% caused the size distribution of microcapsules obtained to become narrower. Size distribution was not appreciably affected by the stirrer speeds employed, provided that adequate Span 85 (5%) was present.

In a subsequent paper, Shigeri et al. [30] reported that other factors, such as decrease in temperature or monomer concentration, which lowered the rate and extent of interfacial polymerization, can increase the size of the polyamide capsules obtained. Also, Ogawa et al. [31] studied the effect of pH of the aqueous phase containing piperazine and variable concentration of Span 85 on the formation of polyterephthalamide microcapsules. Span 85 aided the formation of microcapsules increasingly as the pH rose from 9 to 12 by increasing the transfer of undissociated diamine molecules into the organic phase, whose number increased with rise in pH over this range. Span 85 had no significant effect on the formation of microcapsules outside this pH range, and polyterephthalamide microcapsules were not appreciably formed below pH 9.

Shiba et al. [32] microencapsulated an aqueous solution of bovine serum albumin in various polyamides using 1,6-hexamethylene diamine (A) or piperazine (B) as the monomer soluble in the water phase and sebacoyl chloride (C) or terephthaloyl chloride (D) as the monomer soluble in the organic phase of chloroform and cyclohexane. The order of resistance to rupture by centrifugation was BD > AD > AC > BC. The strength of AD membranes increased when prepared in the presence of the albumin, indicating that amino acid residues of this protein are chemically involved with the acid dichloride in the formation of wall material. This chemical involvement was confirmed by electrophoresis studies on isolated microcapsule membranes whose isoelectric point was reduced presumably by reaction of part of the amino residues of the albumin in the membrane with the acid dichloride. In a subsequent paper by Shiba et al. [33], the electrophoretic behavior of polyphthalamide (BD) microcapsules containing an aqueous solution of bovine serum albumin and of their isolated membranes was investigated. The microcapsules were negatively charged, whereas their membranes were positively charged below pH 3.5 and negatively charged above this pH. These results indicated that the net charge exhibited by the microcapsules is highly dependent on the outer surface charge of the membrane, which probably contains an excess of carboxyl over amino groups derived from the chemical reactivity of the albumin involved in the wall formation.

Takahashi et al. [34] investigated the permeability of polyphthalamide microcapsules prepared from piperazine and terephthaloyl chloride to the entry of various electrolytes. Ion permeability was found to be low and almost independent of temperature, indicating the suitability of this coating material for the design of sustained-release products. In another paper by Takamura et al. [35], the permeability of the electrolytes sodium chloride and sodium sulfate into several kinds of polyamide microcapsules was investigated. It was found that permeability was less when the polyamide contained an aromatic ring rather than a polymethylene chain. Also, increase in apparent viscosity of the microcapsulated aqueous phase by use of polyvinylpyrrolidone (PVP) greatly decreased electrolyte permeability, indicating that diffusion of electrolyte in the interior of the microcapsules was the rate-limiting step during permeation.

Florence and Jenkins [23] microencapsulated trifluoperazine embonate dissolved in polyoxyethylene glycol 400 with poly(piperazine terephthalamide). SEM studies showed the microcapsules not to be well formed, and there was no prolonged drug activity noted when tested in dogs. Kondo and Muramatsu [36] successfully encapsulated the enzyme arginase in the same coating material. However, the specific activity of the microencapsulated enzyme was only about 12% of the native form, probably because of interaction with the terephthaloyl chloride. The presence of 0.1% Span 85 or greater during the production prevented the enzyme from being even further inactivated, probably by forming a protective interfacial layer. Enzyme activity could be increased to 60% if 1% albumin was added to the aqueous phase containing the enzyme, by acting as a competitive source of amino groups. Wood and Whateley [37] microencapsulated aqueous solutions of the enzymes α-chymotrypsin and histidase in nylon 6-10 or poly(piperazine terephthalamide) microcapsules, and maximum activity did not exceed 40%. The maximum of the apparent pH-activity curve for α-chymotrypsin was shifted one unit, indicating that the pH of the microenvironment surrounding the encapsulated enzyme was different from that in bulk solution. This effect may contribute to the reduction in the apparent activity of the microencapsulated enzyme.

5.1.3 Sulfated and Carboxylated Polyphthalamide Microcapsules

Polyphthalamide microcapsules, being uncharged, were found to form aggregates if stored in water for prolonged periods. This tendency could be lessened by addition of Tween 20 as a dispersing agent. Accordingly Koishi et al. [38] prepared negatively charged sulfated polyphthalamide microcapsules by interfacial polymerization between terephthaloyl chloride in the organic phase and a mixture of piperazine and 4,4'-diamino stilbene-2,2'-disulfonic acid in the aqueous phase. The formation of the spherical microcapsules obtained was practically unaffected by the molar ratio of the diamine used. However, in comparison to unsulfated microcapsules, the product was found to be more resistant to centrifugation and did not appreciably aggregate upon storage for 1 month in distilled water, presumably because of electrical repulsion between neighboring microcapsules. The binding efficiency of these sulfated microcapsules for a range of cations was investigated in a subsequent paper by Takahashi et al. [39] with a view to their possible use as ion-exchange material.

Shigeri et al. [40] prepared a range of carboxylated polyphthalamide microcapsules by reacting terephthaloyl chloride in the organic phase with each of the following amino acids in the aqueous phase: L-arginine, L-citrulline, L-cystine, L-histidine, and L-lysine. A typical chemical reaction is

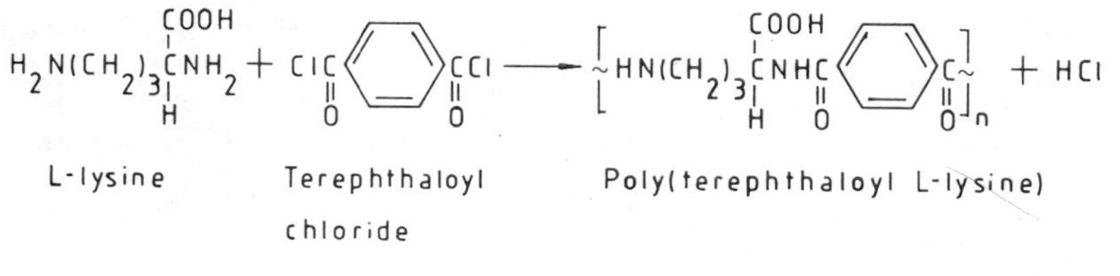

The yield of microcapsules was high when a high concentration of Span 85 and a low temperature were used, as these conditions favored the transfer of amino acids into the organic phase. The resultant microcapsules were negatively charged and did not exhibit an isoelectric point. The charge densities of the microcapsules were comparable to those found on the red blood cells from various animal species, indicating the feasibility of using such microcapsules as models for biological membranes. It was noted that as the valency of a cation increased, its binding to the microcapsules also increased. Further studies on the binding of cations to sulfonated and carboxylated polyphthalamide microcapsules using electrophoresis data were published by Takahashi et al. [34].

Kondo et al. [41] microencapsulated hemoglobin solution from sheep in poly(terephthaloyl L-lysine) microcapsules in an attempt to produce artificial red blood corpuscles. The microcapsules obtained were similar in suspension to dispersions of red blood corpuscles in that they exhibited a mean diameter of about 10 μm, negative surface charge due to dissociation of carboxyl groups of lysine residues, and pseudoplastic flow properties. However, their deformability was lower than that of red blood corpuscles because of their spherical shape, and this might make fine capillary penetration difficult.

Arakawa et al. [42] extensively examined the flow properties of these poly(terephthaloyl L-lysine) microcapsules as a model for blood and concluded from rheological studies using a capillary viscometer that in this regard they were a reasonably good model. In another paper, Arakawa and Kondo [43] reported that the oxygen dissociation equilibrium, zeta potential, and carbonic anhydrase activity of these microcapsules were almost identical to those of sheep erythrocytes but they had greater resistance to flow and lower catalase activity compared to the erythrocytes. Later Suzuki and Kondo [44] studied the interaction of these negatively charged microcapsules with positively charged poly-(diallyldimethyl ammonium) polycations. At low concentration the polycation caused aggregation of the microcapsules and in high concentration, provided that the concentration exceeded a critical value corresponding to exact electrical equivalence of the polyions involved, the polycations caused disintegration of the microcapsules. In another paper, Suzuki et al. [45] showed how such microcapsules undergo disintegration when adequate concentration of alkylpyridinium chloride surfactants was present to solubilize the polymer wall material. Adsorption of lower concentrations of surfactants caused aggregation of the microcapsules. Disintegration of the microcapsules could be prevented by crosslinking the terminal amino groups in the polymer wall material using glutaraldehyde. However, Muramatsu et al. [46] reported that such polyphthalamide microcapsules loaded with sheep hemolysate adsorbed large amounts of fibrinogen and γ-globulin. Platelet adhesion to these artifical red blood cells also occurred, the extent depending on membrane composition when plasma was present. These findings suggest that the microcapsules may cause undesirable thrombosis or angiostenosis if injected into humans or animals.

Recently Arakawa and Kondo [47] attempted to prepare by interfacial polymerization much smaller poly(terephthaloyl L-lysine) capsules in the nanometer size range and loaded with sheep hemolysate that could pass through fine blood capillaries. An electrocapillary emulsification technique was employed whereby the interfacial tension, in the absence or presence of minimal surfactant, at the oil-water interface was reduced to almost zero by an applied potential. This resulted in spontaneous emulsification, forming a monodisperse and stable emulsion with an average droplet diameter less than 1 μm. The ionic strength of the disperse phase must be higher than that of the continuous phase. Details of the theory of the emulsification process have been published by Watanabe et al. [48]. A slow rate of addition of the aqueous hemolysate solution into the organic phase, adequate sorbitan trioleate concentration to just form a monomolecular layer at the interface, and higher temperatures of emulsification tended to favor the formation of optimum nanocapsules having a narrow size range and an average diameter around 500 nm. As the applied potential was increased to about 600 volts, the catalase activity in the hemolysate decreased to about 80% of its unencapsulated activity and the carbonic anhydrase activity decreased to about 40%. Further increase in applied potential did not cause any significant further reduction in activity. Storage in a refrigerator for 40 days caused the microcapsules to lose approximately a further 60% of their residual catalase and carbonic anhydrase activity.

5.1.4 Miscellaneous Biodegradable Polyamide Microcapsules and Their Preparation

Santo and Abend [49], in a patent assigned to Pennwalt Corporation, described the production of polyamide microcapsules formed by interfacial reaction of various amino acids and acid dichlorides to produce biodegradable microcapsules suitable for ingestion. A certain amount of crosslinking appeared to occur between the carboxylic acid groups of the amino acids in the polymer chain and the —COCl groups of the acid dichloride to form anhydride linkages that strengthened the polymer wall material. Reaction of L-lysine with adipyl chloride was particularly recommended. Figure 5.5 shows a scanning electron micrograph of a dried poly(adipyl L-lysine) microcapsule containing vitamin B_{12}, prepared in the author's own laboratories. The smaller particles present are talc used to prevent clumping. The identity of the microcapsule is confirmed by x-ray probe analysis for cobalt. Santo and Vandegaer [50], in another patent assigned to the same company, described an interfacial polymerization process whereby an acid dihalide was reacted with a polyamine to produce a phosphorus-containing

Figure 5.5 Scanning electron micrograph of a dried poly(adipyl L-lysine) microcapsule together with smaller particles of talc.

coating that was claimed to be resistant at acid pH and to dissolve at alkaline pH. In a further patent assigned to Pennwalt by Vandegaer and Meier [51], an apparatus for coating injected droplets with polyamide coating was described.

An interfacial polyamide polymerization procedure for microencapsulation involving interaction between acacia and piperazine was described by Speaker and Lesko [52] in a patent assigned to Temple University. A solution of piperazine in chloroform, containing either a solution or dispersion of nonreactive drug, was emulsified into an aqueous solution of acacia. In the example cited, quinacrine hydrochloride was the drug used. After decanting the aqueous phase and washing to remove unreacted monomers, the separated microcapsules were air-dried to remove the chloroform. The product consisted of drug particles coated with a wrinkled, continuous, shell-like film, and showed sustained-release properties when tested in animals. Related encapsulation procedures were also mentioned in the patent.

Rambourg et al. [53] described the preparation of a wide variety of polyamide microcapsules containing invertase. The activity of the enzyme was reduced by about 40% upon encapsulation. Biodegradable hemoglobin microcapsules [54] were successfully prepared by reacting this polyamine with an acyldichloride followed by crosslinking with glutaraldehyde, and they retained their selective permeability properties even after rehydration from a lyophilized form.

Acryloyl chloride-lysine microcapsules—which are possibly biodegradable—containing the small polypeptide secretin, have been prepared by Bala and Vasudevan [55]. The cumulative percent of drug released from the microcapsules as a function of time was zero-order and decreased as the pH was increased over the range 2 to 10.

5.2 POLYESTER

Polyphenyl ester microcapsules may be prepared by the interfacial reaction of a polyphenol in the aqueous phase with an acid dichloride in the nonaqueous phase. Suzuki et al. [56] prepared such microcapsules by interaction of bisphenol A, i.e., 2,2-bis(4-hydroxyphenyl) propane, and sebacoyl chloride using benzene as the nonaqueous phase and a procedure that was otherwise similar to that employed for nylon 6-10 encapsulation. The interfacial polycondensation reaction was based on the studies of Eareckson [57] and is shown on the next page.

The microcapsules obtained were spherical, probably because of the osmotic pressure of the polyethylene glycol they contained. The membrane thickness was estimated to be of the order of several hundred Angstrom units and was freely permeable to low-molecular-weight ions such as sodium and hydroxyl. In a subsequent paper the effect of chemical structure of the bisphenol, i.e., 2,2-bis(4-hydroxyphenyl) propane or 4,4-dihydroxydiphenyl sulfone, and of the acid dichloride

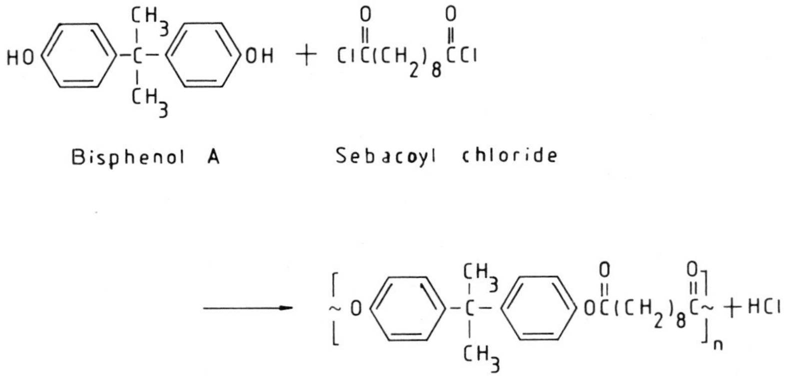

Polyphenol ester

i.e., terephthaloyl chloride or sebacoyl chloride, on the size distribution and mean diameter of microcapsules obtained was investigated by Wakamatsu et al. [58]. Adequate Span 85 concentration favored the transfer of bisphenol into the nonaqueous phase and affected the size distribution of the microcapsules obtained. An increase in the percentage of reacted bisphenol resulted in a narrower size distribution of microcapsules produced.

5.3 POLYURETHANE

Suzuki et al. [56] prepared polyurethane microcapsules using the interfacial polymerization reaction between a diamine in the aqueous phase and a bischloroformate in the nonaqueous phase. The interfacial polycondensation reaction was based on the studies of Wittbecker and Katz [59]. The aqueous phase contained 1,6-hexamethylene diamine, propylene glycol, and sodium carbonate in water, whereas the oil phase contained 2,2-dichlorodiethyl ether and Span 85 in benzene or chloroform: cyclohexane/1:4. The chemical reaction involved in the polymerization process is shown below:

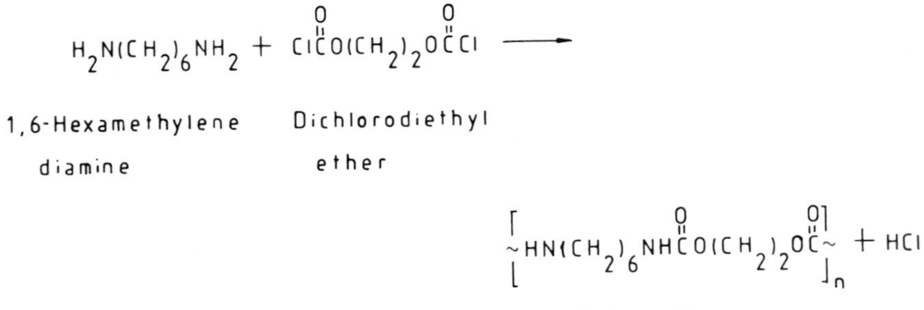

Polyurethane

The microcapsules obtained by centrifugation and washing with further organic solvent containing Span 85 were finally dispersed in a dilute solution of Tween 20 in water. The product contained spherical microcapsules, probably because of the osmotic pressure caused by the polyethylene glycol solution contained within the capsules. Like other types of microcapsules prepared by interfacial polymerization, their membrane thickness was estimated to be of the order of several hundred Angstrom units and was freely permeable to low-molecular-weight sodium and hydroxyl ions. Shigeri and Kondo [60] investigated the permeability of these microcapsules toward the inorganic electrolytes, cesium, sodium, and lithium chlorides, sodium bromide, sodium sulfate, and sodium chromate by electrical conductance measurements. The microcapsules were readily permeable to these cations and anions, the terminal rate being dependent on the hydration of the ions involved.

5.4 MISCELLANEOUS

Many other interfacial polycondensation reactions, involving the formation of polyurea, polycarbonate, polysulfonate, polysulfonamide, polyamide-polyurea, and other types of polymer wall material have been described in the patent literature. A selection of the relevant patents is given in Refs. 61-68. Vandegaer and Meier [69], in a patent assigned to Pennwalt Corporation, described the use of finely divided solid such as silica or talc to lessen aggregation in microcapsules produced by various interfacial polycondensation processes.

REFERENCES

1. H. Nack, Microencapsulation techniques, applications and problems, *J. Soc. Cosmetic Chem. 21*:85–98 (1970).
2. G. O. Fanger, Microencapsulation, a brief history and introduction, in *Microencapsulation: Processes and Applications* (J. E. Vandegaer, ed.), Plenum, New York, 1974, pp. 1–20.
3. P. L. Madan, Microencapsulation II. Interfacial reactions, *Drug Develop. Indust. Pharm. 4*:289–304 (1978).
4. A. Kondo, *Microcapsule Processing and Technology* (J. Wade Van Valkenburg, ed.), Dekker, New York, 1979, pp. 35–45.
5. R. C. Koestler, Microencapsulation by interfacial polymerization techniques—agricultural applications, in *Controlled Release Technologies: Methods, Theory and Applications, Vol. II* (A. F. Kydonieus, ed.), CRC Press, Boca Raton, Fla., 1980, pp. 117-132.
6. E. L. Wittbecker and P. W. Morgan, Interfacial Polycondensation, I, *J. Polym. Sci. 40*:289–297 (1959).
7. T. M. S. Chang, Semipermeable microcapsules, *Science 146*:524–525 (1964).

8. T. M. S. Chang, F. C. MacIntosh, and S. G. Mason, Semipermeable aqueous microcapsules I. Preparation and properties, *Can. J. Physiol. Pharmacol.* *44*:115–128 (1966).
9. T. M. S. Chang, Microcapsules as artificial cells, *Sci. J.* *3*(7): 62–67 (1967).
10. T. M. S. Chang and M. J. Poznansky, Semipermeable aqueous microcapsules (artificial cells). V. Permeability characteristics, *J. Biomed. Mater. Res.* *2*:187–199 (1968).
11. P. W. Morgan and S. L. Kwolek, Interfacial polycondensation II. Fundamentals of polymer formation at liquid interfaces, *J. Polym. Sci.* *40*:299–327 (1959).
12. T. M. S. Chang, Semipermeable aqueous microcapsules ("artificial cells") with emphasis on experiments in an extracorporeal shunt system, *Trans. Amer. Soc. Artif. Int. Organs* *12*:13–19 (1966).
13. T. M. S. Chang, The in vivo effects of semipermeable microcapsules containing L-asparaginase on 6C3 HED lymphosarcoma, *Nature* *229*:117–118 (1971).
14. T. M. S. Chang, Stabilisation of enzymes by microencapsulation with a concentrated protein solution or by microencapsulation followed by cross-linking with glutaraldehyde, *Biochem. Biophys. Res. Commun.* *44*:1531–1536 (1971).
15. E. D. Siu Chang and T. M. S. Chang, In vivo effects of intraperitoneally injected L-asparaginase solution and L-asparaginase immobilized within semipermeable nylon microcapsules with emphasis on blood L-asparaginase, "body" L-asparaginase and plasma L-asparaginase levels, *Enzyme* *18*:218–239 (1974).
16. M. J. Poznansky and T. M. S. Chang, Comparison of the enzyme kinetics and immunological properties of catalase immobilized by microencapsulation and catalase in free solution for enzyme replacement, *Biochim. Biophys. Acta* *334*:103–115 (1974).
17. T. M. S. Chang, *Artifical Cells*, Thomas, Springfield, Ill., 1972.
18. T. M. S. Chang, Artificial cells and microcapsules: comparison of structural and functional differences, in *Microencapsulation: Processes and Applications* (J. E. Vandegaer, ed.), Plenum, New York, 1974, pp. 95–102.
19. T. M. S. Chang, Semipermeable microcapsules as artificial cells: clinical applications and perspectives, in *Microencapsulation* (J. R. Nixon, ed.), Dekker, New York, 1976, pp. 57–65.
20. L. A. Luzzi, M. A. Zoglio, and H. V. Maulding, Preparation and evaluation of the prolonged release properties of nylon microcapsules, *J. Pharm. Sci.* *59*:338–341 (1970).
21. T. Mori, T. Sato, Y. Matsuo, T. Tosa, and I. Chibata, Preparation and characteristics of microcapsules containing asparaginase, *Biotech. Bioeng.* *14*:663–673 (1972).
22. A. W. Jenkins and A. T. Florence, Scanning electron microscopy of nylon microcapsules, *J. Pharm. Pharmacol.* *25*(Suppl.), 57–61 P (1973).

23. A. T. Florence and A. W. Jenkins, In vitro assessment of microencapsulated drug systems as sustained released parenteral dosage forms, in *Microencapsulation* (J. R. Nixon, ed.), Dekker, New York, 1976, pp. 39–55.
24. J. W. McGinity, A. B. Combs, and A. N. Martin, Improved method for microencapsulation of soluble pharmaceuticals, *J. Pharm. Sci. 64*:889–890 (1975).
25. J. W. McGinity, A. Martin, G. W. Cuff, and A. B. Combs, Influences of matrices on nylon-encapsulated pharmaceuticals, *J. Pharm. Sci. 70*:372–375 (1981).
26. M. D. DeGennaro, B. B. Thompson, and L. A. Luzzi, Effect of cross-linking agents on the release of sodium pentobarbital from nylon microcapsules, in *Controlled Release Polymeric Formulations* (D. R. Paul and F. W. Harris, eds.), American Chemical Society, Washington, D.C., 1976, pp. 195–207.
27. M. Koishi, N. Fukuhara, and T. Kondo, Studies on microcapsules. II. Preparation of polyphthalamide microcapsules, *Chem. Pharm. Bull. 17*:804–809 (1969).
28. M. Katz, Interfacial polycondensation. IV. Polyphthalamides, *J. Polym. Sci. 40*:337–342 (1959).
29. V. E. Shashoua and W. M. Eareckson, Interfacial polycondensation, V. Polyterephthalamide from short-chain aliphatic, primary and secondary diamines, *J. Polym. Sci. 40*:343–358 (1959).
30. Y. Shigeri, M. Koishi, T. Kondo, M. Shiba, and S. Tomioka, Studies on microcapsules. VI. Effect of variations in polymerization condition on microcapsule size, *Can. J. Chem. 48*:2047–2051 (1970).
31. T. Ogawa, K. Takamura, M. Koishi, and T. Kondo, Studies on microcapsules. XIII. Effect of span 85 and pH of aqueous phase on the formation of polyamide microcapsules, *Bull. Chem. Soc. Japan 45*:2329–2331 (1972).
32. M. Shiba, S. Tomioka, M. Koishi, and T. Kondo, Studies on microcapsules. V. Preparation of polyamide microcapsules containing aqueous protein solution, *Chem. Pharm. Bull. 18*:803–809 (1970).
33. M. Shiba, Y. Kawano, S. Tomioka, M. Koishi, and T. Kondo, Studies on microcapsules. XI. Electrophoretic behavior of polyphthalamide microcapsules containing aqueous solutions of bovine serum albumen, *Bull. Chem. Sco. Japan 44*:2911–2915 (1971).
34. K. Takahashi, M. Koishi, and T. Kondo, Studies on microcapsules. XVI. Binding of cations to sulfonated and carboxylated polyphthalamide microcapsules as estimated from electrophoresis data, *Kolloid-Z. Polym. 251*:232–235 (1973).
35. K. Takamura, M. Koishi, and T. Kondo, Microcapsules XIV: effects of membrane materials and viscosity of aqueous phase on

permeability of polyamide microcapsules towards electrolytes, *J. Pharm. Sci. 62*:610–612 (1973).
36. T. Kondo and N. Muramatsu, Enzyme inactivation, in *Microencapsulation* (J. R. Nixon, ed.), Dekker, New York, 1976, pp. 67–75.
37. D. A. Wood and T. L. Whateley, A study of enzyme and protein microencapsulation—some factors affecting the low apparent enzymic activity yields, *J. Pharm. Pharmacol. 34*:552–557 (1982).
38. M. Koishi, N. Fukuhara, and T. Kondo, Studies on microcapsules. IV. Preparation and some properties of sulfonated polyphthalamide microcapsules, *Can. J. Chem. 47*:3447–3451 (1969).
39. K. Takahashi, M. Koishi, and T. Kondo, Studies on microcapsules. VII. Further characterization of sulfonated polyphthalamide microcapsules, *Can. J. Chem. 48*:3520–3523 (1970).
40. Y. Shigeri, M. Tomizawa, K. Takahashi, M. Koishi, and T. Kondo, Studies on microcapsules. XII. Preparation and characterization of carboxylated polyphthalamide microcapsules, *Can. J. Chem. 49*:3623–3626 (1971).
41. T. Kondo, M. Arakawa, and B. Tamamushi, Poly(phthaloyl L-lysine) microcapsules containing hemoglobin solution: artificial red blood cells, in *Microencapsulation* (J. R. Nixon, ed.), Dekker, New York, 1976, pp. 163–172.
42. M. Arakawa, T. Kondo, and B. Tamamushi, Flow properties of microcapsule suspensions as a model of blood, *Biorheology 12*:57–66 (1975).
43. M. Arakawa and T. Kondo, Some biophysical and biochemical properties of poly (phthaloyl L-lysine) microcapsules containing hemolysate, *Can. J. Physiol. Pharmacol. 55*:1378–1382 (1977).
44. S. Suzuki and T. Kondo, Disintegration of poly (N^{α},N^{ε}-terephthaloly-L-lysine) microcapsules by poly(diallyldimethylammonium chloride), *J. Colloid Interface Sci. 67*:441–447 (1978).
45. S. Suzuki, T. Nakamura, M. Arakawa, and T. Kondo, Disintegration of poly (N^{α},N^{ε}-terephthaloyl-L-lysine) microcapsules by alkylpyridinium chlorides, *J. Colloid Interface Sci. 71*:141–146 (1979).
46. N. Muramatsu, T. Yoshioka, and T. Kondo, Platelet adhesion to artificial red blood cells having different membrane compositions, *Chem. Pharm. Bull. 30*:257–265 (1982).
47. M. Arakawa and T. Kondo, Preparation of hemolysate-loaded poly (N^{α},N^{ε}-L-lysine-diylterephthaloyl) nanocapsules, *J. Pharm. Sci. 70*:354–357 (1981).
48. A. Watanabe, K. Higashitsuji, and K. Nishizawa, Electric emulsification, in *Colloidal Dispersions and Micellar Behavior* (K. L. Mittal, ed.), American Chemical Society, Washington, D.C., 1975, pp. 97–109.
49. J. E. Santo and P. G. Abend, U.S. Patent 3,607,776 (September 21, 1971).

50. J. E. Santo and J. E. Vandegaer, U.S. Patent 3,492,380 (January 27, 1970).
51. J. E. Vandegaer and F. G. Meier, U.S. Patent 3,464,926 (September 2, 1969).
52. T. J. Speaker and L. J. Lesko, U.S. Patent 3,959,457 (May 25, 1976).
53. P. Rambourg, J. Levy, and M. C. Levy, Microencapsulation III: preparation of invertase microcapsules, *J. Pharm. Sci. 71*:753–758 (1982).
54. M. C. Levy, P. Rambourg, J. Levy, and G. Potron, Microencapsulation IV: cross-linked hemoglobin microcapsules, *J. Pharm. Sci. 71*:759–762 (1982).
55. K. Bala and P. Vasudevan, pH-Sensitive microcapsules for drug release, *J. Pharm. Sci. 71*:960–962 (1982).
56. S. Suzuki, T. Kondo, and S. G. Mason, Studies on microcapsules I. Preparation of polyurethane and polyphenolester microcapsules, *Chem. Pharm. Bull. 16*:1629–1631 (1968).
57. W. M. Eareckson, Interfacial polycondensation. X. Polyphenyl esters, *J. Polym. Sci. 40*:399–406 (1959).
58. Y. Wakamatsu, M. Koishi, and T. Kondo, Studies on microcapsules, XVII. Effect of chemical structure of acid dichlorides and bisphenols on the formation of polyphenyl ester microcapsules, *Chem. Pharm. Bull. 22*:1319–1325 (1974).
59. E. L. Wittbecker and M. Katz, Interfacial polycondensation. VII. Polyurethanes, *J. Polym. Sci. 40*:367–375 (1959).
60. Y. Shigeri and T. Kondo, Studies on microcapsules. III. Permeability of polyurethane microcapsule membranes, *Chem. Pharm. Bull. 17*:1073–1075 (1969).
61. R. L. Hart, D. D. Emrick, and R. G. Bayless, U.S. Patent 3,755,190 (August 28, 1973).
62. C. B. DeSavigny, U.S. Patent 3,959,464 (May 25, 1976).
63. A. E. Vassiliades, U.S. Patent 3,886,084 (May 27, 1975).
64. P. L. Foris, R. W. Brown, and P. S. Phillips, U.S. Patent 4,087,376 (May 2, 1978).
65. P. L. Foris, R. W. Brown, and P. S. Phillips, U.S. Patent 4,089,802 (May 16, 1978).
66. P. L. Foris, R. W. Brown, and P. S. Phillips, U.S. Patent 4,100,103 (July 11, 1978).
67. G. Baatz, M. Dahm, and W. Schäfer, U.S. Patent 4,119,565 (October 10, 1978).
68. H. B. Scher, U.S. Patent 4,140,516 (February 20, 1979).
69. J. E. Vandegaer and F. C. Meier, U.S. Patent 3,575,882 (April 20, 1971).

6

Pan Coating

6.1 INTRODUCTION

Pan coating is among the most widely used procedures for the microencapsulation of drugs. Historical and economic reasons for this are that many pharmaceutical companies possess coating pans for the production of sugar- and film-coated tablets, the technology being adaptable for the coating of the much smaller cores used in microencapsulation. The process may be employed to apply a wide variety of nonenteric and enteric film formers together with other additives such as plasticizers and colorants in organic solvent or aqueous-based systems to cores that have been rounded so as to roll adequately in a coating pan. Unlike many other microencapsulation procedures, the process lends itself to great flexibility in the formulation of the coating. Also, the core may contain a wide range of additives that serve to modify further the release properties of the final dosage form. However, as with tablets, pan coating for microencapsulation is a highly skilled operation, whose subtle complexities are not easily mastered. It is even more time-consuming, as a greater percentage of coating material must be applied to achieve adequate coating thickness and uniformity with smaller cores. Tablet cores typically gain 2 to 5% by weight during film coating, whereas the equivalent gain for microcapsule cores is 10 to 20+%.

Despite the widespread use of pan coating for microencapsulation, there is a surprisingly small published literature relating to aspects of the procedure. Much of what has been published concerns industrial patents, where the finer details of the procedure are generally inadequately described. University and similar institutions appear to have been slow to undertake research in this area because of technical difficulties in obtaining satisfactory coatings and because of the large number of variables, some of which are not easily controlled, that affect coating quality. Lack of consistency and reproducibility are often features of the process. Also, production times of a week or more are not uncommon, to allow for drug loading onto inert core pellets and the subsequent application of a large number of coating fractions, which make the procedure unattractive for fundamental research. Much useful information relating to details of the procedure may be obtained in references relating to the coating of tablets, a number of which will be cited in the course of this chapter. Those not familiar with general aspects of the coating of tablets should refer to the review of Ellis et al. [1] or to the recently published book dealing mainly with the subject edited by Lieberman and Lachman [2]. Many aspects of the formulation of suitable coating materials for application by pan technology have been reviewed in Chap. 2.

6.2 THE PROCESS

Blythe [3], in a patent assigned to Smith, Kline and French Laboratories in 1956, first described how pan coating could be used to apply variable thickness and composition of sprayed wax and/or fat on medicated sugar pellets containing dexamphetamine sulfate to produce a sustained-release "spansule" type dosage form. As the method is the basis on which many subseuqent modifications have been made, it is worth describing first in some detail. Fine spherical sugar pellets (nonpareil seeds), 12–40 mesh, were placed in a rotating 36-in. coating pan. Syrup was poured over the pellets to wet them evenly, and, when tacky, finely divided drug particles were sprinkled onto them. The pellets were then dried using warm air, and the process was repeated three times. A fifth coating of talc was likwise applied using syrup. The medicated pellets produced were screened through a 12-mesh sieve, and three-quarters of the batch obtained was coated with a mixture of glyceryl monostearate and white beeswax in carbon tetrachloride at 70°C. Sufficient coats were applied to the rotating pellets and dried until a 10% coating weight was achieved, whereupon one-third of this batch was removed. Up to 15 or 20% total coating was applied to each half, respectively, of the remaining pellets. The four types of pellets formed were then thoroughly mixed prior to packing at appropriate dosage into outer hard gelatin capsule shells to yield a sustained-release form of the sympathomimetic agent.

In the following sections certain aspects and modifications of the basic approach to the use of pan coating for microencapsulation will be reviewed.

6.2.1 Equipment and Related Facilities

Various coating pans with diameters from 12 to 90 in. and made from stainless steel or occasionally copper are used. A popular size is 36 in. or 1 m. A selection of some of the more common shapes of pans is shown in Fig. 6.1, but many other designs are also used.

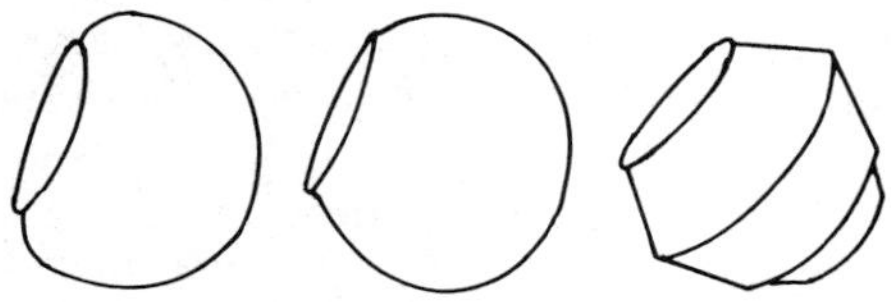

Figure 6.1 Various-shaped coating pans.

The pan is mounted on the drive from a variable-speed motor such that its angle of inclination can be altered. The pan must also be provided with a hot or cold air supply duct and an exhaust duct for the removal of vapor and dust generated during the coating operation. Cores are carried upward by the rotation of the pan until their centrifugal and frictional forces are overcome by gravity, causing individual cores to roll and fall away from the surface of the bed. Figure 6.2 shows a typical arrangement for an operating coating pan. Because of the short lift at the rear and the formation of a vortex at the mouth of a conventional coating pan, cores tend to be relatively stagnant in both of these regions, but this problem can be overcome by periodic manual agitation of the load. Alternatively, a system of properly designed baffles may be fitted to the inside of the pan to improve the uniformity of tumbling or cascading action in the pan.

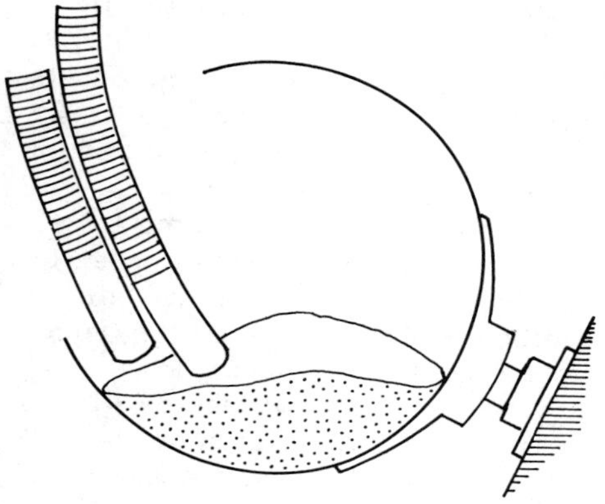

Figure 6.2 Schematic representation of an operating coating pan.

Sutaria [4] has evaluated the usefulness of three basic baffle designs and their placement in a pan. Likewise, Lachman and Cooper [5] have discussed the use of baffles.

Fractions of coating material including duster may be either poured or sprayed onto the moving cores. Spraying is often preferred, as it ensures more uniform coverage of the cores and avoids unwanted localized overwetting. Hydraulic (airless) atomizer systems are usually used in preference to pneumatic ones, as the large volume of air accompanying the latter may cause premature solvent evaporation and excessive rebound, resulting in loss of coating material. Spray pattern, distance and pressure, orifice size, and the nature of the feed are among the important variables that influence spray quality. Details of such variables are discussed in Refs. 5–7. Pickard et al. [8] reported that the undesirable water vapor permeability of free films composed of hydroxypropylmethylcellulose and ethylcellulose with propylene glycol as plasticizer could be reduced if prepared using a poured rather than an airless spray technique. Coatings of superior uniformity were also obtained in the author's own laboratories when applying various enteric coatings to microcapsule cores by pan technology when the coating was poured on rather than delivered from a hydraulic atomizer.

In order to reduce the so-called art of pan coating, various partial or fully automated systems have been developed. Such systems usually involve a programmable unit that controls the periodic delivery of coating solution and duster from atomizers, and the switching on and off of the drying air supply to the revolving cores. Some of the advantages claimed for these automated coating pans to offset their greater capital cost are shorter batch times, saving in skilled labor, and the formation of more uniform and reproducible coatings, which should be of special benefit in research. Details of the design of a number of automated systems have been discussed in the literature [5, 6, 9, 10]. Lantz et al. [11] have reported various methods that were found useful for monitoring the application of volatilc coating solutions to sugar pellets in a rotation pan. By placing a thermocouple and thermistors in the pellet bed, temperature changes occurring during coating could be recorded and used to control aspects of the procedure.

The coating equipment, which may include several coating pans, should be housed in a fume cupboard type of arrangement in order to reduce health hazards to operators associated with volatile solvents and generated dust. Electrical and other equipment should be explosion-proof and properly grounded to reduce the risk of explosion through accumulation of static charge. Should an explosion or fire occur, it should be contained in the immediate coating area or else be harmlessly vented into an adjacent area. Aspects of safety design and pollution control in the processing area for film coating have been reviewed by Pickard and Rees [12].

6.2.2 Preparation of Cores

Solid-core particles greater than 500 μm are generally considered necessary for satisfactory pan coating. The cores should be approximately spherical in shape so as to roll well in the coating pan. They should be adequately hard and of low friability so as to withstand attrition during the process. Very few drug particles conform to all of these desirable requirements, particularly with regard to shape, which is usually very irregular. Also, drug particles of low aqueous solubility that are dissolution rate-limited in their absorption have to be size-reduced well below 500 μm in order to avoid bioavailability problems. Accordingly, cores composed of pure drug are rarely used unless the drug is amenable to recovery in macro spherical form.

The usual approach to core production involves the use of a spherical substrate such as nonpareil sugar seeds. These core pellets are available commercially from a number of different suppliers. Figure 6.3 shows typical pellets composed of approximately 70% white sugar and 30% corn starch, which are available in a number of size ranges from Hans G. Werner, Hamburg, West Germany.

Enz and King [13], in a patent assigned to The Upjohn Co., described how to produce sucrose pellets. Sieved granular sucrose was moistened by spraying with water in a rotating coating pan, dusted with a mixture of finely powdered starch, sucrose, and talc, and dried. The process was repeated until spherical masses of a suitable size range were obtained by screening. Spherical cores may also be prepared by modified air suspension (see Chap. 7) or by spray-drying (see Chap. 8) processes and by using spheronization apparatus (see Chap. 13).

Various liquid adhesives are used to apply finely divided drug particles onto sugar pellets, the most frequently employed being an alcoholic solution of polyvinylpyrrolidone or a hydroalcoholic solution of gelatin.

A number of alternative approaches to core production have been reported. Heimlich and MacDonnell [14], in a patent assigned to Smith, Kline and French Laboratories, described how large crystals of medicament could be coated with additional fine medicament to produce smooth spherical pellets. As an example, paracetamol crystals, 20-40 mesh, were sprayed with an alcoholic solution of polyvinylpyrrolidone and dusted with finely powdered drug in a coating pan. After allowing to roll-dry using cold air, the process was repeated until spherical masses of the required size and shape were obtained. These were given one or two sealing coats of the polyvinylpyrrolidone solution and again dried. The cores produced were then coated with various sustained-release coatings. A particular advantage of the process over the use of inert sugar pellets is that the core material is composed almost entirely of drug, so enabling the bulk of the final product to be considerably reduced. This is an important consideration with many drugs that have large dosages.

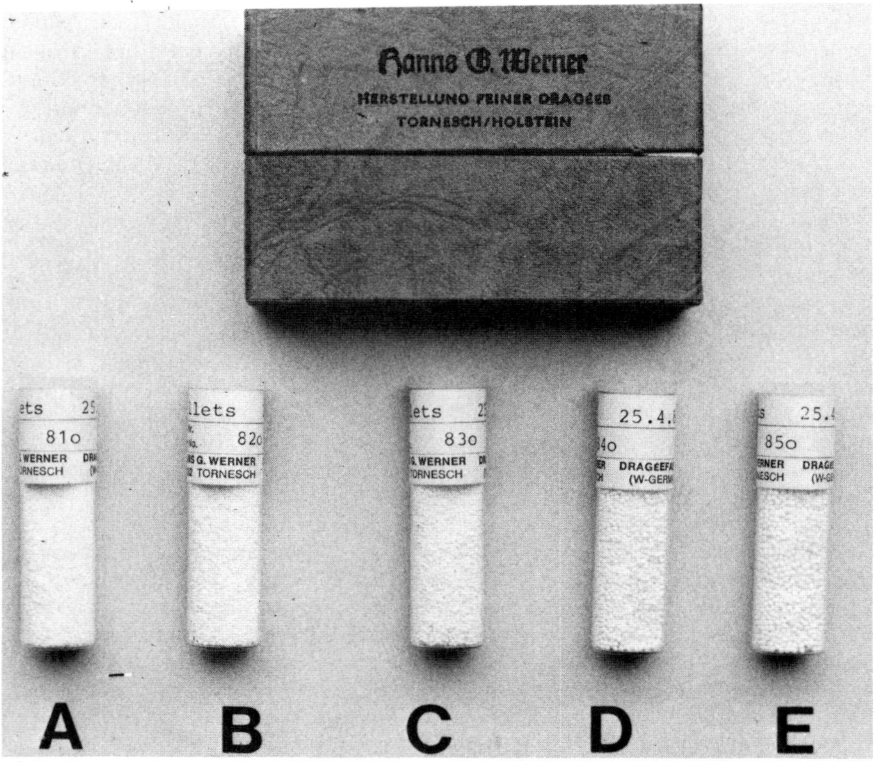

Figure 6.3 Sucrose pellets of varying size range. (Courtesy Werner, Hamburg, West Germany.) A, 0.5–0.6 mm; B, 0.6–0.71 mm; C, 0.71–0.85 mm; D, 0.85–1.00 mm; E, 1.00–1.25 mm.

A different approach to core production involves the incorporation of the drug into suitable matrix materials that will augment the controlled-release properties obtainable upon subsequent coating. For example, Brophy and Deasy [15] used methylene blue as a model drug to form cores with hydrogenated castor oil or polyethylene powder as filler, which markedly retarded dissolution of the drug. Cores prepared with addition of hydroxypropylmethylcellulose, carnauba wax/beeswax 1:1, or carnauba wax exhibited little retardation of drug release. The methylene blue was thoroughly mixed with the selected filler, and the powder bed was sprayed tangentially with a 10% polyvinylpyrrolidone solution in water from an airless spray gun in a coating pan rotated at 30 rpm. As the powders commenced to agglomerate, the rotation speed was gradually reduced to about 15 rpm and the cores were allowed to roll for a further 5 to 10 min. The speed reduction facilitated the growth in size of the cores. The cores were then dried and size fractions were selected by sieving. Controlled-release cores were also prepared by El-Sayed et al. [16], who produced irregular masses containing drug, Eudragit RS (an acrylic polymer) and cellulose acetate phthalate or dibasic calcium phosphate using acetone: isopropanol by granulation through a coarse-mesh sieve. This was followed by rounding of the product in a coating pan by application of sucrose, starch, Aerosil, and talc using water. The dried cores were subsequently pan-coated with Eudragit RS or ethylcellulose and showed sustained-release properties during in vitro dissolution testing.

An interesting possibility for coating small batches of cores that contain drugs that are expensive or in short supply, particularly during research and development work, is suggested in the publication of Rednick et al. [17] on the pan coating of small batches of tablets. Placebo cores of the same dimensions but containing a small amount of reduced iron could be used to bulk the drug-containing cores to a production-size batch for the pan and separated after coating by passing through a magnetic selector. The placebo microcapsules obtained would be useful in clinical trials and for display purposes.

6.2.3 Coating Procedures

A typical approach to the coating of a batch of cores would be as follows. The inner surface of the pan is usually roughened before use if it does not contain baffles. This may be done by coating with an adhesive solution composed of 10% polyvinylpyrrolidone in isopropanol, allowing this to become tacky, sprinkling with talc, and finally dabbing with a damp cloth to produce a stippled surface upon drying. The roughened surface prevents the cores from sliding in the pan. In order to avoid the formation of irregular microcapsules, the cores should be screened initially to remove dust. A suitable load for the coating pan is established, and optimal rolling conditions are determined by varying the speed and angle of rotation.

With the exhaust system switched on, a predetermined amount of coating solution is rapidly but carefully applied to the cores, preferably toward the top of the core bed, where the cores are just starting to fall away. It may be necessary to agitate the cores manually to ensure complete spreading of the coating solution over the entire core bed. Excessive volumes and rate of application will impede tumbling action, unduly prolong drying, and cause the buildup of material on the pan wall.

If the coating solution is sprayed on, then application may be either continuous or intermittent depending on the drying efficiency of the pan. Dusting powders such as talc are often used during the process and are very useful in reducing clumping tendencies. Uniform spreading may be achieved by sprinkling at the top of the bed of cores, but stagnant areas should receive special attention. The powder must be added at the critical moment just as the cores become tacky and in just sufficient quantity to free up any clumps that may be forming. Excessive quantities will interfere with binding between successive layers of the coating material and cause the formation of a friable coating. Also, dust accumulation will cause the development of surface roughness on the cores. If added too soon, the dusting powder may cause premature drying and therefore nonuniform coating. If added too late, the powder may be poorly absorbed by the cores and its effectiveness reduced. Dusting powders are sometimes used in suspension in the coating material in an attempt to standardize the process, but this approach may not produce the best results.

The coating material usually also contains a suspended pigment. Coloring is necessary to visualize the buildup of a uniform coating and to distinguish between microcapsules of a drug that contain different amounts or types of coatings. Different colors allow the uniformity of mixing of different microcapsules filled into outer hard gelatin capsule shells to be readily checked. They also reduce the risk of accidental contamination of one batch of microcapsules with those containing a different drug, particularly in a busy production area.

After applying the first layer of coating and drying, the cores should not be allowed to roll for too long, as they will tend to rub off each other and damage the coating. Dust caused by such abrasion or by the addition of excess dusting powder may give rise to a rough product, as the dust will tend to gather at the rear of the pan and periodically slough off and adhere to the cores. Should roughness develop at any stage, it will be difficult to produce a final product that is smooth, elegant, and of even coating thickness. Rough cores may also result if heat is applied too soon after dusting. It may be necessary to remove the cores repeatedly from the pan during the coating process to free them from excessive dust and to wash off excessive solid deposits that tend to build up on the inside of the pan.

"Jogging" of the pan after each drying cycle may be beneficial in that it allows the bed to turn over and prevents the cores from adhering to each other. This is done by repeatedly switching on for a short period and then off the drive to the coating pan. "Jogging" is particularly necessary if the process has to be shut down at any intermediary stage, because if the cores are left resting against each other while they still retain a significant amount of solvent, they will fuse together and will tend to damage their coatings if subsequently separated. Figure 6.4 shows a photograph of stationary cores partially coated with a cellulose acetate phthalate enteric film in a roughened copper coating pan.

Coating is continued until the required amount has been built up on the cores. This usually involves 20 to 50 applications of coating

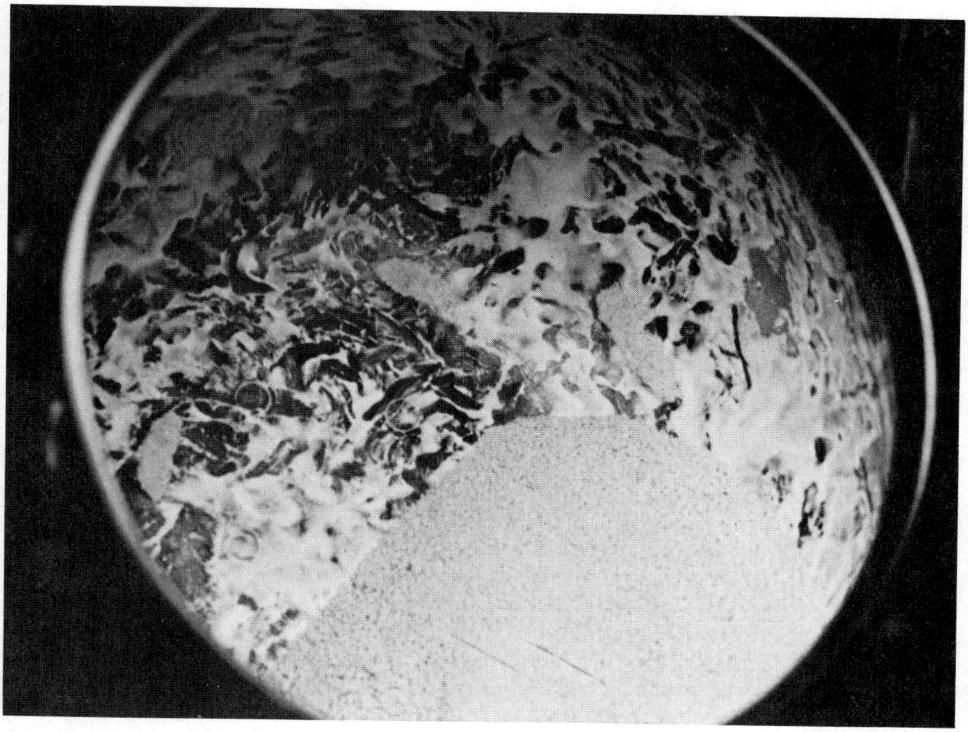

Figure 6.4 Partially coated microcapsule cores in a roughened coating pan.

material if applied discontinuously. The first few layers of coating should not be excessive to avoid cluster formation, and even less coating is applied per occasion as the process proceeds, as at that stage the cores have a partial seal that impedes migration of solvent into them. Increasing difficulty is often experienced in avoiding clumping and achieving a uniform coating as more than 10% of coating is applied. When the coating is complete, the microcapsules produced are spread out evenly on flat trays and dried in an oven at about 40°C for several hours to remove residual solvent and to thermoset the film.

6.3 SIDE-VENTED COATING PAN PROCESS

Side-vented coating pans usch as the Accela-Cota (Manesty Machines, Ltd.) are more efficient than conventional pans for tablet coating. They are normally fitted with a mesh having a 3 mm diameter and are unsuitable for microencapsulation. However, if fitted with a fine laser-perforated mesh of much smaller diameter, they can be used for coating small drug-containing cores. McAinsh and Rowe [18], in a patent assigned to Imperial Chemical Industries, Ltd., England, used such an approach to coat propranol hydrochloride-containing spheres of diameter 0.5 to 2 mm with films of ethylcellulose and hydroxypropyl methylcellulose. The spherical cores contained about 60% drug and 40% microcrystalline cellulose, and were prepared using the spheronization process.

6.4 SOME FURTHER EXAMPLES OF THE MICROENCAPSULATION OF DRUGS BY PAN COATING

In addition to the examples already cited in this chapter, a number of references relating to the use of pan coating for the microencapsulation of drugs will be reviewed in this section. The references have been arranged in chronological order. Any unusual aspect of the coating procedure will be reported. The various coating materials employed will be indicated as representative examples of the potentially enormous number of film formers, plasticizers, and other additives for coating systems that may be employed with this process.

Lowey [19] pan-coated rounded granules containing nitroglycerin, which is a poor candidate for sustained-release formulation because of its extensive metabolism by a first-pass effect, with various cellulose derivatives such as methylcellulose, ethylcellulose, or cellulose acetate phthalate using beeswax or castor oil as plasticizer. Two unusual features of the procedure were that a large number of applications of very dilute coating were applied and the coating pan was rotated at an unusually high speed of 50 to 60 rpm to reduce aggregation and improve the uniformity of deposition of the coating.

Rosen and Switoskey [20] reported that ^{35}S-labeled trimeprazine tartrate, an antipuritic agent, mixed with powdered starch and sugar was applied to the surface of sucrose pellets using a hydroalcoholic gelatin adhesive in a coating pan mounted in a disposable glove-box arrangement. Part of the dried, screened cores produced were then spray-coated with a solution of 11% w/w glyceryl monostearate, 16% w/w glyceryl distearate, and 3% w/w white beeswax in carbon tetrachloride at 40°C. Fractions were removed that had received 14, 15.3, 16.9, or 18.9% of coating material, screened, appropriately mixed, and filled into outer hard gelatin capsule shells. Experiments with animals and subjective clinical studies were used to determine the ratio for mixing the different fractions. Average ^{35}S serum levels for human subjects after oral administration of labeled trimeprazine are shown in Fig. 6.5. The results indicate that a single dose of 15 mg of the sustained-release dosage form produced a similar ^{35}S serum profile as three 5-mg doses of nonsustained release drug given at 0, 4, and 8 hr.

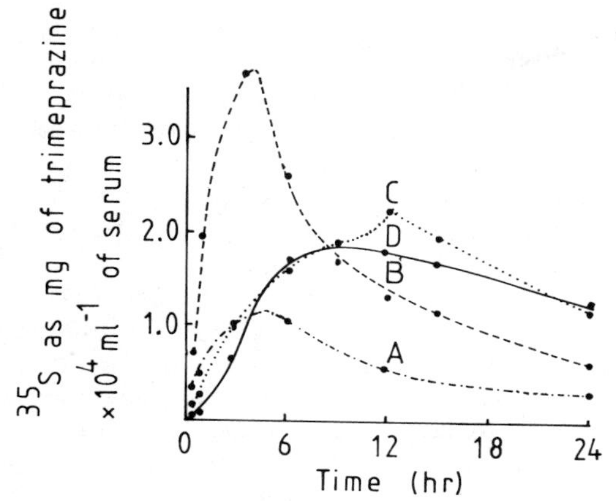

Figure 6.5 Average ^{35}S serum level of adult human subjects after oral administration of labeled trimeprazine: A, 5 mg once daily to 4 subjects; B, 15 mg once daily to 4 subjects; C, 5 mg three times a day every 4 hr (total of 15 mg) to 5 subjects; and D, 15 mg sustained-release capsules to 5 subjects. The drug was administered at 0 hr in the once-a-day regimen and at 0, 4, and 8 hr in the three divided-doses regimen. (From Ref. 20.)

These results were supported by urin 'y excretion data and more extensive clinical trials in 460 human su ,ects receiving unlabeled drug in sustained release capsules having similar in vitro release characteristics.

Cores containing 50% of barium sulfate were prepared by applying a mixture of this radioopaque powder, talc, starch, and cane sugar to sugar crystals in a coating pan using a water spray as described by Wagner et al. [21]. These cores were then coated with a poly(styrene-maleic acid) copolymer using dibutyl phthalate as a plasticizer and a mixture of talc and magnesium stearate as a dusting powder. X-ray studies indicated that the microcapsules resisted disintegration in the stomach of the starved dog for at least 8 hr and were progressively emptied through the pylorus over a period of at least 8 hr. The unusually long clearance time from the stomach was probably due to the high density of the cores as discussed in Chap. 2. The microcapsules disintegrated completely in the proximal part of the small intestine, indicating the suitability of using this or other enteric coatings on microcapsules for sustained-release dosage forms.

A way of achieving variable thickness of coating using the one-coating operation was described by Greif [22] in a patent assigned to American Cyanamid Co. This involved the use of variable size cores in the range 12–80 mesh, the smaller ones achieving thinner coatings and the larger ones thicker coatings. Granules containing tridihexethyl iodide, an anticholinergic agent, were coated by spraying with a mixture of glyceryl monostearate and white beeswax dissolved in 1,1,1-trichloroethane at 60 to 70°C using a coating pan until 44% wax content was achieved. The screened granules produced were then appropriately mixed together and filled into outer hard gelatin capsule shells or tableted to produce a sustained-release dosage form.

Cellulose acetate phthalate in acetone was used by Shepard [23] in a patent assigned to Key Pharmaceuticals, Inc., as the liquid adhesive to apply benzylpenicillin sodium or other drugs to the surface of core pellets. The drug-loaded cores were then coated with a mixture such as glyceryl monostearate, cetyl alcohol, myristyl alcohol, and white beeswax dissolved in chloroform at 65 to 67°C to achieve 7.5 to 67.5% coating. Microcapsules having varying percentages of coating were then further coated with cellulose acetate phthalate to give them enteric properties prior to filling into outer hard gelatin capsule shells or tableting. Although Heimlich and MacDonnell [14] did not use an outer enteric coating, they did claim that the preferred waxy sustained-release coating materials were hydrogenated castor oil, glyceryl monostearate, glyceryl distearate, 12-hydroxystearyl alcohol, and microcrystalline wax sprayed from solution in chloroform at elevated temperature.

To prevent the formation of poorly soluble and poorly absorbable crystalline novobiocin acid in the stomach in the presence of coadministered high concentrations of tetracycline, Schroeter [24], in a patent assigned to The Upjohn Co. partially enteric-coated novobiocin sodium-containing pellets. By delaying the release of the novobiocin sodium for 10 to 20 min, only a small proportion of the tetracycline should remain unabsorbed and consequently not interfere with the release and absorption of the former antibiotic. Enteric film formers, waxy materials, or hydrophobic polymers such as ethylcellulose applied in adequate thickness by air suspension technology were suitable for achieving the desired delay in release. A wetting agent, dioctyl sodium sulfosuccinate, was used in the medicated dusting powder applied to the sugar pellets in the coating pan.

The dog was found by Rosen et al. [25] to act as a satisfactory animal model for evaluating the sustained-release properties in humans of dexamphetamine-^{14}C sulfate from wax-coated microcapsules prepared by pan coating.

An interesting way of overcoming the poor solubility and lack of bioavailability of tetracycline base formed in the small intestine as tetracycline hydrochloride or other derivatives are liberated from enteric-coated microcapsules was described by Corn [26] in a patent assigned to Chemical and Pharmaceutical Patent Holding, Ltd. Tetracycline hydrochloride was granulated with solutions of polyvinylpyrrolidone and shellac in isopropanol together with a poorly soluble organic acid such as fumaric acid or succinic acid and lubricants. The dried granules were then partially enteric-coated in a coating pan with polyvinylpyrrolidone and shellac solution, dusted with further drug, organic acid, and lubricant, and finally coated with further enteric coating. The tetracycline formed a complex with the polyvinylpyrrolidone and when slowly liberated in the small intestine in the presence of the poorly soluble organic acid, the pH of the surrounding medium was not sufficiently high enough to interfere significantly with the solubility of the tetracycline or its absorption.

A baffled coating pan was used by Peters et al. [27] in a patent assigned to Warner Lambert Pharmaceutical Co. to apply sustained-release coatings comprising a mixture of ethylcellulose and waxes such as castor wax, carnauba wax, and paraffin wax from solution in chloroform by pouring or spraying sugar pellets loaded with drugs such as phenylpropanolamine, phenylephrine, or chlorpheniramine maleate, using a 6 to 8% gelatin solution in an acidified hydroalcoholic solvent as liquid adhesive. A lower ratio of ethylcellulose to wax was used for drugs with higher molecular weight and/or lower solubility and a higher ratio of ethylcellulose to wax was used with drugs of low molecular weight and/or high solubility to achieve satisfactory sustained-release properties at about 10% coating level.

A novel way of preventing sustained-release microcapsules containing pepstatin, a pepsin inhibitor used in the treatment of gastric ulcer, from being cleared too rapidly from the stomach when given immediately after a meal was reported by Umezawa [28] in a patent assigned to Zaidan Hojin Biseibutsu Kagaku Kenkyu Kai, Japan. Granules composed of sodium bicarbonate, lactose, and polyvinylpyrrolidone were coated by spraying repeatedly with a solution of hydroxypropylmethylcellulose in methanol in a pan. The product was screened and sprayed with a suspension of finely powdered pepstatin in the coating solution before final drying. The pepstatin could also be incorporated into the cores rather than in the layers of applied coating. Upon ingestion these microcapsules liberated carbon dioxide by the interaction of gastric acid and core sodium bicarbonate, which floated the microcapsules on the stomach content where they liberated pepstatin to suppress pepsin activity for 3 to 5 hr.

Blichare and Jackson [29], in a patent assigned to American Cyanamid Co., heated a mixture of drug and waxy material in a rotating coating pan using an infrared lamp. The heating conditions were adequate to just melt the wax and, as the drug particles were then at a higher temperature, they tended to sink into the molten wax. Spherical granules were obtained upon cooling, and after sieving could be either tableted or filled into capsules.

Hall et al. [30] have discussed the use of an aqueous dispersion of ethylcellulose (Aquacoat, FMC Corporation) to pan-coat tablets. Figure 6.6 shows a photograph of a sample of the coating material which

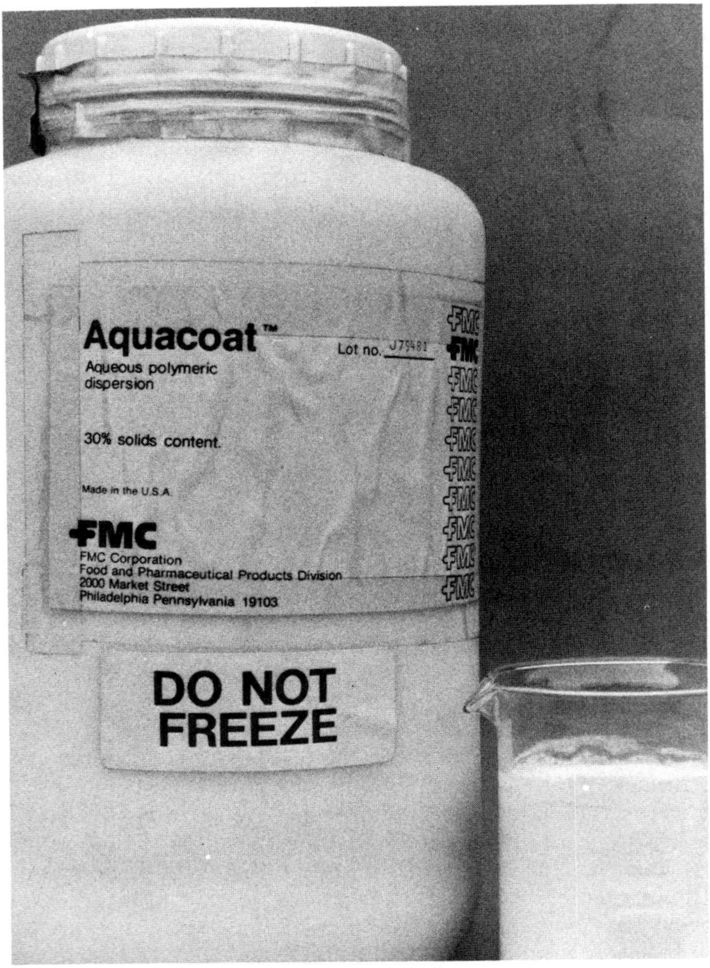

Figure 6.6 Sample of Aquacoat (FMC Corp.), which is an aqueous dispersion of ethylcellulose and surfactants of 30% solids content.

is mixed with a suitable plasticizer before being applied to the cores. This and other aqueous-based coating materials have obvious applications for the coating of microcapsules by pan technology.

Finally, two recent publications relating to the use of pan coating for microencapsulation are worth discussing. In order to coat a water-soluble drug, potassium chloride, using gelatin-acacia complex coacervation (see Chap. 3), Harris [31] initially coated potassium chloride crystals with polymers such as cellulose acetate phthalate, hydroxypropylmethylcellulose, cellulose acetate butyrate, or ethylcellulose and/or a wax composed of a mixture of carnauba wax and stearic acid. Precoating with an ethylcellulose-wax mixture was found to be the most effective at retarding drug dissolution during subsequent rapid complex coacervation coating.

In a publication from the author's own laboratories, Brophy and Deasy [15] pan-coated methylene blue-containing cores with various coatings containing film formers (ethylcellulose and hydroxypropylmethylcellulose) with or without plasticizers (diethyl phthalate, di-*n*-butyl phthalate, or castor oil), waxy sealant (paraffin wax or hydrogenated castor oil), and an extender/antitack agent (talc). Addition of 10% paraffin wax to an ethylcellulose coating was particularly effective at reducing the release of methylene blue. In general, however, modification of the core by incorporation of hydrophobic materials was more effective than coating in retarding drug release, though it tended to produce a coarser product.

REFERENCES

1. J. R. Ellis, E. B. Prillig, and A. H. Amann, Tablet coating, in *The Theory and Practice of Industrial Pharmacy* (L. Lachman, H. A. Lieberman, and J. L. Kanig, eds.), Lea & Febiger, Philadelphia, 1976, pp. 359–388.
2. H. A. Lieberman and L. Lachman (eds.), *Pharmaceutical dosage forms: tablets, Volume 3*, Dekker, New York, 1982.
3. R. H. Blythe, U.S. Patent 2,738,303 (March 13, 1956).
4. R. H. Sutaria, The art and science of tablet coating, *Manuf. Chem. Aerosol News 39*(6):37–42 (1968).
5. L. Lachman and J. Cooper, A programmed automated film-coating process, *J. Pharm. Sci. 52*:490–496 (1963).
6. A. Heyd, Variables involved in an automated tablet-coating system, *J. Pharm. Sci. 62*:818–820 (1973).
7. P. W. Stern, Effects of hydraulic pressure and nozzle orifice size on delivery rates of sprayed materials, *J. Pharm. Sci. 63*:1171–1173 (1974).
8. J. F. Pickard, J. E. Rees, and P. H. Elworthy, Water vapour permeability of poured and sprayed polymer films, *J. Pharm. Pharmacol. 24(Suppl.)*:139P (1972).
9. D. S. Mody, M. W. Scott, and H. A. Lieberman, Development of a simple automated film-coating procedure, *J. Pharm. Sci. 53*:949–952 (1964).
10. A. Heyd and J. L. Kanig, Improved self-programming automated tablet-coating system, *J. Pharm. Sci. 59*:1171–1174 (1970).

11. R. J. Lantz, A. Bailey, and M. J. Robinson, Monitoring volatile coating solution applications in a coating pan, *J. Pharm. Sci. 59*:1174–1177 (1970).
12. J. F. Pickard and J. E. Rees, Film coating: I. Formulation and process considerations, *Manuf. Chem.Aerosol News 45*(4): 19–22 (1974).
13. W. F. Enz and T. E. King, U.S. Patent 3,081,233 (March 12, 1963).
14. K. R. Heimlich and D. R. MacDonnell, U.S. Patent 3,119,742 (January 28, 1964).
15. M. R. Brophy and P. B. Deasy, Influence of coating and core modifications on the in vitro release of methylene blue from ethylcellulose microcapsules produced by pan coating procedure, *J. Pharm. Pharmacol. 33*:495–499 (1981).
16. A. A. El-Sayed, S. A. Said, and A. Sh. Geneidi, Sustaining availability of drugs with Eudragit RS, *Manuf. Chem. Aerosol News 49*(8):52–55 (1978).
17. A. B. Rednick, A. E. Nicholson, and S. J. Tucker, Process for pan coating small batches of tablets, *J. Pharm. Sci. 50*:174 (1961).
18. J. McAinsh and R. C. Rowe, U.S. Patent 4,138,475 (February 6, 1979).
19. H. Lowey, U.S. Patent 2,853,420 (September 23, 1958).
20. E. Rosen and J. V. Swintoskey, Preparation of a ^{35}S labelled trimeprazine tartrate sustained action product for its evaluation in man. *J. Pharm. Pharmacol. 12*:237T–244T (1960).
21. J. G. Wagner, W. Veldkamp, and S. Long, Enteric coatings IV. In vivo testing of granules and tablets coated with styrene-maleic acid copolymer, *J. Amer. Pharm. Ass. Sci. Ed. 49*:128–132 (1960).
22. M. Greif, U.S. Patent 3,078,216 (February 19, 1963).
23. M. Shepard, U.S. Patent 3,080,294 (March 5, 1963).
24. L. C. Schroeter, U.S. Patent 3,220,925 (November 30, 1965).
25. E. Rosen, T. Ellison, P. Tannenbaum, S. M. Free, and A. P. Crosley, Comparative study in man and dog of the absorption and excretion of dextroamphetamine-^{14}C sulfate in sustained-release and nonsustained-release dosage forms, *J. Pharm. Sci. 56*:365–369 (1967).
26. M. E. Corn, U.S. Patent 3,499,959 (March 10, 1070).
27. D. Peters, F. W. Goodpart, and H. A. Lieberman, U.S. Patent 3,492,397 (January 27, 1970).
28. H. Umezawa, U.S. Patent 4,101,650 (July 18, 1978).
29. M. S. Blichare and G. J. Jackson, U.S. Patent 4,132,753 (January 2, 1979).
30. H. S. Hall, K. D. Lillie, and R. E. Pondell, Comparison—aqueous vs solvent based ethylcellulose films, Coating Place, Inc., Verona, Wisc.
31. M. S. Harris, Preparation and release characteristics of potassium chloride microcapsules, *J. Pharm. Sci. 70*:391–394 (1981).

7

Air Suspension Coating

7.1 INTRODUCTION

Much of the microencapuslation of solid cores in the pharmaceitucal industry is done using coating pans or drums as described in the previous chapter. This is often the result of companies having a large investment in this type of equipment together with the necessary expertise in the technology gained because of their need to coat tablets. Such companies are reluctant to invest in novel equipment with minimal advantages for microencapsulation that may be more costly and complex to operate while lacking the general versatility of pan coating. Accordingly, these factors have contributed to a tendency for the pharmaceutical industry to microencapsulate only relatively large spherical and monosize cores with coating materials that do not exhibit excessive tack during the coating operation.

Air suspension coating, which is also widely used in the pharmaceutical industry, is an attractive alternative to pan coating in that it can successfully coat most small solid particles irrespective of size or shape with a wide variety of coating materials. Although the equipment may be more expensive to purchase, the process is much more rapid than pan coating and does not require skilled labor to supervise it, often being fully automated. Coat continuity is superior to that achieved by pan coating, and because of the uniformity of distribution of coating material on and between cores, it is possible to apply less coating material and to dissolve active ingredients in the coating solution. Little or no loss of coating material occurs on the walls of the equipment and, being a closed system, no escape of coating solvent vapor occurs into the working area. Also, the equipment may be used after modification if necessary to carry out other pharmaceutical operations such as mixing, drying, granulation, and the coating of tablets and capsules.

Air suspension coating was originally developed by Dale Wurster and colleagues while at the School of Pharmacy, University of Wisconsin, Madison. It was patented in a series of patents [1–7] describing the process and its related applications assigned to the Wisconsin Alumni Research Foundation (WARF). Hence the process is often referred to as Wurster coating when using equipment conforming with licensees of the patented process. Wurster [8] and Brudney and Toupin [9] presented some preliminary findings on the procedure for the coating of tablets and drug particles.

Figure 7.1 shows a diagram of a typical Wurster coating apparatus, and Fig. 7.2 shows details of its coating chamber. There are many modifications of the equipment available, often incorporating design alterations requested by customers. Common sizes available range from 4 to 46 in. in diameter for air distribution plates with capacities from 0.5 to 450 kg [10] and can be supplied with interchangeable coating chambers to increase output or flexibility. Plant with variable levels of sophistication in the control and monitoring equipment can be supplied.

Using typical Wurster coating equipment, solid cores that must normally be greater than 50 μm in diameter and nonvolatile are placed

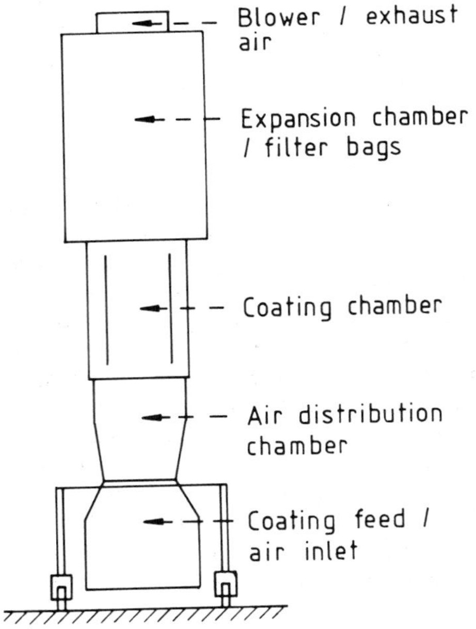

Figure 7.1 Wurster-type coating apparatus.

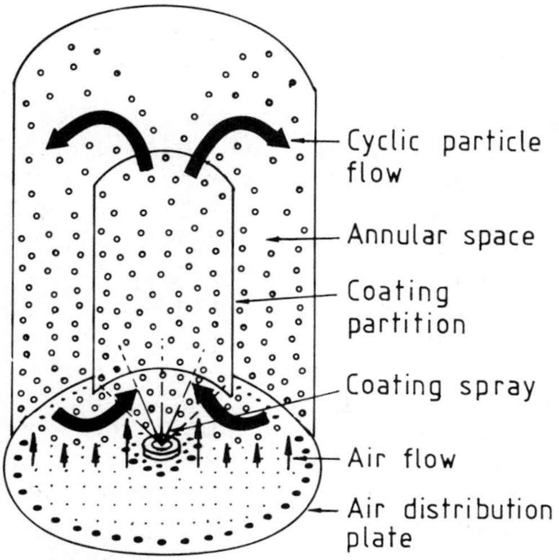

Figure 7.2 Details of the coating chamber of Wurster apparatus.

into the lower part of the coating chamber and fluidized in a nonturbulent manner by a high-velocity, upward-moving airstream as illustrated in Fig. 7.2. The air distribution plate and the coating partition ensure an improved cyclic flow pattern of the cores, which is aided by pressure differential in various regions of the coating chamber. Accordingly, cores make repeated passages about once every 10 s through the coating spray coming from the atomizer nozzle. Collision of cores is reduced, which would otherwise make the coating of small particles in particular very difficult due to aggregation and adherence to the walls of the coating chamber. The process air, which is filtered and preheated as necessary depending on the volatility of the coating solvent used, rapidly dries the product as it rises within the coating partition. Cores as they enter into the expansion chamber can no longer be supported on the fluidizing airstream, which decreases in velocity because of the increase in cross-sectional area of the apparatus. They therefore settle out, descending through the annular space where they receive further drying before reentering the partition to be accelerated upward again and to receive further coating. Exhaust air and vapor are discharged from the top of the apparatus having been previously filtered through a fine-mesh filter to remove suspended solids.

There are several aspects of the equipment and process that deserve further comment. Aspects of the formulation of suitable film coatings have been considered in Chap. 2.

7.2 SOME PROCESS CONSIDERATIONS

7.2.1 Air Distribution and Flow Rate

Design of the proper air distribution plate and partition is very important to the successful microencapsulation of small cores. It is important to set the gap between the plate and the base of the partition correctly. Normally the pressure in the region immediately above the atomizer is less than the region at the base of the annular space, causing particles to be sucked in the direction of the arrows shown in Fig. 7.2. If the design of the plate and partition is faulty or if the gap is too large, the pressure differential may be reduced, resulting in a sluggish or irregular flow of the cores up the coating partition and down the annular space. With gradual increases in velocity of applied air within the partition, the fixed bed of cores is eventually just fluidized. Further increase in air velocity causes the bed of cores to progressively expand up the partition until pneumatic transport occurs at higher air velocities. The ideal air flow rate up the partition for coating cores, which is a function of their size, shape, and density, is between that required to just effect fluidization and pneumatic transport.

7.2.2 Atomizer Type and Feed

Pneumatic atomizers are usually used because their droplet size is finer and hence more desirable for ensuring even wetting of core surfaces than the hydraulic types. Also, the concentration of solids used in the coating feed to the atomizer is usually less than that of pan coating. Both of these factors may contribute to the deposition of a more uniform coat, particularly when thicker coatings are required.

7.2.3 Film Formers

Many different film formers, as indicated in Table 7.1, have been successfully applied to cores using aqueous or organic vehicles as appropriate.

7.2.4 Solvent Selection and Evaporation

It is important to select the proper solvent for any coating application. It should dissolve the film former together with as many of the other components of the coating formulation as possible. Coloring is normally incorporated into the coating to visualize the buildup and uniformity of the coating. Data on relative evaporation rate of solvents as shown in Table 7.2 should be considered, as boiling-point data are not an adequate guide to the rate of solvent loss during drying. The figures quoted in Table 7.2 may vary slightly from one manufacturer to another,

Table 7.1 Some Film Formers Applied by Air Suspension Coating

Acacia
Alginates
Carboxymethylcellulose
Cellulose acetate butyrate
Cellulose acetate phthalate
Ethyl cellulose
Ethyl methacrylate
Gelatin
Hydrogenated castor oil
Hydroxypropylcellulose
Hydroxypropylmethylcellulose
Hydroxypropylmethylcellulose phthalate
Methylcellulose
Methyl methacrylate
Polyethylene glycol
Shellac
Stearic acid
Sugars (various types)
Waxes (various types)

depending on the method used to determine the evaporation rate. The evaporation rate of solvent blends and its effect on polymer solubility and coating deposition are very important and particularly difficult to predict [11]. Data on toxicity and explosion hazards of solvents used should be considered. Evaporating conditions can be adjusted with the aid of a psychrometric chart [12], shown in Fig. 7.3, to be so rapid that even moisture-sensitive core material can be successfully encapsulated using aqeuous-based coatings. Point 1 in Fig. 7.3 represents such a system with an "air" intake temperature of 70°F and 60% relative

Table 7.2 Some Evaporation Rates of Solvents Relative to *n*-Butyl Acetate = 1.0

Acetone	7.7
Ethanol, anhydrous	1.9
Ethyl acetate	4.2
Hexane	3.0
Isoproprопanol, anhydrous	1.7
Methanol, anhydrous	3.5
Toluene	1.9

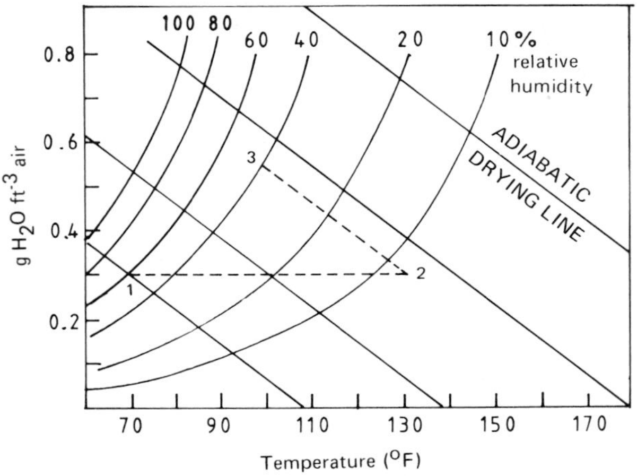

Figure 7.3 Psychrometric chart. (Reprinted with permission from H. S. Hall and R. E. Pondell, The Wurster Process, in *Controlled Release Technologies: Methods, Theory and Applications*, Vol. II (A. F. Kydonieus, ed.), CRC Press, Boca Raton, Fla., 1980, pp. 133-154 [10]).

humidity. When the air is heated to 130°F, its relative humidity decreases to less than 10% (point 2). As the air passes through the coating chamber, it evaporates water from the material being coated, dropping its temperature to 98°F (point 3) and now having a relative humidity of 40%. This air contains 0.55 g of water per cubic foot, compared to 0.30 g ft^{-3} of inlet air. The difference of 0.25 g ft^{-3} means that using these drying conditions one should be able to evaporate 250 g or 250 ml of water per minute for every 100 ft^3min^{-1} of air used. Knowing the concentration of water in the coating system and the flow rate of air to fluidize the bed of cores satisfactorily, the coating application rate can be estimated for particular drying conditions. The evaporation rate of the pure solvent may be decreased by other components in the coating system, which effect may be compensated for by either decreasing the coating application rate or increasing the air inlet temperature, provided that the drug is adequately thermostable. Humidity charts can also be obtained or constructed for other solvents or solvent blends, which normally do not take into account the effect of dissolved substances, such as plasticizers, on solvent evaporation rates. Also, unlike the evaporation of water, the inlet air supply usually contains a negligible amount of such solvents.

7.2.5 Temperature Profile

Hall and Pondell [10] showed data for the temperature profile in an example of an operating Wurster coating chamber. Using process air with an inlet temperature of 170°F, the temperature had dropped to 149°F at the base of the fluidized bed due to the cooling effect of solvent evaporation and had reduced to 132°F above the partition due to further solvent evaporation. Accordingly, it is often possible to coat heat-sensitive materials that are unstable at the temperature of the inlet air but adequately heat-stable at the temperatures achieved in the coating chamber.

7.2.6 Core Size and Shape

Small particles require more coating than large particles to achieve a given film thickness. Hall and Pondell [10] reported that as the particle diameter decreased from 1 mm (U.S. mesh 8) to 0.074 mm (U.S. mesh 200), approximately 17.5 times as much coating was required to produce a film thickness of 0.1 mm. Nonspherical particles of grossly irregular shape require even greater amounts of coating, and difficulties in achieving coat uniformity on angular faces may be experienced.

7.2.7 Particle Interactions

Particle-particle interaction and particle-equipment interaction occur as the cores ascend inside the coating partition and descend in the annular space. The attrition caused by these collisions depends mainly on the durability, velocity, concentration, size, and shape of the particles, and on the properties of the coatings being applied.

Agglomeration problems can often be troublesome during coating. Careful control of coating composition, spray droplet size and delivery rate, and gas flow rate can help prevent them. Aggregation is likely to occur if excessive residual solvent remains on the partially coated cores between passes through the coating spray, causing them to become tacky and adhesive. This will be more pronounced with certain coating formulations and if the coating application rate is excessive for the preset operating conditions of the equipment. Conversely, if the partially coated cores are overdried and the coating lacks bind, it will tend to be excessively abraded. If the exhaust temperature falls excessively during coating, it is indicative of excessive coating application rate leading to aggregation problems. If the exhaust temperature rises excessively, it is indicative of overdrying leading to abrasion problems. For particle separation to occur in a fluidized bed-type device [13, 14], the drag force (F_d) resulting from the interaction among the particle and upward flow gas must equal the sum of the interparticulate adhesive forces (Ψ) and the particle weight W_p, that is

$$F_d = \Psi + W_p \tag{7.1}$$

The superficial gas velocity V_f to provide this drag is

$$V_f = \left(\frac{\Psi + W_p}{\alpha^{1/2} C_d P_g A_p} \right)^{0.5} \tag{7.2}$$

where

α = a constant for the system

C_d = the drag coefficient

ρ_g = the density of the gas

A_p = the effective particle cross-sectional area

Therefore, for a particular system operating at a known gas velocity, the ability to fluidize the bed of particles depends on the values of Ψ and W_p. When Ψ is much larger than W_p, the bed may not fluidize properly because there is insufficient flow to disrupt the channels caused by the upward-moving gas stream; alternatively, an erratic surging bed that is poorly suited to coating may be obtained. If $\Psi >> W_p$, considerable elutriation of particles will occur because the superficial gas velocity required for particle separation must be greater than that required for conveying. Since W_p is a function of particle size, there is a lower limit (50 to 100 μm), depending on the particles and coating, below which cores cannot be successfully coated by the process. Similar considerations apply to flow in coating pans, where the limiting value is around 200 to 300 μm.

7.3 MATHEMATICAL DETERMINATION OF PARAMETERS ASSOCIATED WITH AIR SUSPENSION COATING

In an interesting paper presented by Yum and Eckenhoff [15], a number of empirical equations were presented for obtaining order-of-magnitude estimates of a number of parameters such as particle circulation rate, airflow rate, coating thickness, and coating duration as follows.

7.3.1 Particle Circulation Rate and Airflow Rate Determination

At steady state, the particle circulation rate in the coating partition is given by

$$W_1 = V_1A_1(1 - e_1)\rho_s \tag{7.3}$$

and in the annular space by

$$W_2 = V_2A_2(1 - e_2)\rho_s \tag{7.4}$$

where

W_1 and W_2 = the particle circulation rates for the coating partition and annular space, respectively

V_1 and V_2 = the mean particle velocities in the coating partition and annular space, respectively

A_1 and A_2 = the cross-sectional areas of the partition and annular space, respectively

e_1 and e_2 = the void fractions in the coating partition and annular space, respectively

ρ_s = the mean particle density

By combining Eqs. (7.3) and (7.4), assuming that $W_1 = W_2$,

$$V_1A_1(1 - e_1)\rho_s = V_2A_2(1 - e_2)\rho_s$$

or

$$e_1 = \frac{1 - V_2A_2(1 - e_2)}{V_1A_1} \tag{7.5}$$

By substitution into Eq. (7.5), e_1 can be calculated knowing the values of A_1 and A_2 from the dimensions of the coater and calculating the values of V_1, V_2, and e_2 as follows.

$$V_1 = V_3(1 - 0.179d_s^{0.3}\rho_s^{0.5}) \tag{7.6}$$

where

V_3 = the superficial velocity of the suspending medium (air) in the partition

d_s = the mean equivalent diameter of the particles

Equation (7.6) is an empirical equation after Hinkle [16] that has general application. Likewise, another empirical equation similar to (7.6) was derived from the work of Belden and Kassel [17]

$$V_3 - V_1 = 1.32\left[g'd_s\left(\frac{\rho_s}{\rho_a} - 1\right)\right]^{0.5} \tag{7.7}$$

where

g' = the gravitational constant

ρ_a = the density of air.

By solving the simultaneous equations (7.6) and (7.7), values for V_1 and V_3 can be obtained knowing the values of d_s, ρ_s, and ρ_a. V_2 can be estimated by combining the empirical equation (7.8) after Babu et al. [18] and Eq. (7.9) after Yoon and Kunii [19] as given below:

$$V_4 = \frac{\eta_a}{\rho_a d_s}\{[(25.25)^2 + 0.065G_a]^{0.5} - 25.25\} \tag{7.8}$$

where

V_4 = the superficial air velocity in the annular space,

η_a = the viscosity of air

G_a = the Galileo number = $\dfrac{d_s^3\rho_a(\rho_s - \rho_a)g}{\eta_a^2}$

Thus,

$$\frac{L_1}{V_b} = \frac{L_2}{g}\left[\frac{150\eta_a(V_4 + V_2)(1 - e_2)^2}{d_s^2\gamma^2 e_2^2} + \frac{1.75\eta_a(V_4 + V_2)^2(1 - e_2)}{d_s\gamma e_2}\right] \tag{7.9}$$

where

L_1 = the height of the annular space

L_2 = the height of the coating partition

V_b = the bulk volume of the particles per unit mass

γ = a particle shape factor which may be defined as

$$\gamma = \frac{0.5}{\dfrac{L_{max}}{L_{min}}}$$

where

L_{max} = the maximum dimension of the particles

L_{min} = the minimum dimension of the particles. The value of e_2 can be calculated approximately by the following equation.

$$e_2 = \frac{\bar{\bar{V}}_b - \rho_s^{-1}}{\bar{\bar{V}}_b} \qquad (7.10)$$

7.3.2 Coating Thickness and Duration Determination

Assuming that all of the coating material delivered is applied to the particles, the following equation is applicable:

$$Q_1C_1 = \frac{'N d('At_m\rho_m)}{d_t} \qquad (7.11)$$

or

$$\frac{Q_1C_1\,dt}{'N} = d('At_m\rho_m) \qquad (7.12)$$

Integrating both sides from $t = 0 \rightarrow t$, we obtain

$$\frac{Q_1C_1t}{'N} = 'At_m\rho_m \qquad (7.13)$$

or

$$t_m = \frac{Q_1C_1t}{'N\,'A\rho_m} \qquad (7.14)$$

or

$$t = \frac{'At_m\rho_m\,'N}{Q_1C_1} \qquad (7.15)$$

where

t_m = the thicknesses of the dry coating

t = the coating duration

Q_1 = the solution delivery rate

C_1 = the concentration of solids in the coating solution

$'N$ = the total number of particle circulations

$'A$ = the surface area of the particles

ρ_m = the density of the dry coating

$'N$ may be obtained from the expression $'N = t/f$, where f is the frequency of particle circulation, and is given by $f = W_2/W$, where W is the total weight of particles.

Reasonably good agreement was found by Yum and Eckenhoff [15] between calculated values and those obtained experimentally when using a 6 in., 12 in., or 18 in. Wurster coater for the coating of a mixed bed of cylindrical particles and disklike tablets. The mixed cores produced sufficient bulk density and voidage to optimize fluidization and hence deposition of uniform coating, which would otherwise have been very difficult to achieve on the lightweight, cylindrical particles alone.

7.4 AIR SUSPENSION EQUIPMENT AND ITS OPERATION

Various modifications of air suspension equipment for the coating of particles have been described in the literature. Robinson et al. [14] described equipment that worked on a fluidization principle but that employed a truncated partition and an atomizer spraying downward in a countercurrent fashion on the fluidized cores. Air activation vibrators were fitted to the coating and expansion chambers to reduce the tendency of finer particles to cling to their surfaces during the coating operation. Because of its improved control of particle separation and transport, it was possible to use the apparatus to coat particles from 10 to 1000 μm in diameter with a variety of coating materials. Kondo [20] also discussed various modifications of the Wurster process.

Friedman and Donbrow [21] described the construction of an inexpensive fluid bed coating apparatus suitable for use with very small batches (50 to 500 g) of cores. The apparatus was used to apply ethylcellulose with polyethylene glycol to granules prepared by a moist granulation process and containing the model drugs salicylic acid and caffeine. Used under optimal conditions of temperature, concentration of coating solution, and spray pressure, the apparatus was capable of producing excellent batch homogeneity and reproducibility together with acceptable granule friability.

Recently Higashide et al. [22] described a new device employing a pneumatic transport system for very rapidly applying ethylcellulose to mask the taste of granules containing the bitter drug dicloxacillin sodium. Whereas the effect of applying 5% or more ethylcellulose was adequate for taste-masking purposes, surface coverage of the granules was only 65 to 70%, which may approach 100% when optimal working conditions for the coating of the granules are eventually developed.

A number of commercial companies [23–25] market Wurster and other air suspension-type equipment suitable for microencapsulation. In many cases the equipment is purposely built to suit the coating and output requirements of the customer's product. It is always advisable to run the product initially on pilot-scale equipment to ascertain the design features of the equipment needed for a particular coating application. Modifications to equipment after installation are usually very costly and can be avoided by proper planning at the development stage. If one is interested in coating a number of different products using the same equipment, it will normally result in the design of more elaborate and costly plant, often using interchangeable coating chambers.

With all types of fluidization equipment, safety precautions against solvent and dust explosion using explosion relief flaps, suppression systems, or inertization methods need to be considered in the design. These features are particularly required when working with potentially explosive substances such as organic solvents and powders such as sugars and starch. With inflammable or explosive solvents, operating conditions may be set to provide a very low evaporation rate in order to minimize the risk of these hazards occurring. All electrical equipment should be explosion-proofed, and care should be taken to ensure proper grounding of equipment to prevent the buildup of static electricity causing sparking. Also, precautions against air pollution caused by venting toxic solvents and other materials into the atmosphere may be required. Simon [26] has recently reviewed aspects of the containment of hazards in fluid bed technology. The equipment can be instrumented with a wide range of controls, including those to monitor inlet and exhaust temperature, so causing corrective changes to be made to air or coating flow rate as necessary to prevent under- or overdrying.

7.5 SOME APPLICATIONS OF AIR SUSPENSION COATING

The references in this section are arranged roughly in chronological order and give a good indication of the wide application of the process for the coating of drugs.

Coletta and Rubin [27] described the coating of aspirin crystals of various mesh sizes with mixtures of ethylcellulose and methylcellulose sprayed from methylene chloride:isopropyl alcohol 1:1 using Wurster air suspension apparatus. A dye was used in the coating to visualize its application, and it was noted that finer material required more coating to obtain approximately the same final film thickness. Coating was complete in 30 min, and the microencapsulated aspirin produced was then tableted. Dissolution studies confirmed that as the content of ethylcellulose in the coating was increased, there was a progressive reduction in the rate of release of aspirin. In a subsequent related paper, Wood and Syarto [28] showed that the microencapsulated aspirin produced a sustained-release effect when tested in human subjects. Incomplete drug absorption was noted with products containing a high content of ethylcellulose in the coating. Also, successful correlation of the results of in vitro and in vivo studies was found to be very dependent on the composition of the coating.

Caldwell and Rosen [29] used an air suspension coater to successfully apply dexamphetamine sulfate onto 20–25 mesh sugar pellets using a solution of gelatin in a hydroalcoholic solvent as adhesive. About 60 to 70% of the powder charge applied was taken up by the cores. These were then coated with various materials, including melted microcrystalline wax, glyceryl monostearate, glyceryl distearate, beeswax, and 12-hydroxystearyl alcohol, to obtain lipid-coated microcapsules with sustained-release properties. Enteric-coated pellets were also prepared using cellulose acetate phthalate that released only 1 to 3% of drug after 1.5 hr exposure to U.S.P. gastric fluid, while rapidly releasing drug in U.S.P. intestinal fluid. Excellent batch-to-batch reproducibility was obtained with the equipment, and no scale-up problems were encountered.

Robinson et al. [14] successfully microencapsulated for controlled-release purposes spherical sugar pellets with about 10.5% ethylcellulose, aspirin crystals of irregular form with about 37% cellulose acetate phthalate, sodium bicarbonate particles with 33% and 50% hydrogenated castor oil, and both paracetamol and saccharin particles with 12 hydroxystearyl alcohol using an air suspension coater.

A number of patents mention the use of air suspension coating for the microencapsulation of drugs. Thus Brown [30] patented a process whereby a drug was deposited on very small corn/potato starch or similar sorbable-core particles. The resultant product was then encapsulated using air suspension equipment with suitable coating material such as methylcellulose prior to formulation as part of a suspension or emulsion for injectable use. The drug should slowly diffuse away from the site of injection. The process was particularly recommended for the administration of antiallergic agents such as pollen extracts for hay fever suffers.

Milosovich [31], in a patent assigned to Parke, Davis and Co., described a process involving the extrusion of gelatin, glycerin, and the antihistamine drug diphenhydramine hydrochloride in water at 70°C through a bank of no. 22 gauge hypodermic needles into a coolant liquid immiscible with the gel. Core beads of 590 to 840 μm diameter were obtained, which were then coated in a Wurster coater with ethylcellulose films of varying thickness to produce a sustained-release product.

Michaels [32], in a patent assigned to Alza Corp., used a Wurster air suspension technique to encapsulate chloramphenicol particles of size 50 μm with polylactic acid polymer of molecular weight approximately 50,000 dissolved in chloroform. The resultant microcapsules were dispersed in an aqueous solution of sodium alginate, which was then dried to form a film. This film was immersed in a zinc chloride solution, which formed a crosslinked anionic polyelectrolyte with the alginate that was biodegradable as a result of polyvalent ion displacement. This washed film was dried and cut into elliptically shaped ovals suitable for insertion and retention in the eye sac. The ocular device formed gave sustained administration of the antibiotic to the eye at a rate controlled by the polylactic acid microcapsule coating, while concurrently the alginate matrix completely biodegraded in the environment of the eye.

Benedikt [33], in a patent assigned to Byk Gulden Chemischi Fabrik GmbH, applied sodium salicylate by pan coating onto sugar pellets using a 10% solution of ethylcellulose in ethanol as adhesive until they were 1 to 2 mm in size and contained about 84% by weight of drug. These cores were then sprayed in a fluidized bed-type apparatus with a solution of either ethylcellulose and shellac in denatured ethanol:dichloromethane 1:1 or various methacrylic acid-methacrylic acid ester copolymers dissolved in the same solvent to obtain a product with enteric and sustained-release properties.

Juslin and Puumalainen [34] spheronized pellets containing phenazone, microcrystalline cellulose, and potato starch using spheronization apparatus (see Chap. 13). The product was coated with Eudragit RS (an acrylic copolymer) containing triacetin as plasticizer and dissolved in isopropanol:acetone 1:1, using an air suspension coater. The coated pellets together with suitable additives were then tableted. In vitro dissolution studies indicated that the pellets had a slower release rate of drug than the tablets, whose microcapsules were presumably partially ruptured upon compression. As the compression force was markedly increased, drug release diminished due to greater compaction in the tablets formed.

Various acrylic resins (Eudragit range) were used by Lehmann et al. [35] to encapsulate spherical pellets and compact granules using an air suspenson technique. Even drug crystals with a size range of 100 to 300 μm having sharp edges and angles were successfully coated. Figure 7.4 shows methaqualone crystals before and after coating. If talc or magnesium stearate was suspended in the coating solution and if operating conditions were optimized, agglomeration of the particles was avoided.

In the first of a series of papers, Friedman et al. [36] used a fluidized coating procedure to apply ethylcellulose/polyethylene glycol 4000 mixtures onto granules spheronised by pan rotation and containing salicylic acid and caffeine as model drugs to produce a prolonged-release product. They [37] also used the same equipment to coat commercially available placebo lactose pellets with ethylcellulose films containing the drugs. The pellets were found to have satisfactory friability (1 to 1.5%) for the coating process.

Hall et al. [38] used air suspension coating and also pan coating to compare aqueous and organic solvent-based ethylcellulose films. Coatings based on organic solvents are undesirable because of possible fire, explosion, or pollution problems, toxicity of residual solvents in the product, and the high cost of such solvents. Ethylcellulose, which is water-insoluble, has become available recently as a pseudolatex containing surfactants under the trade name Aquacoat [39], and was used with a suitable plasticizer such as diethyl phthalate. Samples of a drug in granular form were coated with this pseudolatex, which has a high solids content, and the product exhibited sustained-release properties when tested in vitro. The release rate decreased with increasing amount of coating applied over the range 5 to 20%, but was not as effective as delaying drug release as an ethylcellulose film applied from a methanol/methylene chloride solvent. However, change of plasticizer in the pseudolatex to tributyrin appeared to give better film formation with slower drug

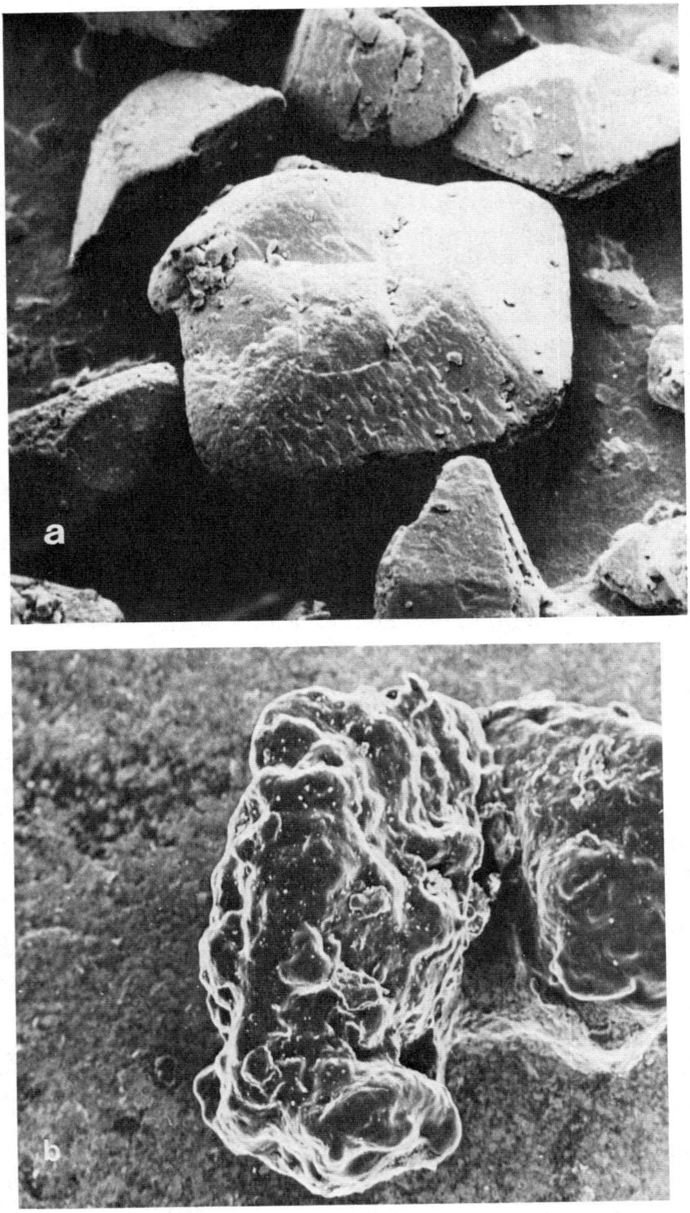

Figure 7.4 Methaqualone crystals before (a) and after (b) coating with Eudragit retard. (Reproduced with permission from Ref. 35 and courtesy of Dr. K. Lehmann, Rohm Pharma GmbH, Darmstadt, West Germany.)

release than that obtained with the diethyl phthalate-containing coating. Cores that were highly water-soluble or that absorb water very rapidly needed hydrophobic precoats to prevent loss of water into the core before it had fused the coating droplets into a continuous film.

Raghunathan et al. [40] reported the use of Wurster equipment to apply ethylcellulose-vegetable oil in methylene chloride:acetone 10:1 onto cores consisting of an ion-exchange resin-drug complex to make drug release diffusion-controlled. Results reported related mainly to release of phenylpropanolamine and to a lesser extent dextromethorphan. It was necessary to pretreat the ion-exchange resin-drug complex with an agent such as polyethylene glycol 4000, which has hydrophilic, nonvolatile, solvent properties. The pretreatment prevented swelling of the cores of the microcapsules upon immersion in water from rupturing the coating. The phenylpropanolamine-containing microcapsules were shown to have enhanced delayed-release properties in vitro when compared to the uncoated form. In vivo tests indicated that mixtures of both could be used to obtain varying drug release profiles.

Recently, Bogentoft et al. [41] enterically coated granules containing digoxin or aspirin with carboxylic polymers using a fluidized bed technique. The digoxin-containing product showed delayed excretion of degradation products in the urine of healthy subjects. The aspirin-containing product showed less damage to the gastric mucosa in comparison to a commercially available sustained-release product when determined by gastroscopic evaluation in healthy subjects. The emptying of ^{51}Cr-labeled aspirin-containing coated granules from the stomach was followed by scintigraphy and was found to occur over a period of several hours with $t_{90\%}$ for emptying ranging from 2 to 5 hr. In comparison to enterically coated tablets, the coated granules were a more reliable dosage form for ensuring a reproducible absorption profile. Also, the plasma levels of aspirin achieved were more uniform with the enteric-coated granules.

REFERENCES

1. D. E. Wurster, U.S. Patent 2,648,609 (August 11, 1953).
2. D. E. Wurster, U.S. Patent 2,799,241 (July 16, 1957).
3. D. E. Wurster, U.S. Patent 3,089,824 (May 14, 1963).
4. D. E. Wurster and J. A. Lindlof, U.S. Patent 3,196,827 (July 27, 1965).
5. D. E. Wurster, J. V. Battista, and J. A. Lindlof, U.S. Patent 3,207,824 (September 21, 1965).
6. D. E. Wurster, U.S. Patent 3,241,520 (March 22, 1966).
7. D. E. Wurster, U.S. Patent 3,253,944 (May 31, 1966).
8. D. E. Wurster, Air-suspension technique of coating drug particles: a preliminary report, *J. Amer. Pharm. Assoc. Sci. Ed. 48*:451–454 (1959).

9. N. Brudney and P. Y. Toupin, Air suspension coating, *Can. Pharm. J. 94*:18–19 (1962).
10. H. S. Hall and R. E. Pondell, The Wurster process, in *Controlled Release Technologies: Methods, Theory and Applications*, Vol. II (A. F. Kydonieus, ed.), CRC Press, Boca Raton, Fla., 1980, pp. 133–154.
11. A. A. Sarnotsky, Evaporation of solvents from paint films, *J. Paint Technol. 41*:692–701 (1969).
12. H. A. Lieberman and A. Rankell, Drying, in *The Theory and Practice of Industrial Pharmacy* (L. Lachman, H. A. Lieberman, and J. L. Kanig, eds.), Lea & Febiger, Philadelphia, 1970, pp. 24–48.
13. M. Baerns, Effect of interparticle adhesive forces on fluidization of fine particles, *Ind. Eng. Chem. Fundamentals 5*:508–516 (1966).
14. M. J. Robinson, G. M. Grass, and R. J. Lantz, An apparatus and method for the coating of solid particles, *J. Pharm. Sci. 57*:1983–1988 (1968).
15. S. I. Yum and J. B. Eckenhoff, Development of fluidized-bed spray coating process for axisymmetrical particles, *Drug Develop. Indust. Pharm. 7*:27–61 (1981).
16. B. L. Hinkle, Acceleration of particles and pressure drops encountered in horizontal pneumatic conveying, Ph.D. thesis, Georgia Institute of Technology, Atlanta, 1953.
17. D. H. Belden and L. S. Kassel, Pressure drops encountered in conveying particles of large diameter in vertical transfer lines, *Ind. Eng. Chem. 41*:1174–1178 (1949).
18. S. Babu, B. Shah, and A. Talwalker, Fluidization characteristics of coal gasification materials. Presented at the A.I.Ch.E. 69th Annual Meeting, Chicago, November 28–December 2, 1976.
19. S. M. Yoon and D. Kunii, Gas flow and pressure drop through moving beds, *Ind. Eng. Chem. Process Des. Develop. 9*:559–565 (1970).
20. A. Kondo, Air-suspension coating process, in *Microcapsule Processing and Technology* (J. Wade Van Valkenburg, ed), Dekker, New York, 1979, pp. 142–153.
21. M. Friedman and M. Donbrow, Fluidized bed coating technique for production of sustained release granules, *Drug Develop. Indust. Pharm. 4*:319–331 (1978).
22. F. Higashide, A. Miyagishima, and Y. Omura, A new method for film coating granules, *J. Pharm. Pharmacol. 32*:55–56 (1980).
23. Lakso Division, Package Machinery Co., East Longmeadow, Mass.
24. Glatt Maschinen-und Apparatebau AG, Prattein, Switzerland.
25. Aeromatic AG, Muttenz, Switzerland.
26. E. Simon, Containment of hazards in fluid-bed technology, *Manuf. Chem. Aersol News 49*(1):23–24, 28, 31–32 (1978).

27. V. Coletta and H. Rubin, Wurster coated aspirin I. Film-coating techniques, *J. Pharm. Sci.* 53:953–955 (1964).
28. J. H. Wood and J. Syarto, Wurster coated aspirin II. An in vitro and in vivo correlation of rate from sustained-release preparations, *J. Pharm. Sci.* 53:877–881 (1964).
29. H. C. Caldwell and E. Rosen, New air suspension apparatus for coating discrete solids, *J. Pharm. Sci.* 53:1387–1391 (1964).
30. E. A. Brown, U.S. Patent 3,185,625 (May 25, 1965).
31. G. Milosovich, U.S. Patent 3,247,066 (April 19, 1966).
32. A. S. Michaels, U.S. Patent 3,962,414 (June 8, 1976).
33. G. Benedikt, U.S. Patent 4,083,949 (April 11, 1978).
34. M. Juslin and P. Puumalainen, Pellets coated with acrylate plastic in tabletting, in *F.I.P. Abstracts, 37th International Congress of Pharmaceutical Sciences, The Hague, September,* 1977, p. 48.
35. K. O. R. Lehmann, H. M. Bossler, and D. K. Dreher, Controlled drug release from small particles encapsulated with acrylic resins, in *Polymeric Delivery Systems* (R. J. Kostelnik, ed.), Gordon & Breach, New York, 1978, pp. 111–119.
36. M. Friedman, M. Donbrow, and Y. Samuelov, Release rate of drugs from ethyl cellulose coated granules containing caffeine and salicylic acid, *Drug Develop. Indust. Pharm.* 5:407–424 (1979).
37. M. Friedman, M. Donbrow, and Y. Samuelov, Placebo granules as cores for timed release drug delivery systems, *J. Pharm. Pharmacol.* 31:396–399 (1979).
38. H. S. Hall, K. D. Lillie, and R. E. Pondell, *Comparison—aqueous vs. solvent based ethylcellulose films,* Coating Place, Inc., Verona, Wisc.
39. *Aquacoat Application Bulletin,* FMC Corp., Philadelphia.
40. Y. Raghunathan, L. Amsel, O. Hinsvark, and W. Bryant, Sustained-release drug delivery system I: coated ion-exchange resin system for phenylpropanolamine and other drugs, *J. Pharm. Sci.* 70:379–384 (1981).
41. C. Bogentoft, G. Ekenved, and U. E. Jonsson, Controlled release of drugs to the small intestine, in *Proceedings of the 8th International Symposium on Controlled Release of Bioactive Materials, Ft. Lauderdale, July 26–29,* 1981, pp. 42–45.

8

Spray Drying, Spray Congealing, Spray Embedding, and Spray Polycondensation

8.1 SOME BASIC PRINCIPLES OF SPRAY DRYING AND SPRAY CONGEALING

Applications of the use of spray drying to aid in the recovery of microcapsules produced by other processes, such as coacervation or polymerization procedures, have been mentioned in previous chapters. However, spray drying and spray congealing have been used as the primary means of microencapsulation in a wide range of drugs and flavors. In the former process the core substance is dispersed in a solution of the coating material, which is then atomized and the solvent dried off using heated air in a spray dryer. The latter process is similar except that no solvent is used for the coating material, which has the property of melting at elevated temperature when being atomized and congealing when the droplets formed meet cool air in a spray dryer. Congealing can also be accomplished by spraying the dispersion of core material in the coating into a chilled organic solvent, desolvating liquid, or sorptive particles.

Both processes have the advantage over many other microencapsulation procedures of being rapid, single-stage operations suitable for batch or continuous production of large quantities of product. Heat-sensitive core substances can be coated by spray drying because exposure to elevated temperature is very short, normally ranging from 5 to 30 s. Moisture-sensitive drugs can be encapsulated by using nonaqueous coating systems. However, the coating produced by spray drying tend to be rather porous which may make them adequate for taste-masking and other purposes but not for controlled release. Also, the capital and running costs of a spray dryer are high, which makes the procedure expensive unless large production runs are required or the equipment can be used for other applications. Figure 8.1 shows a

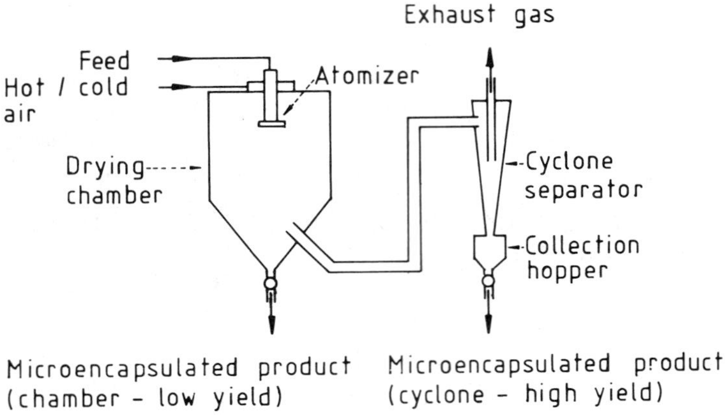

Figure 8.1 Schematic diagram of a cocurrent spray dryer.

diagram of a typical spray dryer. There are many different types of spray dryers available, varying in design to suit the requirements of the feed material and product. The coating material should have an adequate viscosity to ensure that it contains a uniform dispersion of core substance at room temperature for spray drying and at elevated temperature for spray congealing. The feed is circulated to the atomizer from a thermostated mixing tank by a suitable pumping system. A centrifugal or spinning-disk atomizer is normally employed whereby the feed is delivered to the center of a high-speed rotating disk, which causes it to break into droplets. Various designs of such rotary atomizers are available to suit feed and product requirements and may be driven electrically or by means of an air turbine. In spray congealing the air supply to the turbine can be preheated to prevent the feed from congealing in the atomizer and blocking it. Nozzle-type atomizers are used less frequently, as they tend to clog.

Air is blown into the drying chamber, having been preheated if necessary by passage over a heat exchanger. Normally, air flow is cocurrent with the direction of atomization for simplicity in design, but some spray dryers have countercurrent or mixed flow patterns. It is essential to obtain good mixing of droplets and air, which is also influenced by the design of the coating chamber.

When spray drying, heat and mass transfer occur rapidly between the droplets and the surrounding hot air because of the large surface area available for evaporation. The rate of drying is a complex function of feed rate, droplet size and distribution, coating solvent, inlet/outlet temperature, humidity, gas velocity, and other factors. The residence time for a droplet in the drying chamber is only a few seconds. High inlet gas temperatures of 150 to 200°C can be employed even with thermolabile substances, as the product rises above the wet-bulb temperature only at the end of the process. As the solvent evaporates, it tends to deposit a spherical coating of solids as a skin around one or more core particles. Further evaporation of solvent may be hindered, depending on its rate of diffusion through the coating, possibly resulting in the formation of blow holes in the final coating. When spray congealing using a spray dryer, the coating is applied as a hot melt and solidification of the coating occurs upon spraying into cold air. Much higher concentrations of coatings are required for spray congealing than for spray drying, because only the molten coating forms the coating liquid phase. Mechanical rakes or air sweepers may be used to prevent the product from adhering to the walls and base of the drying chamber.

Most of the microencapsulated product is recovered from the collection hopper after it has been separated from the effluent gases by passage through a cyclone separator. Waste gases, from which dust has been removed, usually by filtration, are vented with regard to avoiding environmental pollution. A smaller amount of product may also be recovered from the base of the drying chamber, which if produced by spray drying will have been exposed to elevated temperature for an extended period and may have been adversely affected in physical and chemical characteristics.

The interested reader should refer to the book by Masters [1] on spray drying for extensive details of the construction and operation of this type of equipment. The importance of pilot-plant studies for developing dryer design and specification is stressed. Precautions against fire and explosion are discussed. A large number of mathematical equations are presented for quantifying aspects of the process. The section on spray-drying equipment by Belcher et al. [2] in the book edited by Mead should also be consulted. Newton [3] has reviewed spray drying and its uses in pharmacy. Ganderton [4], Lieberman and Rankell [5], and Nielsen [6] have published on the design and pharmaceutical applications of spray dryers. Kondo [7] has an interesting chapter in his book on the use of spray drying for microencapsulation where a number of nonpharmaceutical applications of the process are discussed. Recently, Nolen and Kool [8] microencapsulated thermonuclear fuel pellets by spray drying using various polymers. Of the coating materials examined of pharmaceutical interest, polymethyl methacrylate delivered from a low-boiling-point organic solvent formed a good coating, whereas sodium carboxymethylcellulose was difficult to encapsulate with because, being water-soluble, its solution was very viscous and difficult to evaporate.

8.2 SOME EXAMPLES OF THE MICROENCAPSULATION OF EXCIPIENTS AND DRUGS BY SPRAY DRYING

8.2.1 Using Aqueous Coating Solutions

Many water-soluble coatings have been used for the microencapsulation of dispersed core material by spray drying at elevated temperature. For example, various aromatic oils used as flavorings in the formulation of pharmaceuticals may be microencapsulated in this manner to produce a free-flowing powder with reduced volatility [9]. The oil is emulsified as an oil-in-water (o/w) system into a concentrated aqueous acacia solution, which is then preheated to about 50°C prior to atomization into a hot air blast. The very rapid evaporation of the water produces a cooling effect, which minimizes loss and thermal degradation of the volatile oil in the final product.

Marotta et al. [10], in a patent assigned to the National Starch and Chemical Corporation, used an aqueous solution of acid ester dextrins to encapsulate an emulsified dispersion of lemon oil by spray drying. The product did not have the characteristic odor, color, or taste associated with ordinary dextrins, and it also had superior volatile oil-retention properties.

In a patent assigned to Beatrice Foods Co., Noznick and Tatter [11] used an ammoniacal aqueous solution of gluten with an oil phase containing peppermint oil to form an o/w emulsion with the aid of a suitable emulsifier. The emulsion was then spray dried, and as the ammonia dried off under the influence of the heat treatment, the gluten formed a coating around the core droplets that was insoluble in neutral aqueous solutions. The aqueous permeability of the coating could be varied by altering the percentage gluten used and by incorporating dextrin and/or acacia into the aqueous phase of the emulsions whose solids formed the coating.

Various penicillins that are water-insoluble or poorly soluble have been microencapsulated to reduce the incidence of sensitivity reactions occurring among personnel involved in subsequent production procedures. Seager [12], in a patent assigned to Beecham Group Ltd. (G.B.), described a spray-drying process suitable for such a purpose that was unusual in that the content of coating on the final product was normally less than 6% of its total weight. This requirement was necessary in order to avoid reducing the bioavailability of the antibiotic. As an example of the process, a high concentration of ampicillin trihydrate particles were suspended in a dilute aqueous solution of sodium carboxymethylcellulose. Foaming was controlled by adding octanol to the concentrated slurry obtained, which was filtered to remove clumps of poorly dispersed medicament. The stirred slurry was then spray dried using a nozzle atomizer and an inlet temperature of about 160°C and an outlet temperature of about 84°C. Product less than 75 μm was recycled into the feed material, as otherwise it would tend to be very dusty and difficult to handle during subsequent processing into dosage forms. Other adequately water-soluble coating materials used included hydroxypropylmethylcellulose, hydroxypropylcellulose, gelatin, polyvinylpyrrolidone, and polyvinyl alcohol. With the exception of the latter two materials, the coatings formed tended to have blow holes.

8.2.2 Using Nonaqueous Coating Solutions

Robinson and Svedres [13], in a patent assigned to Smith, Kline and French Laboratories, suspended finely divided sulfamethylthiadiazole particles in a solution of hydrogenated castor oil in chloroform which was then spray dried using an inlet temperature of 90°C and an outlet temperature of 40°C. The product was used in the formulation of an aqueous suspension of the medicament having sustained-release properties. The suspension formed sedimented slowly, partially because of the very small size of the microcapsules (20 μm) and their low density.

As another example of the use of a nonaqueous vehicle in a spray drying process, Granatek et al. [14], in a patent assigned to Bristol-Meyers Co., coated a dispersion of dicloxacillin sodium particles with a mixture of ethylcellulose and spermaceti that had been previously dissolved in methylene chloride. Other penicillins were similarly microencapsulated for taste-masking purposes and used to prepare either capsules or tablets. In a subsequent patent [15], their use for the preparation of suspensions was described.

8.3 SOME EXAMPLES OF THE MICROENCAPSULATION OF EXCIPIENTS AND DRUGS BY SPRAY CONGEALING

Robinson and Swintosky [16] microencapsulated particles of sulfaethylthiadiazole by mixing them with molten hydrogenated castor oil at 110°C; the suspension was then spray congealed into an air-cooled chamber using a centrifugal-wheel atomizer. The spherical microcapsules obtained were observed to consist of finely divided drug particles uniformly dispersed throughout a matrix of hydrogenated castor oil and to have a uniform film of the oil over the surface of each microcapsule. The product was formulated into an aqueous suspension and was shown to have sustained-release properties when tested by in vitro and in vivo procedures. Scott et al. [17], in a subsequent paper from the same company, Smith, Kline and French, reported that waxes, fatty acids or alcohols, sugars, and other coating materials that melt without decomposition at high temperature and solidify at normal storage temperatures are suitable for spray congealing. They used molten stearic acid thickened with dissolved ethylcellulose at 70°C without any core substance in a modified laboratory Niro spray dryer fitted with a centrifugal-wheel atomizer and having an ambient air inlet temperature of 25 to 27°C.

The particle size of the spray-congealed coating material was 17 to 40 μm for mean surface volume diameter and was found to increase within this range with decreasing wheel speed, increasing feed rate, and decreasing viscosity of the feed. Variation in wheel speed was more important in controlling particle size than variation in feed rate, whereas viscosity of the feed, which surprisingly was reported as Newtonian, had relatively little influence. Obviously, if a concentrated dispersion of core particles in the molten wax was used, the system would exhibit non-Newtonian behavior and its apparent viscosity would depend on wheel speed, which might significantly influence the particle size of a product. A correlating equation was developed for the three variables that could be of assistance in scaling-up operations for production.

In a series of papers by John and Becker [18] and Cusimano and Becker [19], certain details of a spray congealing process for sulfaethylthiadiazole were investigated, such as the composition of the coating wax, the nozzle size of the atomizer, and the presence of surfactant in the wax matrix. Both decreasing nozzle size and increasing surfactant concentration tended to produce products with a faster rate of drug release in acid pepsin or alkaline pancreatin media. However, dissolution behavior was most affected by the type of waxy coating material used. White beeswax U.S.P. glyceryl tristearate, carnauba wax, hydrogenated castor oil, cetyl alcohol, and glyceryl monostearate were examined. In a subsequent paper by Hamid and Becker [20], the in vitro dissolution patterns of some spray-congealed sulfaethylthiadiazole-wax products in a tablet form were studied. Tableting caused a decrease in the rate of drug release, as is shown in Fig. 8.2.

Two other examples of spray congealing of pharmaceutical interest are as follows. Koff [21] dispersed thiamine mononitrate in a molten mixture of mono- and diglycerides of palmitic and stearic acids at 74°C and spray-congealed the suspension into ambient air at 20°C using a centrifugal atomizer. The average size of the coated particles obtained was about 60 μm, and the process was reported to be suitable for the encapsulation of other vitamins of the B group for taste-masking purposes.

Smith and Lambrou [22], in a patent assigned to Bush, Boake Allen, Ltd., described a multiple encapsulation process whereby primary capsules were prepared by spray drying an emulsion of lemon oil in water containing a modified gelatin as emulsifying agent and coating material. The gelatin-coated microcapsules were then dispersed in molten non-self-emulsifying glyceryl monostearate and spray-cooled. The microcapsules obtained contained one or more primary capsules embedded in the hydrophobic matrix and were suitable for use as a flavoring agent in powder form.

Waxy beads of average diameter 0.1 to 2.5 mm were prepared by Ross and Reul [23] in a patent assigned to Farbwerke Hoechst AG vormals Meister Lucius & Bruning, Germany. A glycol ester of a wax and a surfactant were used as the carrier for a dispersion of fine drug

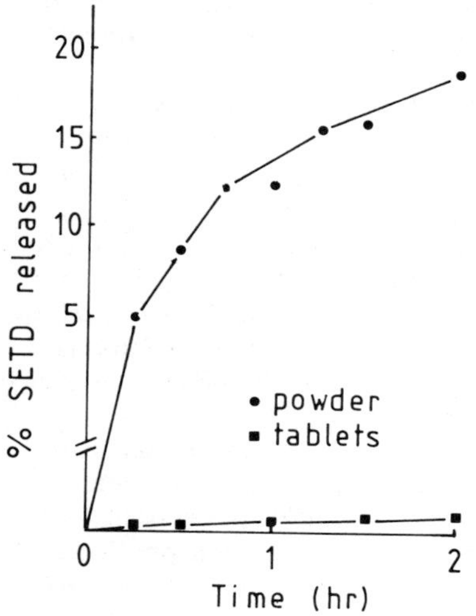

Figure 8.2 Plot of percent sulfaethylthiadiazole (SETB) released as a function of time for dissolution from SETD-synthetic waxlike ester-white beeswax (1:1) tablets and powder in acid pepsin medium. (Reproduced with permission of the copyright owner, American Pharmaceutical Association; from Ref. 20.)

particles. Increasing the content of surfactant gave faster release rate of drug. The melt was passed through a spray head and the droplets allowed to fall into an 8-m-deep tunnel against an upward-moving cold airstream, which solidified the beads and removed upward fine particles for reprocessing. The product collected at the bottom of the tunnel was dosed into capsules.

8.4 SPRAYING INTO CHILLED ORGANIC SOLVENT, DEHYDRATING LIQUID, OR SORPTIVE SOLID PARTICLES

Some examples of spraying into vehicles other than air are given as follows.

Hecker and Hawks [24], in a patent assigned to Eastman Kodak Co., Rochester, N.Y., emulsified vitamin A palmitate and antioxidants into a continuous phase composed of gelatin, glucose, and water at 65°C. The glucose acted as a plasticizer for the coating. The emulsion was then further heated to 75°C and sprayed into agitated hexane cooled to 25°C, which was a nonsolvent for the gelatin coating and which contained corn starch to reduce agglomeration. The microcapsules formed were filtered off and were dried in warm air to free them from residual water and hexane. The final product had a particle size of 250 to 600 μm.

Spraying into polyglycols has been used to dehydrate droplets of hydrophilic colloid containing dispersed core material so as to form microcapsules. Pasin [25], in a patent assigned to Balchem Corporation, emulsified lemon oil into an aqueous solution of dextrin and sprayed the emulsion into agitated liquid polyethylene glycol having a preferred molecular weight in the range 190 to 630. The polyethylene glycol dehydrated the microcapsules formed, which were then separated by centrifugation. Adhering polyethylene could be washed off using a suitable solvent or retained to reduce gaseous environmental penetration into the microcapsules.

Palmer [26] microencapsulated orange oil by initially emulsifying it with acacia and water, then made the emulsion formed more tacky by adding corn syrup solids and more acacia, and finally sprayed this system onto an agitated bed of acacia particles. The acacia particles absorbed the volatile oil and water, leaving the other solids to form a substantially dry continuous coating over the acacia particles. The product was sieved and formed a free-flowing powder that showed no loss or decomposition of flavor after 2 years storage at room temperature.

8.5 SPRAY EMBEDDING

In spray embedding an aqueous or organic solvent containing a solution of both a drug and a dissolution-retardant polymer are spray-dried. For example, Asker and Becker [27] prepared either an aqueous ammoniacal solution of shellac or cellulose acetate phthalate and sulfaethylthiadiazole, or an alcohol:chloroform solution of Glycowax S-932, castor wax MP80, aluminum monostearate, or glyceryl monostearate and the drug prior to spray drying. The products are best described as microparticles. They tended to clump and to have a porous appearance. The product containing the glycowax had the most prolonged release when tested by in vitro dissolution studies. It also had the lowest cumulative urinary excretion, as shown in Fig. 8.3, which may have been due to its reduced bioavailability.

More recently Takenaka et al. [28] spray-dried ammoniacal solutions of sulfamethoxazole and cellulose acetate phthalate using a rotary atomizer. Additives such as colloidal silica or talc in the feed for spray drying greatly improved the flow properties of the products, which could then be compressed into tablets wtih enteric properties. X-ray analysis (see Fig. 8.4) and infrared spectroscopy (see Fig. 8.5) confirmed that

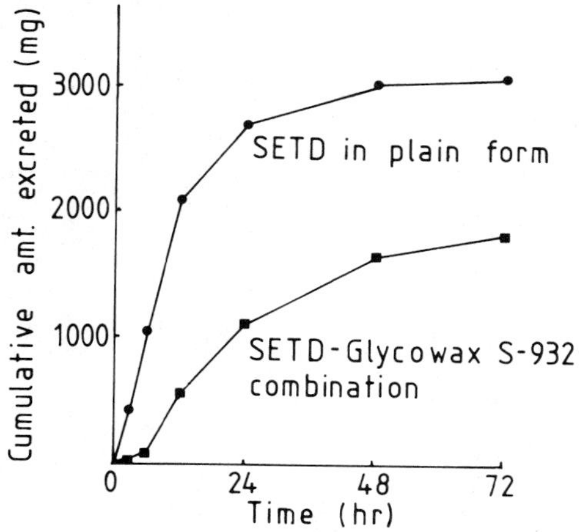

Figure 8.3 Average cumulative urinary excretion of free SETD for 4 humans receiving a 3.9-g oral dose of SETD in plain form and SETD-Glycowax S-932 combination. (Reproduced with permission of the copyright owner, American Pharmaceutical Association; from Ref. 27.)

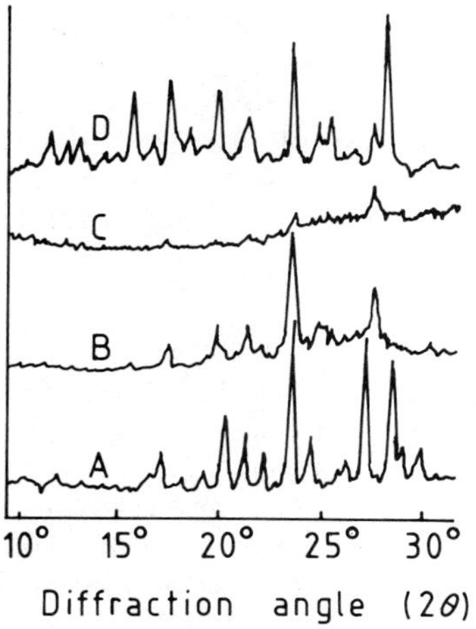

Figure 8.4 X-ray diffraction patterns of original and spray-dried sulfamethoxazole. Key: A, form I, original sulfamethoxazole; B, form II, sulfamethoxazole recrystallized in water at dry ice-acetone temperature; C, spray-dried products prepared from formulations containing cellulose acetate phthalate (50 g) and colloidal silica (50 g); and D, spray-dried products prepared from formulations containing talc (50 g). (Reproduced with permission of the copyright owner, American Pharmaceutical Association; from Ref. 28.)

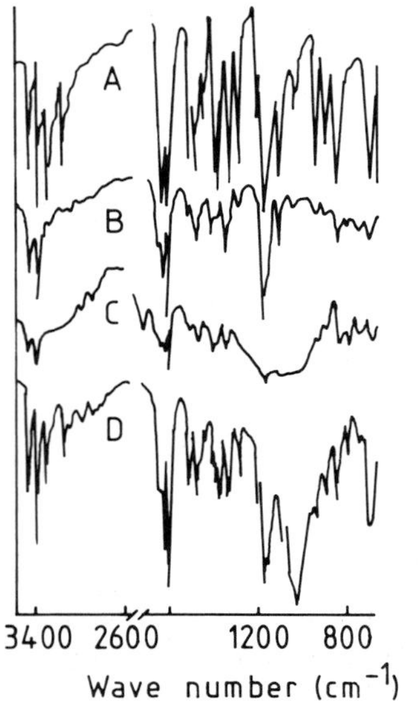

Figure 8.5 Infrared spectra of original and spray-dried sulfamethoxazole. Key: A, form I, original sulfamethoxazole; B, Form II, sulfamethoxazole recrystallized in water at dry ice-acetone temperature; C, spray-dried products prepared from formulations containing cellulose acetate phthalate (50 g) and colloidal silica (50 g); and D, spray-dried products prepared from formulations containing talc (50 g). (Reproduced with permission of the copyright owner, American Pharmaceutical Association; from Ref. 28.)

some of the crystalline drug was converted from its form I to form II. Further studies on the effect of cellulose acetate phthalate and excipients on the polymorphism of spray-embedded sulfamethoxazole were reported by Takenaka et al. [29]. Also, Wise et al. [30] spray dried a 1,1,1,3,3,3-hexafluoro-2-propanol and benzene solution of the antimalarial drug, 2,4-diamino-6-(d-naphthylsulfonyl)-quinazoline and a copolymer of dl-lactic acid and glycolic acid. The fine particles (less than 125 μm) produced were suspended in carboxymethylcellulose solution and implanted into mice to give prolonged protection against rodent malaria, *Plasmodium berghei*. Recovery of excreted radioactivity indicated that the drug-polymer material had a sustained release through 14 weeks and confirmed the potential of the drug delivery system for use against human malaria.

Of course, spray embedding with hydrophilic polymers can be used to enhance the dissolution of poorly soluble drugs. For example, Kala et al. [31] prepared phenobarbitone-polyvinyl alcohol and digoxin-polyvinylpyrrolidone compounds in fine powder form that showed enhanced in vitro dissolution due to the amorphous form of the drug produced.

8.6 SPRAY POLYCONDENSATION

Spray polycondensation is based on spray drying whereby a dispersion of core material in a continuous phase containing reactive monomer and precondensate with catalyst, in addition to other film-forming agents, is caused to polymerize at the high temperature of the process by vaporization of water. Voellmy et al. [32] used this process to microencapsulate phenobarbitone particles. Polyvinyl alcohol was used as a film-forming agent. Hexamethylolmelamine hexamethyl ether, trimethylolamine trimethyl ether, or dimethylourea dimethyl ether was used as monomer and a cationic reactive tenside on a melamine-formaldehyde base was used as precondensate. Hydrochloric acid was used as catalyst. The drug was uniformly dispersed in a solution of these materials in distilled water and the pH adjusted to 2.5 with hydrochloric acid prior to spray drying using an inlet temperature of 220°C and an outlet temperature of 90°C using a rotary atomizer. The microcapsules formed were normally cured by further drying for 1 hr at 100°C. They were essentially spherical and exhibited a sustained-release effect in vitro. Factors influencing drug release from the microcapsules, such as monomer composition and pH of the coating solution, were investigated.

Another example of the use of spray polycondensation was reported by Takenaka et al. [33]. L-ascorbyl monostearate was microencapsulated by the process at 140°C using polyvinyl alcohol, carboxymethylcellulose, or polyvinylpyrrolidone as film-forming polymer, trimethylolmelamin trimethyl ether as precondensate, and glutaraldehyde as hardening agent. An organic amine salt was used to catalyze the polymerization of the precondensate. The mechanical strength of the microcapsules and both the rate of release and oxidation of their core material tended to decrease with increasing trimethylolmelamin trimethyl ether content.

REFERENCES

1. K. Masters, *Spray Drying Handbook*, 3rd ed., George Godwin, London, 1979.
2. D. W. Belcher, D. A. Smith, and E. M. Cook, Spray drying equipment, in *The Encyclopedia of Chemical Process Equipment* (W. J. Mead, ed.), Reinhold, New York, 1964, pp. 861–874.
3. J. M. Newton, Spray drying and its application to pharmaceuticals, *Manuf. Chem. Aerosol News* *33*(4):33–36, 55 (1966).
4. D. Ganderton, in *Unit Processes in Pharmacy*, William Heinemann Medical Books, London, 1968, pp. 110–113.

5. H. A. Lieberman and A. Rankell, Drying, in *The Theory and Practice of Industrial Pharmacy* (L. Lachman, H. A. Lieberman, and J. L. Kanig, eds.), Lea & Febiger, Philadelphia, 1970, pp. 22–48.
6. F. Nielsen, Spray drying pharmaceuticals, *Manuf. Chem. 53*(7): 38–39, 41 (1982).
7. A. Kondo, Microencapsulation by spray-drying process, in *Microcapsule Processing and Technology* (J. Wade Van Valkenburg, ed.), Dekker, New York, 1979, pp. 154–165.
8. R. L. Nolen and L. B. Kool, Microencapsulation and fabrication of fuel pellets for inertial confinement fusion, *J. Pharm. Sci. 70*:364–367 (1981).
9. J. Merory, *Food Flavorings—Composition, Manufacture and Use*, Avi, Westport, Conn., 1960, pp. 274–277.
10. N. G. Marotta, R. M. Boettger, B. H. Nappen, and C. D. Szymanski, U.S. patent 3,455,838 (July 15, 1969).
11. P. P. Noznick and C. W. Tatter, U.S. Patent 3,351,531 (November 7, 1967).
12. H. Seager, U.S. Patent 4,016,254 (April 5, 1977).
13. M. J. Robinson and E. V. Svedres, U.S. Patent 2,805,977 (September 10, 1957).
14. A. P. Granatek, B. C. Nunning, N. G. Athanas, R. L. Dana, E. S. Granatek, and R. G. Daoust, U.S. Patent 3,549,746 (December 22, 1970).
15. A. P. Granatek, B. C. Nunning, N. G. Athanas, R. L. Dana, E. S. Granatek, and R. G. Daoust, U.S. Patent 3,626,056 (December 7, 1971).
16. M. J. Robinson and J. V. Swintosky, Sulfaethylthiadiazole V. Design and study of an oral sustained release dosage form, *J. Amer. Pharm. Assoc. Sci. Ed. 48*:473–478 (1959).
17. M. W. Scott, M. J. Robinson, J. F. Pauls, and R. J. Lentz, Spray congealing: particle size relationships using a centrifugal wheel atomizer, *J. Pharm. Sci. 53*:670–675 (1964).
18. P. M. John and C. H. Becker, Surfactant effects on spray-congealed formulations of sulfaethylthiadiazole-wax, *J. Pharm. Sci. 57*:584–589 (1968).
19. A. G. Cusimano and C. H. Becker, Spray-congealed formulations of sulfaethylthiadiazole (SETD) and waxes for prolonged-release medication. Effect of wax, *J. Pharm. Sci.* 57:1104–1112 (1968).
20. I. S. Hamid and C. H. Becker, Release study of sulfaethylthiadiazole (STED) from a tablet dosage form prepared from spray-congealed formulations of SETD and wax, *J. Pharm. Sci. 59*:511–514 (1970).
21. A. Koff, U.S. Patent 3,080,292 (March 5, 1963).
22. R. A. Smith and A. Lambrou, U.S. Patent 3,819,838 (June 25, 1974).

23. G. Ross and B. Reul, U.S. Patent 3,857,933 (December 31, 1974).
24. J. C. Hecker and O. D. Hawks, U.S. Patent 3,137,630 (June 16, 1964).
25. J. Z. Pasin, U.S. Patent 3,664,963 (May 23, 1972).
26. E. Palmer, U.S. Patent 3,903,295 (September 2, 1975).
27. A. F. Asker and C. H. Becker, Some spray-dried formulations of sulfaethylthiadiazole for prolonged-release medication, *J. Pharm. Sci.* *55*:90–94 (1966).
28. H. Takenaka, Y. Kawashima, and S. Y. Lin, Preparation of enteric-coated microcapsules for tableting by spray-drying technique and in vitro simulation of drug release from the tablet in GI tract, *J. Pharm. Sci.* *69*:1388–1392 (1980).
29. H. Takenaka, Y. Kawashima, and S. Y. Lin, Polymorphism of spray-dried microencapsulated sulfamethoxazole with cellulose acetate phthalate and colloidal silica, montmorillonite, or talc, *J. Pharm. Sci.* *70*:1256–1260 (1981).
30. D. L. Wise, G. J. McCormick, G. P. Willet, and L. C. Anderson, Sustained release of an antimalarial drug using a copolymer of glycolic/lactic acid, *Life Sci.* *19*:867–873 (1976).
31. H. Kala, J. Traue, H. Moldenhauer, and G. Zessin, On the production and the characterization of spray embeddings, *Pharmazie* *36*:106–111 (1981).
32. C. Voellmy, P. Speiser, and M. Soliva, Microencapsulation of phenobarbital by spray polycondensation, *J. Pharm. Sci.* *66*:631–634 (1977).
33. H. Takenaka, Y. Kawashima, Y. Chikamatsu, and Y. Ando, Mechanical properties, dissolution behavior and stability to oxidation of l-ascorbylmonostearate microcapsules prepared by a spray-drying polycondensation technique, *Chem. Pharm. Bull.* *30*:2189–2195 (1982).

9

Polymerization Procedures for Nonbiodegradable Micro- and Nanocapsules and Particles

9.1 INTRODUCTION

In recent years considerable interest has been shown in various polymerization techniques for the containment of biologically active materials in polymers. This chapter discusses the principal polymerization procedures employed and their application for the production of biodegradable capsules and particles in the micrometer (μm) and nanometer (nm = 1/1000 μm) size range. The next chapter considers applications of the polymerization techniques where the polymer formed is biodegradable.

Polymerization techniques of pharmaceutical interest are normally carried out in the liquid phase by bulk, suspension, emulsion, or micelle processes. The nonreactive, biologically active material may be incorporated into the polymer during polymerization if adequately stable or may be subsequently incorporated into the preformed polymer. The monomers normally chosen are those whose polymers have already gained wide medical acceptance, often for purposes other than encapsulation, such as polymethacrylates for prostheses and contact lenses or polyglycolic acid for absorbable suture material. Other nontoxic polymers can also be used, provided that they are freed from harmful unreacted monomer, catalyst/initiator, and other materials used in their preparation.

9.1.1 Bulk Polymerization

In bulk polymerization only a monomer or mixture of monomers and possibly the biologically active material are usually heated, often in the presence of a catalyst/initiator to increase the reaction rate. As the polymer forms through reaction of functional groups in a stepwise manner, there is a progressive increase in the amount and molecular weight of polymer formed. This is accompanied by a progressive rise in apparent viscosity of the system, which may be maintained in the fluid state by keeping the temperature high. The process has the advantage of forming a relatively pure polymer, which solidifies as a block on cooling. However, the process has a number of disadvantages for pharmaceutical applications. It is difficult to dissipate the high exothermic heat of reaction, which if present may have an adverse effect on a thermolabile drug or other material. The polymer block formed may need to be mechanically fragmented to produce fine particles having irregular size, shape, and release properties.

9.1.2 Suspension Polymerization

Suspension polymerization is often referred to as bead polymerization or pearl polymerization. Typically it involves heating a water-insoluble liquid monomer or monomers and possibly a biologically active material as a dispersion of droplets (usually 100 to 5000 μm in diameter) in a continuous aqueous phase. The droplets may also contain an initiator and are formed by suitable mechanical agitation. The aqueous phase may contain stabilizers such as thickening agents to increase the viscosity of the continuous phase, electrolytes to increase the interfacial tension between the phases, and finely divided insoluble filler to interfere mechanically with agglomeration. Minor amounts of emulsifiers and buffers may also be needed. Adhesion between droplets often becomes a particularly troublesome problem when polymerization has proceeded to a critical point where the surface of polymer beads have become tacky. The kinetics of polymerization are the same as those of bulk polymerization, again resulting in the formation of a relatively pure polymer. On completion of polymerization, the product is washed to free it from stabilizers and is dried as beads or pearls. The major advantage of the process is that the continuous phase absorbs the heat of the polymerization reaction and, by acting as a coolant, prevents excessive temperature rise. Also, the product is in the form of spherical beads of relatively uniform size and release characteristics. The major disadvantages associated with the process are difficulty in freeing the product of unwanted stabilizers and other additives, and coalescence problems with soft polymer beads.

9.1.3 Emulsion Polymerization

Emulsion polymerization differs from the superficially similar suspension polymerization procedure in three important respects. The initiator is initially located in the aqueous phase. More vigorous agitation is employed, as a result of which the droplet size is usually below 100 μm and is often less than 1 μm, i.e., in the nanometer range. The surfactant concentration employed is much higher, being usually well in excess of its critical micelle concentration (CMC). This results in an altered mechanism of polymerization, which may be summarized as follows. Excess surfactant molecules form micelles whose hydrophobic interior take up part of the available monomer, causing them to swell. Initiator radicals generated in the aqueous medium, often by heating or irradiation, diffuse into these swollen micelles to start polymerization. As the monomer is consumed, it is replaced by progressive diffusion of the remaining monomer from its location as emulsified droplets into the micelles, which continue to enlarge as polymerization proceeds. The associated enlarging surfaces compete for available surfactant, thus influencing the number of available micelles that can form polymer as a result of their fortuitous gain of initiator radicals. Smith and Eward [1] were able to quantitatively predict the number of particles that would form when the supply of available surfactant was depleted in a system. Upon completion of polymerization, a latex is often obtained that consists of a colloidal dispersion of polymer particles with a size of approximately 100 nm. Because of the minute particle size, sedimentation does not occur. Polymerization can also occur in systems containing surfactant below its CMC by homogeneous nucleation onto polymer-radical complexes. This mechanism may also be partially responsible for particle formation in systems containing surfactant above its CMC. Nonaqueous-based emulsion polymerization procedures, whereby an aqueous solution of hydrophilic monomer such as acrylic acid or acrylamide is emulsified in a continuous oil phase using a water-in-oil (w/o) emulgent/surfactant and an oil-soluble initiator, have been reported.

The main advantages of emulsion polymerization are that higher-molecular-weight polymer is usually formed at a faster rate and a lower temperature. The heat of reaction is readily dissipated. The process is particularly suitable for the formation of minute spherical particles for injectable and other use. However, a major disadvantage of the process is usually high associated concentration of unreacted monomer, many of which are quite toxic and difficult to free the product of. There are also increased difficulties in recovering the particles from the dispersion because of their minute size.

9.1.4 Micelle Polymerization

Micelle polymerization differs from emulsion polymerization in that all of the monomer, and possibly the biologically active material if added, is contained within the micelles formed by a suitable concentration of a surfactant in excess of its CMC before polymerization is commenced. The diffusion of monomer from the micelles is prevented by the nonsolvent properties of the outer phase used. Therefore, unlike emulsion polymerization, negligible increase in particle size of the product occurs

Table 9.1 Typical Polymerization Techniques with Medical Applications

Bulk polymerization	Suspension (bead or pearl) polymerization	Emulsion and micelle polymerization (see text for difference)
Preparation		
Monomer ± bioactive material + catalyst/initiator if required	Monomer ± bioactive material + initiator—disperse as droplets in water phase + stabilizers, i.e., thickening agents, electrolyte filler	Monomer ± bioactive material—disperse as droplets in aqueous or nonaqueous phase + initiator + surfactant above CMC + stabilizers
No agitation required	Continuous vigorous agitation required	No further agitation may be required after emulsification
Heat or irradiate to assist polymerization	Heat or irradiate to assist polymerization	Heat or irradiate to assist polymerization
↓ Polymer block solidifies upon cooling—mold or mechanically fragment	↓ Spherical polymeric particles formed > 100 μm in diameter by bulk polymerization kinetics—separate and dry	↓ Minute spherical polymeric particles formed often ∿100 nm in diameter by emulsion polymerization kinetics—separate and dry
Advantages/disadvantages		
Relatively pure polymer formed—difficult to dissipate high exothermic heat of reaction—unsuitable for thermolabile materials unless added to preformed polymer	Heat of polymerization dissipated—spherical, uniform-sized particles obtained having more reproducible release characteristics—difficult to free product of stabilizers, etc.—soft polymer beads tend to adhere	Different kinetics gives polymer of higher molecular weight—heat of reaction easily dissipated—small spherical particles suitable for injection—difficult to free product from large amount of unreacted monomer, stabilizers, etc.—difficult to isolate product because of its small size

as polymerization proceeds. The particles obtained remain extremely small, which is particularly useful when preparing nanoparticles.

Table 9.1 summarizes the main features of the four polymerization techniques discussed. Solution polymerization, whereby the monomer is dissolved in a solvent prior to polymerization, is not used for pharmaceutical applications other than for modifications of emulsion and micelle polymerization. The solvent normally used should be a nonsolvent for the polymer to facilitate its separation. In the following sections, applications of these polymerization techniques are discussed grouped according to the nature of the polymer formed.

9.2 ACRYLIC PRODUCTS

9.2.1 Some Basic Chemistry

A wide variety of monomers have been used for the production of various acrylic polymers. Aspects of the chemistry of some of the more important compounds are briefly discussed below. Acrylic acid and methacrylic acid are colorless liquid monomers that may be polymerized to

$$CH_2{=}CH{-}COOH$$

Acrylic acid

$$CH_2{=}C(CH_3){-}COOH$$

Methacrylic acid

form their corresponding polymers, polyacrylic acid and polymethacrylic acid, as follows:

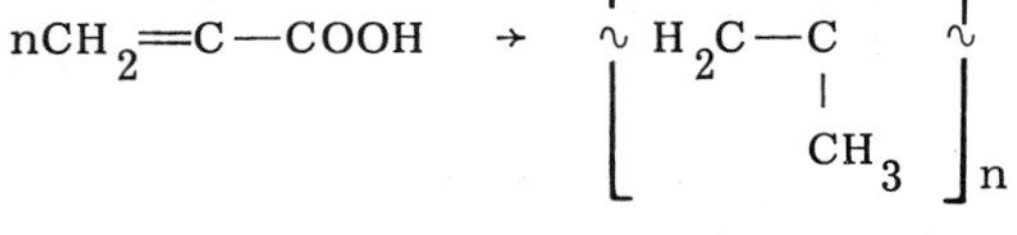

Polyacrylic acid (PAA)

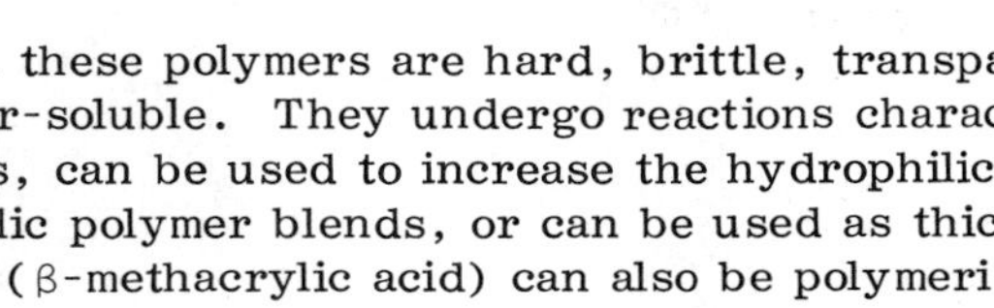

Polymethacrylic acid (PMMA)

Both these polymers are hard, brittle, transparent materials that are water-soluble. They undergo reactions characteristic of carboxylic acids, can be used to increase the hydrophilic character of various acrylic polymer blends, or can be used as thickening agents. Crotonic acid (β-methacrylic acid) can also be polymerized to form a polymer

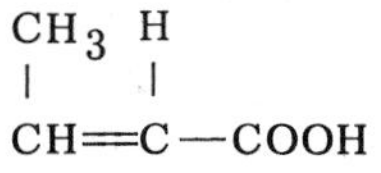

Crotonic acid

with similar properties. Acrylamide can be polymerized to form polyacrylamide. This polymer undergoes reactions characteristic of its free amino groups and is also very water-soluble.

$$CH_2{=}CH{-}CO{-}NH_2$$

Acrylamide

In order to form water-insoluble polymers, various esters of acrylic and methacrylic acid, such as ethyl acrylate, methyl methacrylate, and 2-hydroxyethyl methacrylate (HEMA), may be polymerized. For

$$CH_2{=}CH{-}COOC_2H_5$$

Ethyl acrylate

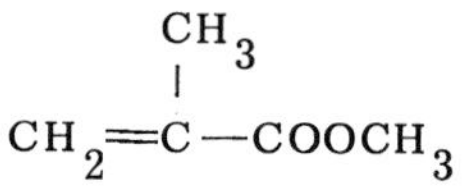

Methyl methacrylate

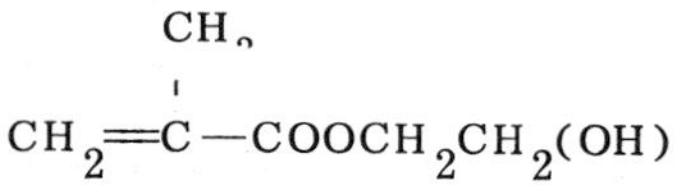

2-Hydroxyethyl methacrylate
(HEMA)

example, methyl methacrylate is a colorless liquid that may need to be initially freed of added inhibitors such as hydroquinone or *p*-methoxyphenol and polymerized as follows:

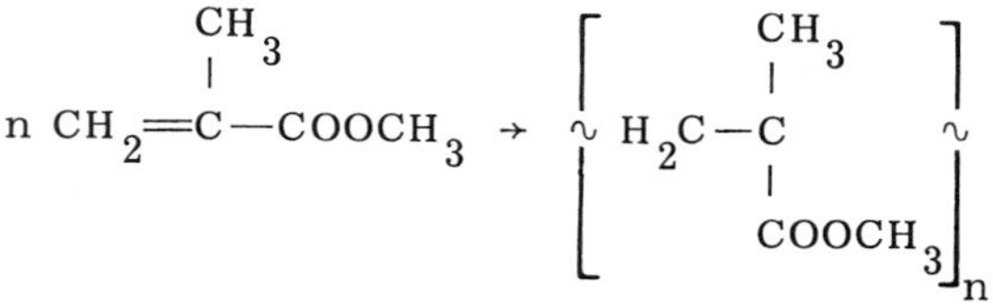

Polymethyl methacrylate (PMMA)

The free radical polymerization may be initiated by heat or irradiation. However, it is more commonly achieved by adding an initiator such as benzoyl peroxide or azobisisobutyronitrile that decomposes on heating into free radicals (i.e., organic molecules containing an unpaired electron) as follows:

Benzoyl peroxide

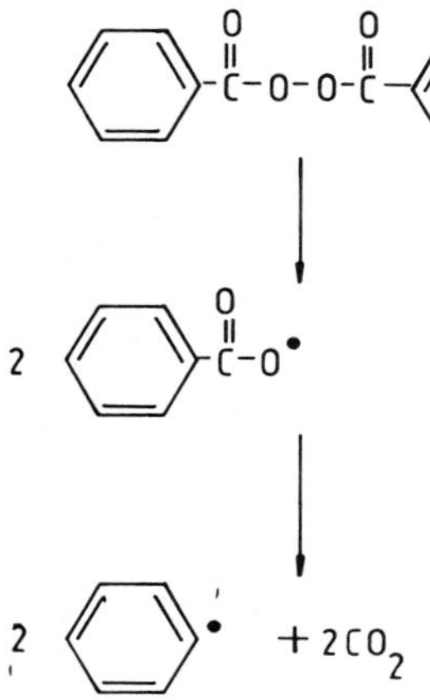

Azobisisobutyronitrile

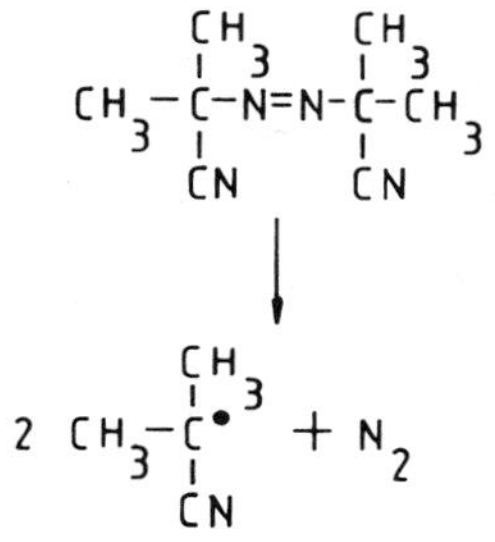

The initiator fragment free radical ($I^\bullet$) formed is then added to a vinyl monomer radical:

$$I^\bullet + CH_2{=}C(R_1){-}R_2 \longrightarrow I{-}CH_2{-}\overset{\bullet}{C}(R_1){-}R_2$$

This new radical then rapidly adds on further monomer molecules to form a polymer chain:

$$I{-}CH_2{-}\overset{\bullet}{C}(R_1){-}R_2 + CH_2{=}C(R_1){-}R_2 \longrightarrow I{-}CH_2{-}C(R_1)R_2{-}CH_2{-}\overset{\bullet}{C}(R_1){-}R_2 +$$

$$n\ CH_2{-}C(R_1){-}R_2 \longrightarrow \left[I{-}CH_2{-}C(R_1){-}R_2 \right]_n{-}CH_2{-}\overset{\bullet}{C}(R_1){-}R_2$$

Termination commonly occurs by reaction with another long-chain radical or by transfer. PMMA is a hard, water-resistant, rigid, transparent material used to produce dentures and artificial bone material. It and related polymers may be crosslinked with N,N'-methylene-bis(acrylamide) as follows:

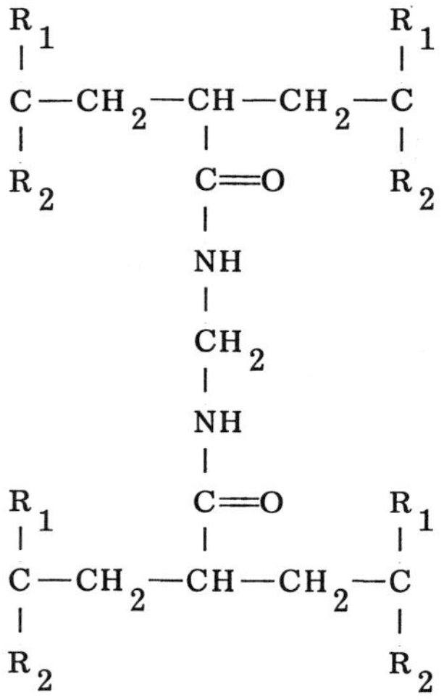

PolyHEMA can be crosslinked with polyglycol dimethacrylate to form the material from which hydrophilic contact lenses are formed.

Various copolymers can also be formed, such as by reaction between methacrylic acid and methyl methacrylate as follows:

$$nCH_2{=}C(CH_3){-}COOH + nCH_2{=}C(CH_3){-}COOCH_3 \rightarrow$$

Methacrylic acid   Methyl methacrylate

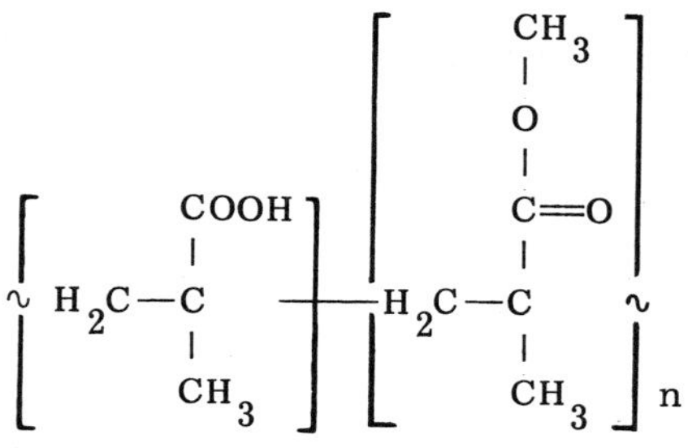

Copolymer

Acrylic Products

Branching and crosslinking of these types of polymers occurs due to the ease with which the tertiary hydrogen can be extracted from the polymer chain:

$$\begin{array}{ccc} \text{CH}_3 & \text{CH}_3 & \text{CH}_3 \\ | & | & | \\ \text{O} \quad \text{I}^{\bullet} \rightarrow & \text{O} \quad \leftrightarrow & \text{O} \\ | & | & | \\ \text{CO} & \text{CO} & \text{C}-\text{O}^{\bullet} \\ | & |^{\bullet} & \| \\ \text{H}_2\text{C}-\text{CH} & -\text{H}_2\text{C}- & -\text{H}_2\text{C}- \end{array}$$

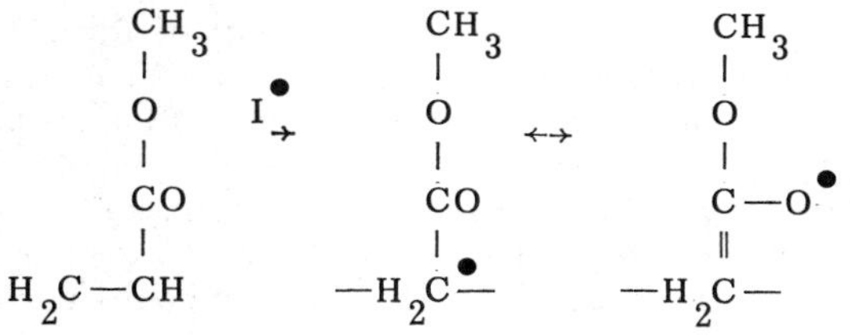

9.2.2 Acrylic Microparticles

Among the earliest reported applications of acrylic polymers for medical use was the paper by Hjerten [2] describing how crosslinked polyacrylamide in chromatographic columns could be adapted for the separation of substances with a wide range of molecular weight from a few hundred to about 1 million. The method was used to separate human serum proteins.

Because of the selective permeability of this type of polymer and its freedom from toxicity, interest quickly shifted to a study of its use for the entrapment of biological material, particularly those of high molecular weight. Thus Bernfeld and Wan [3] entrapped antigens and enzymes in crosslinked, water-insoluble polyacrylamide particles and showed that the entrapped material retained biological activity. Wada and Kishizaki [4] described a chromatographic device in which polynucleotide was immobilized into a polyacrylamide gel matrix prepared in the form of beads by a suspension polymerization process. The residues of the immobilized chains were accessible to globular proteins of molecular weight up to 60,000.

Very considerable interest has been shown in the immobilization of various enzymes in acrylic polymers and copolymers [5]. For example, Nilsson et al. [6] used a suspension polymerization technique to immobilize trypsin. In a typical production run the enzyme was dissolved with acrylamide (monomer) and N,N'-methylene-bis-(acrylamide) (crosslinking agent) in an aqueous triethanolamine-HCl buffer pH 7. After addition of the free radical initiator ammonium persulfate and the accelerator N,N,N',N'-tetramethylethylenediamine, the aqueous phase was added to the organic phase (toulene:chloroform 290:110) containing the emulgent sorbitan sesquioleate and stirred vigorously to form a w/o dispersion. Polymerization was carried out at 4°C under nitrogen for 30 min and the beads formed were filtered off and washed with toluene and aqueous media to remove residual chloroform and enzyme not properly trapped. The average diameter of the beads was 100 to 250 μm, and as they retained about 67% of the activity of free trypsin, denaturation of the enzyme during the process was considered low. Enzyme immobilized by a block polymerization procedure followed by mechanical fragmentation continued to leak trypsin during the enzyme assay. By substituting potassium acrylate for half the acrylamide, subsequent adjustment of the pH to 7.0 using HCl followed by addition of 1-cyclohexyl-3-(2-morpholinoethyl)-carbodiimide, and change in the ratio of toluene:chloroform to 265:135 in the organic phase, covalent bonding of the enzyme occurred with enhanced enzymic activity being retained in the product. The spherical beads formed were suitable for column processes giving high flow rates. Smaller particles down to 10 μm in diameter could be produced by increasing the amount of emulgent used.

In a subsequent paper, Johansson and Mosbach [7] successfully immobilized other enzymes, such as ribonuclease, β-glucosidase, yeast alcohol dehydrogenase, and urease using entrapment or covalent bonding to acrylic polymers and copolymers in bead form. Covalent bonding to polymers of polyacrylamide was accomplished using glutaraldehyde, to copolymers of acrylamide and 2-hydroxyethyl methacrylate using cyanogen bromide, or to copolymers of acrylamide and acrylic acid using a water-soluble carbodiimide.

In a series of papers by Ekman, Sjöholm, and co-workers [8-11], incorporation of macromolecules into spherical microparticles, preferably around 1 to 3 μm, was reported. The crosslinked polyacrylamide particles were prepared by a procedure similar to that described by Nilsson et al. Their use for macromolecular ligand interactions in radioimmunoassays, in binding studies of drugs such as salicylic acid and warfarin with entrapped albumin, and to form immunofluorescent complexes with specific cells or components to facilitate their detection and separation by sedimentation or centrifugation techniques because of altered density, was discussed.

The microparticles were observed to have a macroporous structure that retained macromolecules by either chemically binding them or entrapping them in the polyacrylamide matrix. The amount entrapped increased with increase in the amount of crosslinking agents used and with increasing size of the macromolecule used. Low-molecular-weight substrate could freely penetrate the microparticles. For example, immobilized lactate dehydrogenase was shown to be active in the analysis of lactate and pyruvate. As some of the macromolecules were exposed on the surface of the microparticles, they could react with cells that could not penetrate the microparticles.

In a subsequent paper, Ekman and Sjöholm [12] reported that carbonic anhydrase and other proteins immobilized in spherical microparticles of crosslinked polyacrylamide exhibited improved stability against heat denaturation and proteolytic enzymes. A modified production procedure was employed whereby the accelerator was added to the emulsion only when it had been homogenized to the required particle size. This prevented undesirable premature polymerization giving broken spheres of irregular size. However, a major disadvantage of the process was the low amount of protein (3 to 10%) immobilized into the microparticles. Ljungstedt et al. [13] used horse anti(human lymphocyte) globulin immobilized in spherical polyacrylamide particles 1 to 5 μm in diameter together with fluorescein in dextran to detect and separate human lymphocytes with specific surface receptors (antigens). Also, Edman and Sjöholm [14] immobilized asparaginase in polyacrylamide microparticles and showed it to be effective in enzyme replacement therapy but also to be immunogenic, particularly upon intraperitoneal and intravenous injection into mice. Morehouse and Bolton [15, 16], in two patents assigned to Dow Chemical Co., described the production of microparticles by use, for example, of a dispersed organic phase containing 2-hydroxyethyl acrylate and methyl methacrylate in an aqueous phase containing a copolymer of 2-sulfoethyl methacrylate and butyl acrylate as a coalescing acid. The emulsion polymerization process used may have pharmaceutical applications.

During recent years considerable interest has been shown in the entrapment of drug molecules of lower molecular weight in various acrylic polymers and copolymers. In 1970 Khanna et al. [17] reported the first application of suspension polymerization for the production of sustained-release dosage forms whereby a drug was embedded during the polymerization of one or more monomers. Other polymerization techniques were also tried. Emulsion and solution polymerization were considered undesirable because of the high concentration of unreacted monomers left in the product, as was the bulk polymerization technique because the poorly dissipated exothermic heat of reaction liberated might cause degradation with thermolabile drugs. Methacrylic acid, methyl methacrylate, and crotonic acid were used as monomers, benzoyl peroxide or azobisisobutyronitrile as initiator, and carboxyvinyl polymer (Carbopol) or polyvinylpyrrolidone was used as stabilizer to increase the viscosity of the continuous water phase. A high concentration of sodium sulfate was used to decrease the solubility of the monomers in the water phase by a salting-out effect which, however, must not flocculate the stabilizers. Sulfuric acid was added to prevent the ready hydrolysis of monomers under basic conditions. The drugs selected included chloramphenicol, chlorothiazide, pentobarbitone, and its sodium salt, papaverine base, hydrocortisone, and gluthemide.

Polymerization was carried out in a three-necked flask provided with a reflux condenser, inert gas supply, and stirrer similar to that shown in Fig. 9.1. It is necessary to exclude oxygen because its quenching effect on free radicals produced would tend to slow the rate of polymerization.

To the water phase and stabilizers previously heated to the required temperature of up to 75°C, the monomer and drug were added and stirred to produce droplets of the desired size. The initiator was then added and polymerization was allowed to proceed for periods up to 24 hr to produce solid beads, which were removed, washed with

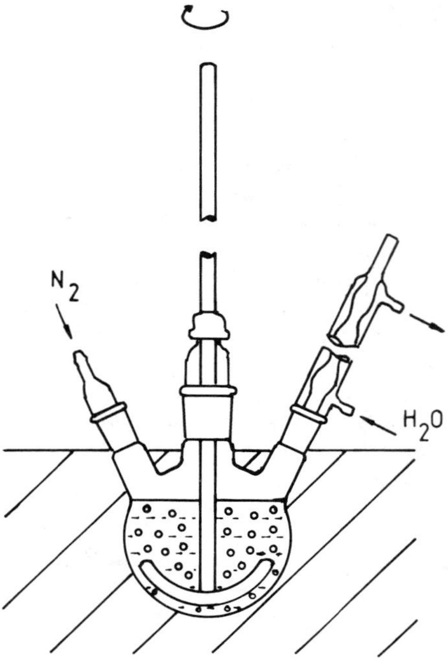

Figure 9.1 Apparatus for suspension polymerization.

water or dilute hydrochloric acid to remove impurities from their surfaces, and dried. The product was composed mainly of individual spherical particles with a range in size from 300 to 1000 μm, together with a small amount of deformed particles and agglomerates. The beads were transparent if the drug was soluble in the polymer and opaque if the drug was dispersed. Drugs having no affinity for the monomer phase were not successfully embedded in the beads.

In a subsequent paper, Khanna and Speiser [18] investigated the release of chloramphenicol from beads polymerized from methacrylic acid and methyl methacrylate using dissolution studies. As the percentage of the more water-soluble monomer, methacrylic acid, in the copolymer was increased, the lower was the pH of the dissolution medium required to allow drug release to commence. Release of drug was also promoted by decrease in particle size of the beads.

In a series of papers by Kala and co-workers [19, 20], the production of spherical acrylic particles containing drugs such as benzodiazepines, indomethacin, meprobamate, and phenobarbitone were discussed using thermally mild conditions. They used methyl methacrylate and acrylic acid as monomer and benzoyl peroxide or azobisisobutyronitrile as initiator, thus allowing the temperature to be reduced to 60°C even though up to 200 min were required for polymerization. The time of polymerization could be reduced by the use of copolymers of vinyl acetate and crotonic acid with the acrylic monomers and use of the peroxide but not the azo initiator. Use as initiators of unstable peroxides that are potentially explosive, such as diisopropyl peroxiddicarbonate, allowed polymerization of beads containing phenobarbitone at 30°C and meprobamate at 50°C. When ultraviolet light was used with the initiators to induce free radical formation, the temperature of polymerization could be further reduced. By use of redox systems such as benzoyl peroxide as oxidizing agent and dimethylanaline as reducing agent, the polymerization temperature could be reduced to 20°C, though the beads formed showed an orange color due to the formation of byproducts.

Wahlig et al. [21] reported that gentamicin was slowly released from commercially available PMMA beads [22] (Septopal, BDH Pharmaceuticals, Ltd.) used for implantation. The beads have a diameter of 7 mm, contain 7.5 mg of gentamicin sulfate, and 20 mg zirconium dioxide as a radiocontrast medium in a copolymer formed from methyl methacrylate and methyl acrylate. The beads are threaded on a multifilament surgical wire. In a clinical trial, 41 patients with infection of either bone or soft tissue mainly of the lower limb were shown to be effectively treated by local implantation of the beads. Excellent tissue tolerance was observed, though the 14 to 180 beads used were usually extracted after 10 to 14 days, as the formation of new connective tissue tended to fix the chain and cause breakage of the wire upon withdrawal. In some weak patients, when a second operation was considered undesirable, the beads were left in situ for considerable periods without any observable adverse effect. As systemic levels of this normally toxic antibiotic detected in the serum and urine were extremely low, the normal side effects were not observed. The product was originally developed by E. Merck (Darmstadt, West Germany). The proceedings of a recent symposium on the treatment of chronic bone infections with Septopal gives greater detail of the clinical evaluation of the product [23].

Yoshida et al. [24] dissolved polymethyl methacrylate or polystyrene in a large excess of various acrylic monomers to which were dispersed the drug potassium chloride. Such systems were then allowed to fall as droplets into a cold (-78°C) precipitation medium such as *n*-hexane and the spherical particles formed were then polymerized by γ-irradiation with 2 Mrad from a ^{60}Co source. After irradiation the polymerized spheres were removed from the hexane and dried. Depending on the composition of the polymers used, various in vitro release profiles for the potassium chloride could be obtained. Despite polymerization being induced at low temperature, the large dose of irradiation involved may limit the usefulness of this procedure for the encapsulation of many drugs.

Recently El-Samaligy and Rohdewald [25] prepared polyacrylamide microbeads with diameters of about 100 to 600 μm containing tetracycline hydrochloride or theophylline. A suspension polymerization process was used similar to that described by Nilsson et al. [6]. The presence of gelatin during the polymerization process led to the formation of larger microspheres and to slower drug release.

9.2.3 Acrylic Nanoparticles

A number of papers discussed in the previous section could give particles in the nanometer size range. However, this section will not refer to them again but will concentrate on the literature relating to the production of ultrafine acrylic particles for medical use.

A medical application involving use of various acrylic nanoparticles containing entrapped antibodies and prepared by aqueous emulsion copolymerization of methacrylate derivatives has been reported [26-29]. Figure 9.2 shows a scanning electron micrograph of the spheres, which usually ranged in size from 30 to 340 nm. These small immunological nanoparticles were employed as visual markers for the detection and localization of surface antigens on cells by scanning electron microscopy. The primary amino groups of the antibodies were linked covalently to the hydroxyl and carboxyl groups on the surface of the acrylic spheres using glutaraldehyde, cyanogen bromide, or aqueous carbodiimide techniques. Larger spheres were prepared for use with light microscopy by a ^{60}Co γ-irradiation technique that polymerized 2-hydroxyethyl methacrylate with or without comonomers.

Birrenbach and Speiser [30] described a process for the production of nanoparticles using polymerized acrylic acid, acrylamide, and/or their derivatives and other monomers. Particles in the size range 20 to 200 nm and preferably around 80 nm were obtained. A micelle polymerization process was employed, as an example of which acrylamide and N,N'-methylene-bis(acrylamide) were solubilized in aqueous solution into *n*-hexane using a blend of anionic (Aerosol OT) and nonionic (tenside LA 55-4) surfactants. The soluble protein was filtered through a membrane filter to form a sterile system to which was added aseptically tetanus toxin solution. Other bioactive materials such as urease or ^{131}I-labeled human immunoglobulin (IgG) could be added using slight modifications of the procedure. These systems were then polymerized by exposure to 0.3 Mrad from a ^{60}Co γ-irradiation source while sealed under nitrogen at 20 to 30°C. Polymerization was judged to be complete by the aid of an acidimetric color titration to determine unsaturated compounds by reaction with morpholine. Alternatively, ultraviolet light or visible light and the initiator riboflavin 5-sodium phosphate and potassium peroxodisulfate were used to cause polymerization. Figure 9.3a shows a schematic representation of the preparation of the nanoparticles.

Figure 9.2 Scanning electron micrograph of copolymer latex spheres synthesized by aqueous emulsion polymerization. (Reproduced from *The Journal of Cell Biology 64*:75–88 (1975) by copyright permission of The Rockefeller University Press; from Ref. 26.)

After polymerization the nanoparticles were separated by various centrifugation or ultrafiltration procedures and could be isolated as a freeze-dried, free-flowing powder that readily formed a colloidal solution upon dispersion in water. The nanoparticles formed were probably tiny spherical polymeric matrices rather than nanocapsules, as the water-soluble monomers used would be unlikely to show any preferential film formation at the surface of the micelles. Embedded human IgG was released only to the extent of 25% over 48 days, indicating that it was mainly immobilized within the nanoparticles. Studies on the adjuvant effect of nanoparticles as carriers for human IgG in the immunization of guinea pigs showed that higher antibody titers were obtained

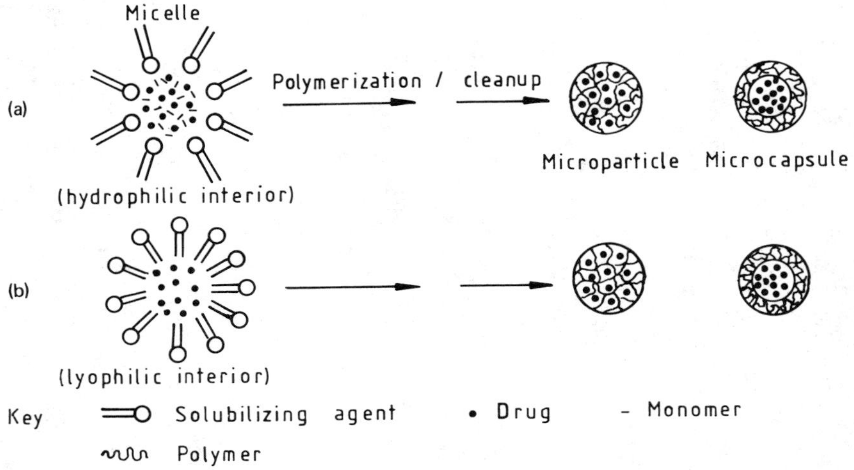

Figure 9.3 Schematic diagram of preparation of nanoparticles.

compared to using aluminum oxide as the adjuvant, the levels being comparable to those obtained with Freund's adjuvant. However, nanoparticles containing tetanus toxoid produced slightly lower antibody titers than the standard aluminium phosphate adjuvant.

The micelle polymerization procedure was patented by Speiser and Birrenbach [31]. Water-insoluble material could also be entrapped in nanoparticles by solubilizing it together with a water-insoluble monomer such as an acrylic acid ester in liquid paraffin using a suitable surfactant such as polyoxyethylene sorbitan monooleate (Tween 80) and an aqueous outer phase as shown diagramatically in Fig. 9.3b. Some properties of nanoparticles have been reviewed by Speiser [32]. Scanning electron micrographs in a paper by Kopf et al. [33] showed them to be composed of aggregates of cylindrical tubes or spherical particles of regular size. There was no significant increase in viscosity when the particles were suspended in water, indicating lack of hydration of the particles. These workers also claimed that washing was adequate to reduce the residual monomer concentration in the polyacrylamide nanoparticles to an acceptable level.

There may be some problems associated with the process, however. These nanoparticles should be suitable for injectionable use, so particular care must be taken to ensure that the polymerization reaction is complete. Dillingham et al. [34] claimed that monomer residues associated with polymethyl methacrylate prosthetic devices exhibited adverse reactions in the body. Kreuter and Zehnder [35] reported that polymethyl methacrylate nanoparticles contained not more than 1% of extraneous material such as unreacted monomer, a level that might be toxicologically unacceptable.

The process is less likely to be suitable for the containment of low-molecular-weight drugs that would readily escape from the porous polymer matrix formed. However, Kopf et al. [36] successfully immobilized phenylpropanolamine hydrochloride in the nanoparticles, because they showed that this relatively low-molecular-weight drug was covalently bound to the polymer. Of course, the bioactive material may be adsorbed into the previously prepared blank PMMA nanoparticles, as reported by Kreuter and Speiser [37] for influenza virions. Likewise, larger molecules with molecular weights greater than 150,000 were claimed by Speiser and Birrenbach [31] to be unsuitable for incorporation into these nanoparticles, presumably because they are too bulky, as was noted by Kreuter and Speiser [37] for virions that were only partly enclosed by PMMA nanoparticles.

Another difficulty is the lack of biodegradability of the acrylic polymer materials used. This may excessively retard the liberation of the entrapped material and may give rise to long-term accumulation problems. Obviously, accumulation is a less serious problem with a dosage form such as a vaccine, which would be administered infrequently, and more serious in enzyme replacement therapy, where frequent dosing might be necessary. Also, Kreuter et al. [38] investigated the distribution and elimination of labeled PMMA nanoparticles after intravenous injection into rats. Thirty minutes after injection the nanoparticles were found in high concentration in the lungs and mainly in the liver. The liver and spleen, and to a lesser extent the bone marrow, continued to accumulate the nanoparticles while the lungs lost nanoparticles over 7 days. During this time there was only a small loss ($<6\%$) of nanoparticle-associated radioactivity from the body via the urine, feces, and breath. Upon intramuscular injection into mice, all of the radioactivity persisted at the site of injection for up to 70 days. Kreuter et al. [39] reported that unlabeled nanoparticles did not induce granulomas within 1 year around the injection site in guinea pigs. Collectively, these results suggest the possibility of producing unwanted capillary obstruction in the lungs and elsewhere with high doses of these nanoparticles and the difficulty of obtaining effective local concentrations of drugs in other than certain organs upon intravenous injection unless injected locally into a specific body site. Of course, for many drugs such as anticancer agents and antiinfective agents, accumulation in a diseased liver would be desirable. Couvreur et al. [40] investigated the uptake of polyacrylamide nanoparticles labeled with fluorescein by cultured rat fibroblasts. After 12 hr incubation, the nanoparticles appeared to be associated mainly with the lysosomes of cells, probably as a result of endocytosis. Thus these lysomotropic carriers may be useful for promoting cellular uptake of drugs that would not otherwise diffuse through biological membranes.

The use of radiation to induce polymerization can cause appreciable radiolytic decomposition of many drugs, particularly in macromolecules such as enzymes, serum proteins, and vaccines whose biological activity may be reduced. Doses up to 0.5 Mrad were reported by Kreuter and Zehnder [35] to be necessary for the production of PMMA nanoparticles, and dose rate used was important.

Despite these reservations, a number of applications of these nanoparticles have been investigated. Lubbers et al. [41] and Opitz et al. [42] entrapped β-methylumbelliferone to indicate pH, and pyrene butyric acid to indicate pO_2, using nanoparticles. Upon injection into blood vessels or tissues, the indicators contained in the nanoparticles had reduced toxicity and allowed the related physiological factors to be determined in minute regions of the body.

Kreuter and Speiser [43] polymerized methyl methacrylate in the presence of influenza virus by simply dissolving the monomer together with acrylamide in some cases to make the product less hydrophobic in the virus suspension. This was then polymerized by 0.4 Mrad γ-irradiation under nitrogen. The product tended to be composed of agglomerates of tiny round particles ranging from 50 to 300 nm. Virus was also added to previously prepared PMMA/PAA nanoparticles. Those prepared using 0.5% methyl methacrylate in the presence of the virus had optimum effect in promoting antibody response upon injection into mice and guinea pigs. Their adjuvant effect was considerably more effective than virus added to prepared nanoparticles or to aluminum hydroxide, though not as good as Freund's complete adjuvant (w/o emulsion + mycobacteria cell wall; proved too toxic for vaccination). Later Kreuter and Haenzel [44] reported the adjuvant effect of various acrylic polymers and monomers also using influenza vaccine as a model antigen. The adjuvant effect in mice of 100 to 200 nm virion-containing nanoparticles composed of PMMA was found to be much more reproducible than that of aluminium hydroxide.

Kreuter and Liehl [45] reported that the adjuvant effect of nanoparticles composed of a copolymer of PMMA/PAA and containing adsorbed or embedded inactivated influenza virus was comparable to that of aluminum hydroxide in mouse protection experiments against infectious virus. These results contrasted with those previously reported by Kreuter and Speiser [43], but in that case the proportion of virus antigen to adjuvant was much higher. In a subsequent paper by these workers [46], long-term experiments showed prolonged antibody response in mice to the adjuvant with entrapped or adsorbed influenza virus. Also, the polymer vaccines were more heat-stable at 40°C than those prepared using aluminium hydroxide or no adjuvant. The molecular weights of polymethyl methacrylate nanoparticles were determined by Bentele et al. [47] with HPLC-gel permeation chromatography.

In an attempt to improve the biodegradability of acrylic particles, Edman et al. [48] prepared polyacryldextran spheres containing immobilized protein by an emulsion copolymerization procedure involving bisacrylamide with acryl dextran. They claimed that the hydrophilic molecules formed part of the hydrocarbon chain in the polyacrylamide and hence the polymer would be more readily metabolized. The yield of immobilized protein and its heat stability was greater than with polyacrylamide particles, but leakage of protein was greater.

9.3 POLYSTYRENE PRODUCTS

Styrene is a colorless liquid, sparingly soluble in water, which may be polymerized by bulk, suspension, or emulsion techniques. Spherical particles of polystyrene can be prepared by heating to about 90°C a dispersion of the monomer in water in the presence of a free radical initiator such as benzoyl peroxide and suitable stabilizers [49]. Alternatively, ionizing radiation may be used to initiate the polymerization by its formation of free radicals [50]. The typical head-to-tail structure of polystyrene produced is shown below.

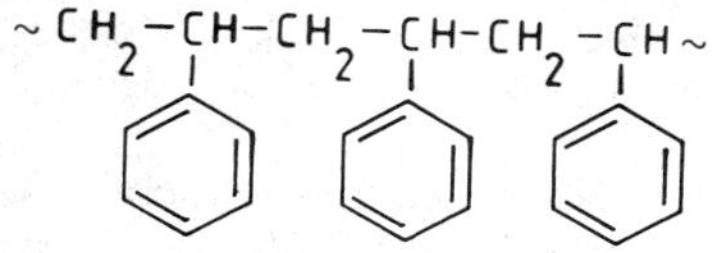

In the preparation of expandable polystyrene spheres, the styrene is usually polymerized in the presence of a low-boiling-point hydrocarbon such as *n*-pentane (termed a blowing agent), which becomes entrapped in the polystyrene spheres formed. When these are then heated with steam, the resultant combination of polymer softening, volatilization of hydrocarbon, and diffusion of steam into the spheres causes them to expand to about 40 times their original size.

Because of the uniformity of size and the range of sizes available, the earliest interest in the medical uses of polystyrene spheres was to calibrate measuring instruments such as the electron microscope and the Coulter counter. Later these monodisperse particles were used in various serological tests used in the diagnosis of rheumatoid arthritis, disseminated lupus erythematous, human pregnancy, etc. In these tests gamma globulin is adsorbed onto the surface of small (0.8 to 1.2 μm) polystyrene particles and mixed with body fluid from the patient containing characteristic factors, which react with the sensitized latex to produce visible agglutination. Covalently bound antibodies on polystyrene spheres have also been used to locate cell surface antigens using scanning electron microscopy. These and other medical applications that do not involve controlled release of drugs are discussed in greater detail by Vanderhoff [51], Singer [52], and Quash et al. [53].

Crosswell and Becker [54] used a suspension technique to polymerize droplets of styrene monomer dispersed in an aqueous medium at 84°C under nitrogen alone or in the presence of drug to produce spherical beads approximately 30 mesh in size. The initiators used were benzoyl peroxide and azobisisobutyronitrile, and the suspension stabilizers were calcium phosphate and polyvinyl alcohol. To prepare drug-loaded expanded beads, initially blank nonexpanded beads were immersed in *n*-pentane and then added to boiling water. The irregularly shaped expanded beads obtained were loaded by agitation in an alcoholic solution of paracetamol and dried. Dissolution studies showed that the nonexpanded beads released 64% of drug within 30 min, which was presumably located on or near the surface of the beads. The remaining 36% was not released even after 24 hr of dissolution treatment, as presumably it was bound within the plastic matrix. The expanded beads released nearly all their drug within 24 hr of the dissolution treatment. Actual drug incorporation into the nonexpanded beads was only 1.3%, and because the initiators used at high temperature were strong oxidizing agents, decomposition of other drugs was a major problem. In contrast, the expanded beads could be loaded with around 20% drug, depending on how porous they were made, and the process did not give rise to drug decomposition problems. Attempts to polymerize the terpolymer, styrene-divinylbenzene-ethylvinylbenzene did not yield a satisfactory product due to the polymer agglomeration and degradation of medicament.

Reyes [55], in a patent assigned to International Business Machines Corp., described a process whereby a suitable hydrophobic monomer or polymer such as polystyrene in a suitable organic solvent is graft-polymerized as a coating onto a gelled hydrophilic polymer core such as agar or an alginic acid derivative in water using ionizing radiation in a dose from 1 to 5 Mrad. The hydrophilic polymer solution was dispersed as droplets in a solution of the hydrophobic monomer or polymer as a w/o emulsion, usually prior to initiating the irradiation. No drug was reported as being dissolved or dispersed in the internal phase of the emulsion, and the influence of the large dose of radiation on drug stability would have to be investigated. However, the process could have pharmaceutical applications for obtaining a tightly adhering hydrophobic coating around a gelled hydrophilic core.

9.4 POLYSILOXANE PRODUCTS

Very considerable interest has been shown in the formation of polysiloxane implantable devices containing contraceptive steroids for fertility control because the polymer material is reported to be biocompatible, though not biodegradable. Chang [56] described a process for forming polydimethylsiloxane microcapsules containing erythrocyte hemolysate. An aqueous solution of hemolysate was dispersed by adding dropwise with constant agitation into Silastic (Dow Corning). While maintaining stirring, stannous octoate was added as catalyst, followed within 2 sec by 50% aqueous Tween 20 solution containing sodium hydrosulfite (to prevent degradation of the hemolysate). Stirring of the w/o/w emulsion was then continued at slower speed for at least 20 min until the polymerized polydimethylsiloxane coating was no longer tacky. The microcapsules formed exhibited semipermeable properties, being able to combine reversibly with oxygen for several days.

Drug particles were dispersed in polyalkoxy- or polyorganoalkoxysiloxanes having adequate viscosity by Unger et al. [57] in a patent assigned to Merck Patent GmbH, Germany. When polycondensation was complete, the product was dried and size-reduced to produce drug-containing particles that had sustained-release properties.

REFERENCES

1. W. V. Smith and R. H. Eward, Kinetics of emulsion polymerization, *J. Chem. Phys. 16*:592–599 (1948).
2. S. Hjerten, "Molecular sieve" chromatography on polyacrylamide gels prepared according to a simplified method, *Arch. Biochem. Biophys. Suppl. 1*:147–151 (1962).
3. P. Bernfeld and J. Wan, Antigens and enzymes made insoluble by entrapping them into latices of synthetic polymers, *Science 142*:678–679 (1963).
4. A. Wada and A. Kishizaki, Chromatographic studies with immobilized polynucleotide in acrylamide gel matrix, *Biochim. Biophys. Acta 166*:29–39 (1968).
5. O. R. Zaborsky, *Immobilized Enzymes*, CRC Press, Cleveland, Ohio, 1973.
6. H. Nilsson, R. Mosbach, and K. Mosbach, The use of bead polymerization of acrylic monomers for immobilization of enzymes, *Biochim. Biophys. Acta 268*:253–256 (1972).
7. A. C. Johansson and K. Mosbach, Acrylic copolymers as matrices for the immobilization of enzymes. Covalent binding or entrapping of various enzymes to bead-formed acrylic copolymers, *Biochim. Biophys. Acta 370*:339–347 (1974).
8. B. Ekman and I. Sjöholm, Use of macromolecules in microparticles, *Nature 257*:825–826 (1975).
9. B. Ekman, C. Lofter, and I. Sjöholm, Incorporation of macromolecules in microparticles: preparation and characteristics, *Biochemistry 15*:5115–5120 (1976).
10. S. E. Brolin, A. Agren, B. Ekman, and I. Sjöholm, The analytical use of lactate dehydrogenase entrapped in microparticles, *Anal. Biochem. 78*:577–581 (1977).
11. I. Ljungstedt, B. Ekman, and I. Sjöholm, Detection and separation of lymphocytes with specific surface receptors by using microparticles, *Biochem. J. 170*:161–165 (1978).

12. B. Ekman and I. Sjöholm, Improved stability of proteins immobilized in microparticles prepared by a modified emulsion polymerization technique, *J. Pharm. Sci. 67*:693–696 (1978).
13. I. Ljungstedt, B. Ekman, and I. Sjöholm, Detection and separation of lymphocytes with specific surface receptors by using microparticles, *Biochem. J. 170*:161–165 (1978).
14. P. Edman and I. Sjöholm, Acrylic microspheres in vivo V: Immunological properties of immobilized asparaginase in microparticles, *J. Pharm. Sci. 71*:576–580 (1982).
15. D. S. Morehouse and F. H. Bolton, U.S. Patent 4,075,134 (February 21, 1978).
16. D. S. Morehouse and F. H. Bolton, U.S. Patent 4,049,604 (September 20, 1977).
17. S. C. Khanna, T. Jecklin, and P. Speiser, Bead polymerization technique for sustained-release dosage form, *J. Pharm. Sci. 59*: 614–618 (1970).
18. S. C. Khanna and P. Speiser, In vitro release of chloramphenicol from polymer beads of α-methacrylic acid and methylmethacrylate, *J. Pharm. Sci. 59*:1398–1401 (1970).
19. H. Kala, M. Dittgen, and W. Schmollack, Über die herstellung arzneistoffhaltiger perlopolymerisate auf polyacrylatbasis, *Pharmazie 31*:793–799 (1976).
20. H. Kala and R. Mank, Preparation of bead polymers under thermically mild conditions, *Drug Develop. Indust. Pharm. 7*: 453–460 (1981).
21. H. Wahlig, E. Dingeldein, R. Bergmann, and K. Reuss, The release of gentamicin from polymethylmethacrylate beads, *J. Bone Joint Surg. 60-B*:270–275 (1978).
22. *Septopal—Product Particulars*, BDH Pharmaceuticals, Ltd., Lenten House, Alton, Hants., England.
23. *Treatment of Chronic Bone Infections with Septopal*, Proceedings of a symposium held at the Royal College of Surgeons, London, June 26, 1980, BDH Pharmaceuticals Ltd., Lenten House, Alton, Hants, England.
24. M. Yoshida, M. Kumakura, and I. Kaetsu, Drug entrapment for controlled release in radiation-polymerized beads, *J. Pharm. Sci. 68*:628–631 (1979).
25. M. El-Samaligy and P. Rohdewald, Polyacrylamide microbeads, a sustained release drug delivery system, *Int. J. Pharm. 13*:23–34 (1982).
26. R. S. Molday, W. J. Dreyer, A. Rembaum, and S. P. S. Yen, New immunological spheres: visual markers of antigens on lymphocytes for scanning electron microscopy, *J. Cell Biol. 64*: 75–88 (1975).
27. S. P. S. Yen, A. Rembaum, R. S. Molday, and W. Dreyer, Functional colloid particles for immunoresearch, in *Emulsion*

Polymerization (I. Piirma and J. L. Gardon, eds.), ACS Symposium Series 24, Washington, D.C., 1976, pp. 236–257.

28. S. P. S. Yen, A. Rembaum, and R. S. Molday, U.S. Patent 4,035,316 (July 12, 1977).
29. A. Rembaum, S. P. S. Yen, and W. J. Dreyer, U.S. Patent 4,138,383 (February 6, 1979).
30. G. Birrenbach and P. P. Speiser, Polymerized micelles and their use as adjuvants in immunology, *J. Pharm. Sci.* *65*:1763–1766 (1976).
31. P. Speiser and G. Birrenbach, U.S. Patent 4,021,364 (May 3, 1977).
32. P. Speiser, Microencapsulation by coacervation, spray encapsulation and nanoencapsulation, in *Microencapsulation* (J. R. Nixon, ed.), Dekker, New York, 1976, pp. 1–14.
33. H. Kopf, R. Joshi, M. Soliva, and P. Speiser, Production and isolation of polyacrylamide nanoparticles, *Pharm. Ind.* *38*:281–284 (1976).
34. E. O. Dillingham, N. Webb, W. H. Lawrence, and J. Autian, Biological evaluation of polymers, I. Poly(methyl methacrylate), *J. Biomed. Mater. Res.* *9*:569–596 (1975).
35. J. Kreuter and H. J. Zehnder, The use of ^{60}Co γ-irradiation for the production of vaccines, *Radiat. Eff.* *35*:161–166 (1978).
36. H. Kopf, R. K. Joshi, M. Soliva, and P. Speiser, Studium der mizellpolymerisation in gegenwart niedermolekularer arzneistoffe, *Pharm. Ind.* *39*:993–997 (1977).
37. J. Kreuter and P. P. Speiser, In vitro studies of poly(methyl methacrylate) adjuvants, *J. Pharm. Sci.* *65*:1624–1627 (1976).
38. J. Kreuter, U. Tauber, and V. Illi, Distribution and elimination of poly (methyl-2-^{14}C-methacrylate) nanoparticle radioactivity after injection in rats and mice, *J. Pharm. Sci.* *68*:1443–1447 (1979).
39. J. Kreuter, R. Mauler, H. Gruschkau, and P. P. Speiser, The use of new polymethylmethacrylate adjuvants for split influenza vaccines, *Exp. Cell Biol.* *44*:12–19 (1976).
40. P. Couvreur, P. Tulkens, M. Roland, A. Trouet, and P. Speiser, Nanocapsules: a new type of lysosomotropic carrier, *FEBS lett.* *84*:323–326 (1977).
41. D. W. Lubbers, N. Opitz, P. P. Speiser, and H. J. Bisson, Nanoencapsulated fluorescence indicator molecules measuring pH and pO_2 down to submicroscopical regions on the basis of the optode-principle, *Z. Naturforsch.* *32c*:133–134 (1977).
42. N. Opitz, D. W. Lubbers, P. P. Speiser, and H. J. Bisson, Determination of hydrogen ion activities via nanoencapsulated fluorescing pH-indicator molecules, *Pflug. Arch.* *368,* Suppl., R49, No. 194 (1977).

43. J. Kreuter and P. P. Speiser, New adjuvants on a polymethylmethacrylate base, *Infect. Immunol.* *13*:204–210 (1976).
44. J. Kreuter and I. Haenzel, Mode of action of immunological adjuvants: some physicochemical factors influencing the effectivity of polyacrylic adjuvants, *Infect. Immunol.* *19*:667–675 (1978).
45. J. Kreuter and E. Liehl, Protection induced by inactivated influenza virus vaccines with polymethylmethacrylate adjuvants, *Med. Microbiol. Immunol.* *165*:111–117 (1978).
46. J. Kreuter and E. Liehl, Long-term studies of microencapsulated and adsorbed influenza vaccine nanoparticles, *J. Pharm. Sci.* *70*:367–371 (1981).
47. V. Bentele, U. E. Berg, and J. Kreuter, Molecular weights of poly(methyl methacrylate) nanoparticles, *Int. J. Pharm.* *13*:109–113 (1982).
48. P. Edman, B. Ekman, and I. Sjöholm, Immobilization of proteins in microspheres of biodegradable polyacryldextran, *J. Pharm. Sci.* *69*:838–842 (1980).
49. K. J. Saunders, *Organic Polymer Chemistry*, Chapman and Hall, London, 1973, pp. 71–83.
50. V. Stannett, H. Shiota, H. Garreau, and J. L. Williams, Kinetics of the radiation induced polymerization of styrene in emulsions, *J. Colloid Interface Sci.* *71*:130–140 (1979).
51. J. W. Vanderhoff, The use of monodisperse latex particles in medical research, *ACS Div. Org. Coating Plastics Chem. Reprints* *24*:223–232 (1964).
52. J. M. Singer, The latex fixation test in rheumatic diseases, *Amer. J. Med.*, *31*:766–779 (1961).
53. G. A. Quash, A. Niveleau, M. Aupoix, and T. Greenland, Immunolatex visualisation of cell surface Forssman and polyamine antigens, *Exp. Cell Res.* *98*:253–261 (1976).
54. R. W. Crosswell and C. H. Becker, Suspension polymerization for preparation of timed-release dosage forms, *J. Pharm. Sci.* *63*:440–442 (1974).
55. Z. Reyes, U.S. Patent 3,405,071 (October 8, 1968).
56. T. M. S. Chang, Semipermeable aqueous microcapsules ("artificial cells") with emphasis on experiments in an extracorporeal shunt system, *Trans. Amer. Soc. Artif. Int. Organs* *12*:13–19 (1966).
57. K. Unger, H. Kramer, H. Rupprecht, and W. Kircher, U.S. Patent 4,169,069 (September 25, 1979).

10

Polymerization Procedures for Biodegradable Micro- and Nanocapsules and Particles

In the previous chapter various polymerization techniques were described together with some pharmaceutical applications involving the use of essentially nonbiodegradable polymers. In this chapter other applications are considered that involve the use of biodegradable polymeric materials. Interest in this type of polymer has been very considerable in the last 15 years because of its obvious advantage for administration to patients, particularly in parenteral dosage forms. As the monomers and other agents employed are usually nontoxic and rarely bioincompatible, they can be safely used to develop controlled-release and other dosage forms. Wood [1] has reviewed aspects of biodegradable drug delivery systems.

10.1 POLYMERS AND COPOLYMERS OF LACTIC/GLYCOLIC ACIDS—OTHER ALIPHATIC POLYESTERS

Both polylactic acid and polyglycolic acid are used as absorbable sutures, implants, and various prosthetic devices. Since the early 1970s considerable interest has been shown in lactic and glycolic acid polymers as erodible drug carrier materials. Being tissue-compatible, they are particularly suitable for the production of controlled-release formulations for parenteral and implantation use. Lactic acid has an asymmetric carbon atom and therefore has two optical isomers,

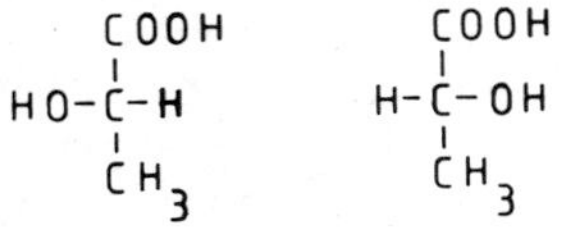

(L+) (D-)

Lactic acid

whereas glycolic acid,

$$\begin{array}{c} COOH \\ | \\ HO-C-H \\ | \\ H \end{array}$$

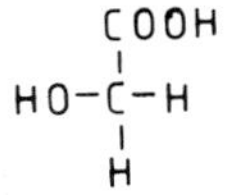

Glycolic acid

has no optical isomers. The hydroxy acids are widely distributed in plants and animals and, being bifunctional, may be condensed to form polyesters. In concentrated aqueous solution lactic acid forms linear polymer of the following general formula:

$$HO-\underset{\underset{CH_3}{|}}{CH}-CO\left[O-\underset{\underset{CH_3}{|}}{CH}-CO\right]_n O-\underset{\underset{CH_3}{|}}{CH}-COOH$$

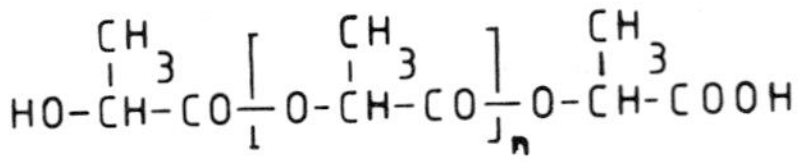

Polylactic acid

Six-membered cyclic dimers of lactic or glycolic acid may be formed by removal of the free and much of the bound water together with the water produced by dimerization from aqueous solutions of the corresponding acids by vacuum distillation as follows:

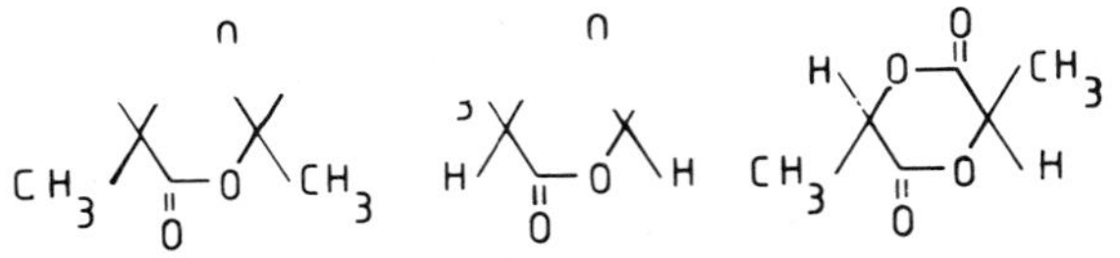

L(+) Lactide   D(-) Lactide   meso-Lactide

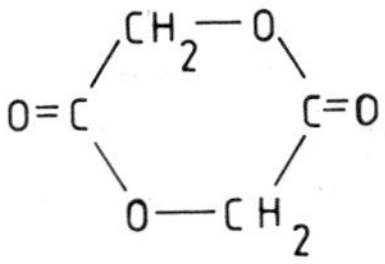

Glycolide

The α crystalline form of glycolide is more stable hydrolytically than the β form and is therefore preferred for the preparation of higher polymers.

Polymers of lactic acid and glycolic acid can be prepared by a bulk polymerization technique either by direct polymerization or via cyclic dimers. In direct polymerization a low-molecular-weight polymer (number average <10,000) is formed by boiling off water from the reaction mixture or by azeotropic distillation with an aromatic hydrocarbon. An acid catalyst is beneficial to increase reaction rate at temperatures below 120°C. In order to prepare lactic and glycolic acid polymers and copolymers of a higher molecular weight, it is necessary to employ purified lactide and/or glycolide as cyclic monomer units. Acid catalysts recommended include compounds of antimony, cadmium, lead, tin, titanium, or zinc, and a variety of amines. As each mole of catalyst contains a mole of water, the number of moles of polymer formed will depend on the number of moles of catalyst added.

In order to form polymers with number average molecular weights in excess of 40,000 for biomedical applications, the organometallic catalyst trialkyl aluminum (TEAL) should preferably be used. Because of the high temperatures involved for long duration in these bulk polymerization procedures, drugs are normally incorporated into prepared polymer. Drug may be incorporated into the hot polymer melt formed; being thermoplastic, the polymers can be readily molded. Use of polyglycolic acid tends to be limited because of difficulties in fabricating composites due to its low solubility in common solvents.

Polylactic acid (sometimes termed polylactide) has been recommended for the production of woven grafts, absorbable staples, and surgical mesh apart from its use in sutures, implants, and prosthetic devices. It slowly undergoes hydrolytic decomposiiton in the body to lactic acid, which is a normal byproduct of muscle metabolism. The polymer degrades in vivo at about 12 to 14% in 3 months, taking about 80 days to halve its average molecular weight. Obviously, the higher the molecular weight and the more crystalline as prepared from a pure optical isomer, the longer is the biological half-life of the polymer. Polyglycolic acid (PGA) is likewise metabolized to a harmless byproduct at somewhat faster rate than polylactic acid (PLA). Most studies indicate that the half-life of PLA/PGA copolymers increases with PLA content.

Further details of the production and properties of these polymers and copolymers are contained in the reviews of Wise et al. [2], Yolles and Sartori [3], and Yolles et al. [4]. Apart from their use in such preparations as slow-release pesticides [5, 6], such polymeric materials have been investigated for the production of controlled-release dosage forms of various drugs such as steroids, narcotic antagonists, anticancer agents, and antimalarial agents as discussed in the following sections.

10.1.1 Steroids

The first application of the use of these polymers as carrier systems for contraceptive steroids was reported by Jackanicz et al. [7]. They initially prepared poly-L(+)-lactic acid by the polymerization of the corresponding L(+)-lactide using tetraphenyl tin as a catalyst and elevated temperature under vacuum. The polymer was used to prepare films to minimize changes in surface area in an attempt to obtain zero-order release. This was done by suspending the drug d-norgestrel in a solution of the polymer in benzene and methanol, which was spread as a film and dried to remove the solvent. The in vitro release of drug from the film into 1% bovine serum albumin solution was relatively constant over 80 days. Almost uniform drug release was observed when the film was implanted into rats, it being broken down with minimal tissue reaction at a slower rate than the rate of drug release.

Later studies [2] on cylinders prepared from drug and copolymers of PLA/PGA showed that release of steroid in animals was affected by the optical isomer of lactic acid used, its content in the copolymer, the molecular weight of the polymers, and the drug loading. Their usefulness for long-term fertility control was confirmed.

An injectable delivery device was then developed by Anderson et al. [8] by cryogenically grinding cylinders of poly-L(+)-lactic acid (210,000 molecular weight) containing 20% norethindrone into particles of 90 to 190 μm, which were suspended in a 1% w/v methylcellulose aqueous vehicle. Upon intramuscular injection into rats, apparent zero-order release of steroid was observed for most of 90 days with no evidence of tissue irritation or encapsulation at the site of injection. However, despite overcoating the particles with poly-DL-lactic acid, an initial burst of steroid release was observed, probably due to poorly entrapped drug present on the surface of the particles as a result of their preparation by mechanical fragmentation (see Fig. 10.1). Also, a final burst of drug release was observed around day 88, presumably resulting from the breakup of the remaining polymer.

The preparation of solid solutions of d-norgestrel in PLA having sustained-release properties was described by Sadek [9] in a patent assigned to Dynatech Corporation.

Schindler et al. [10] reported that another polyester, poly(ε-caprolactone), which has a longer biological half-life in the body than

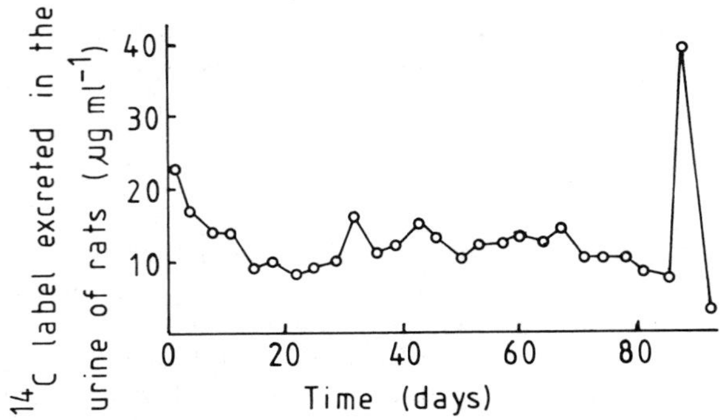

Figure 10.1 ^{14}C label from norethindrone excreted in vivo in the urine of rats from 90 to 180 μm particles of poly-L(+)-lactic acid containing 20% by weight norethisterone. (From Ref. 8.)

PLA, was particularly suitable for the design of biodegradable contraceptive steroid devices with a useful lifespan of about 1 year. This polymer is also more permeable than PLA, having a much lower glass transition temperature. Polymerization of ε-caprolactone was generally performed in bulk using stannous octoate at 130°C to give a polymer of the chemical structure

$$\left[\sim OCH_2CH_2CH_2CH_2CH_2\overset{\overset{O}{\|}}{C}\sim\right]_n$$

Poly(ε-caprolactone)

Copolymerization of lactide and caprolactone was performed using stannous octoate or other catalyst at 130 or 180°C. The stability of the copolymer at ambient temperature decreased with increasing lactide content until the glass transition temperature exceeded ambient. Further details of the properties of these and other related biodegradable polyesters used for the construction of monolithic and reservoir-type devices for the controlled release of steroids and a narcotic antagonist are described in two recent publications of Pitt et al. [11, 12]. Their production was also the subject of a patent assigned to Pitt and Schindler [13].

10.1.2 Narcotic Antagonists

Woodland et al. [14] used polylactic acid to prepare a composite containing cyclazocine as discrete small particles and as a reservoir device enclosed in a film of the pure polymer. In vivo studies in rats showed that lower-molecular-weight polymer was metabolized faster than higher-molecular-weight polymer giving quicker drug release. In a subsequent related paper, Yolles et al. [15] used naltrexone as the narcotic antagonist, adding it to prepared polylactic acid; after evaporation of the solvent, methylene chloride, the film formed upon melt pressing was ground into a powder. Particles in the size range 500 to 710 μm were suspended in carboxymethylcellulose solution, injected into various animals, and showed a significant blocking action against the effect of morphine for up to 20 days.

In later studies Schwope et al. [16, 17] and Wise et al. [18] investigated the effect of composition of PLA/PGA copolymer on the controlled release of naltrexone base and naltrexone pamoate molded in bead form. Injectable forms of narcotic antagonists in such polymers were considered unsuitable, because it would be difficult to reverse the effect in cases where narcotic addicts needed a morphinelike analgesic for the relief of serious pain, perhaps following injury. Thus beads about 1.5 mm in diameter suitable for trochar implantation of about 20 were developed for such drug-polymer combinations. Unlike cylindrical implants, which could be self-removed by addicts, the beads could only be removed in emergency by minor surgery. The researchers found that naltrexone base was actually soluble in the polymer material [with the exception of some copolymers of L(+)-lactic acid], enabling high drug loadings to be achieved. In general, increased drug release was observed in vitro and in vivo using animals when the glycolic acid content was increased, the drug loading was increased, and the polarity of the drug was increased by use of the more water-soluble base. The beads containing the drug base showed approximately zero-order drug release with a duration of action of 20 to 70 days depending on the formulation used. One formulation could be extended in duration of delivery by approximately 20 days further by dip coating with pure copolymer. If the naltrexone pamoate was used in the polymer matrix, the duration of action could be extended to 6 months. Similar results in animals for a naltrexone-PLA/PGA delivery system in bead form were reported by Reuning et al. [19].

Thies [20] and Mason et al. [21] reported on the in vitro and in vivo release of various narcotic antagonists from DL-polylactic acid microcapsules. Details of the production procedure were not given. Microcapsules were prepared in size fractions less than 300 μm and contained between 50 and 75% of drug. Because some of the microcapsules contained wall defects, they tended to release drug in vitro with an initial burst effect. The microcapsules had an effective antagonism of morphine analgesic effect in the rat or mouse following injection of about 14 days for cyclazocine free base and naltrexone free base, and of about 28 days for naltrexone pamoate. In vivo studies using mice and monkeys on these microcapsules containing naltrexone pamoate were reported by Harrigan et al. [22].

10.1.3 Anticancer Agents

Sustained release of cyclophosphamide and *cis*-dichlorodiammineplatinum from polylactic acid composites containing the plasticizer tributyl citrate in film form was reported by Yolles et al. [23]. In a subsequent paper, Yolles et al. [24] also reported results with the anticancer agent doxorubicin, the three separate drugs being prepared from melt-pressed composites with polylactic acid in particle form. In vivo tests showed a considerable improvement in the lifespan of mice with ascites sarcoma when treated by injection with the composites compared to free drug.

10.1.4 Antimalarial Agents

As mentioned in Chap. 8, Wise et al. spray-embedded a copolymer of DL-lactic acid and glycolic acid and the antimalarial drug 2,4-diamino-6-(2-naphthylsulfonyl)-quinazoline to produce coated drug particles that upon injection had a sustained effect over 14 weeks against rodent malaria, *Plasmodium berghei*. In a later publication of Wise et al. [25], sulfadiazine was chosen as the antimalarial drug and incorporated with composites of (1) L(+)-polylactic acid/DL-polylactic acid, (2) L(+)-polylactic acid/polyglycolic acid, or (3) L(+)-polylactic acid. The products were formed into spherical beads 1.5 mm in diameter, rods 0.75 mm in diameter, or cryogenically ground into fine particles (90 to 180 μm). The greater the drug loading employed, the lower the molecular weight of the polymer, and the smaller the particle size of the product, the more rapidly was the drug released. When mice were implanted with drug containing beads composed of poly 90% L(+)-lactic acid/10% DL-lactic acid at dosages of drug equivalent to at least 57 mg kg^{-1}, they were effectively protected against infection by *P. berghei* for 21 weeks.

10.2 ALBUMIN PRODUCTS

Another process involving a type of polymerization of a disperse phase containing usually either human, bovine, or egg albumin has attracted considerable interest for the entrapment of drugs in micro- and nanospheres. An aqueous solution of the albumin, containing either dissolved or dispersed drug, is emulsified by mechanical means as droplets into a continuous phase composed of a suitable oil and emulgent if necessary at room temperature. On increasing the temperature of the water-in-oil (w/o) emulsion formed to 60 to 70°C the albumin crosslinks and polymerizes to form solid drug-loaded spheres. These are then separated from the continuous phase by centrifugation, washed with organic solvent to remove traces of oil adhering to their surfaces, and dried.

In 1968 Kosar and Arkins [26], in a patent assigned to Keuffel and Esser Company, described a different microencapsulation process whereby droplets of oil phase were coated with polymerized albumin. Cotton seed oil was emulsified as an o/w system into an aqueous solution of egg albumin. Upon heating to 70°C while maintaining stirring, the albumin at the interface polymerized to form a coating around the oil droplets. A silicone agent was used to control foaming of the albumin solution. To reduce polymerization of albumin in the bulk of the continuous phase, an infrared-absorbing compound was added to the oil phase. Upon subsequent exposure to infrared radiation, the droplets became hot enough to polymerize a coating of albumin around their surface, while the albumin in the bulk of the continuous phase did not polymerize as its temperature reached only about 40°C. Obviously the process could have pharmaceutical applications if a drug was added to the oil phase.

An example of microencapsulation using heat-polymerized albumin was reported by Farhadieh [27] in a patent assigned to Abbott Laboratories. As an example of the process, erythromycin ethyl succinate, which is poorly water-soluble, has an objectionable taste, and is inactivated by acid in the stomach, was dispersed in an aqueous solution of egg albumin at pH 8. This was emulsified into a continuous phase of isooctane using sorbitan trioleate as a w/o emulgent. While maintaining stirring the flask containing the emulsion was immersed in a water bath at 74°C to coagulate the albumin as a coating around the drug particles. After cooling the microcapsules were washed by repeated suspension and centrifugation from a dilute aqueous solution of polyethylene sorbitan monolaurate and then from water before being recovered as a free-flowing powder by freeze-drying. The final microcapsules were about 50 to 80 μm in size and had similar or better bioavailability than the unencapsulated drug particles.

Interest in polymerization of human serum albumin involving formation of a w/o emulsion was reported by Zolle, Rhodes, and colleagues [28-30]. Albumin microcapsules were typically prepared by emulsifying a 25% aqueous solution of human serum albumin added dropwise into cottonseed oil using an electrically driven stirrer, which for faster speeds gave smaller microspheres, and heating the resultant emulsion up to 118, 146, or 165°C over 45 min and then maintaining at the temperature for another 30 min to ensure complete evaporation of the water. After cooling and separation by filtration or centrifugation, the microspheres were washed with heptane and acetone or with diethyl ether to remove residual oil and then air-dried. The spheres were radioactively labeled with a gamma-emitting radioisotope added to the starting albumin solution or combined with preformed spheres containing additives such as ferric hydroxide if necessary to facilitate labeling. The size distribution and quantity of the microspheres formed were also affected by the type of oil used and by the volume of human serum albumin solution, its concentration, and pH. Clusters of microspheres were frequently obtained. When suspended for 5 min in normal saline, the microspheres swelled about 50% if prepared at 118°C and about 20% if prepared at the higher temperatures. The higher the temperature of preparation, the slower was the rate of the degradation of the microspheres in the lung.

These studies have lead to the production of ^{99m}Tc-labeled human albumin microspheres of a narrow size range from approximately 12 to 44 μm, which are frequently used nowadays by injection upstream of the lungs to cause a temporary occlusion of some of the capillaries in the lungs of patients. The γ-rays coming from the isotope can be visualized using a gamma camera, allowing the efficiency of lung perfusion to be studied by this noninvasive procedure. This facilitates the diagnosis of pulmonary embolism and other lung disorders. The microspheres are considered nonantigenic and are quickly metabolized, so blood flow through the lungs is impeded for only a short period of time. However, caution must always be exercised in the injection of this potentially antigenic material into humans. Littenberg [31] reported a case of an anaphylactoid reaction in a 22-year-old woman to the intravenous administration of ^{99m}Tc-labeled human albumin microspheres. Also, one case of a fatal reaction to ^{131}I-labeled macroaggregates of human serum albumin immediately following injection has been reported by Atkins et al. [32].

In a subsequent paper, Scheffel et al. [33] prepared much smaller labeled human albumin spheres, 0.3 to 1 μm in size, for investigation of the reticuloendothelial system. The albumin-containing droplets were formed by homogenization in the oil phase and heated at 175 to 185°C for 10 min. They were then cooled and diluted with diethyl ether to reduce the viscosity of the oil phase to permit separation by centrifugation. The separated nanospheres were further washed with ether to remove residual oil and repeatedly suspended in absolute alcohol using sonication to permit contaminants and clumps to settle out and be removed. The nanospheres were then suspended in ether and passed through a filter of pore size 14 μm and dried. The resultant spheres were finally labeled with ^{99m}Tc and used successfully to visualize the liver and spleen.

Evans [34], in a patent assigned to the Minnesota Mining and Manufacturing (3M) Company, described the production of microspheres made from human albumin, hemoglobin, or parenterally acceptable gelatin that were crosslinked either thermally or chemically and subsequently labeled with radioactive isotope for scanning applications. Also, in a patent assigned to Zolle [35], preparation of human albumin microspheres was described that could be labeled with radionuclides. Zolle also effectively incorporated insulin into albumin microspheres without significant reduction in biological activity so as to form a sustained-release product.

Human albumin microspheres were investigated by Kramer [36] as vehicles for cell specific delivery of the anticancer agent, mercaptopurine. The microspheres were prepared by a method similar to that of Scheffel et al., the labeled drug being added to the aqueous albumin solution containing the solvent dimethylformamide prior to emulsification. If an equivalent amount of mercaptopurine was added to prepared blank albumin microspheres, about 40% less drug was associated with the product. Larger quantities of the more water-soluble anticancer agent daunomycin hydrochloride could be entrapped in the albumin spheres without using mixed solvent systems. The microspheres containing daunomycin hydrochloride, which were not sonicated, were about 0.2 to 1.2 μm in diameter and showed aggregation. Figure 10.2 shows a scanning electron micrograph of albumin microspheres containing sulfamethizole prepared in the author's laboratory, which likewise show extensive clustering before sonication. By being phagocytozed after injection, such microspheres could be useful in the treatment of cancer and bacterial infections of the reticuloendothelial system and for the delivery of immunosuppressive agents. Kramer and Burnstein [37] later reported that human albumin microspheres containing mercaptopurine were phagocytozed by several tumor cell lines.

Sugibayashi and co-workers [38-41] also investigated the use of bovine serum albumin nanospheres of average diameter 660 nm containing 5-fluorouracil for cancer chemotherapy. They were prepared by a modification of the method of Scheffel et al. The nanospheres showed a tendency to form clusters, and the particle size was influenced by the

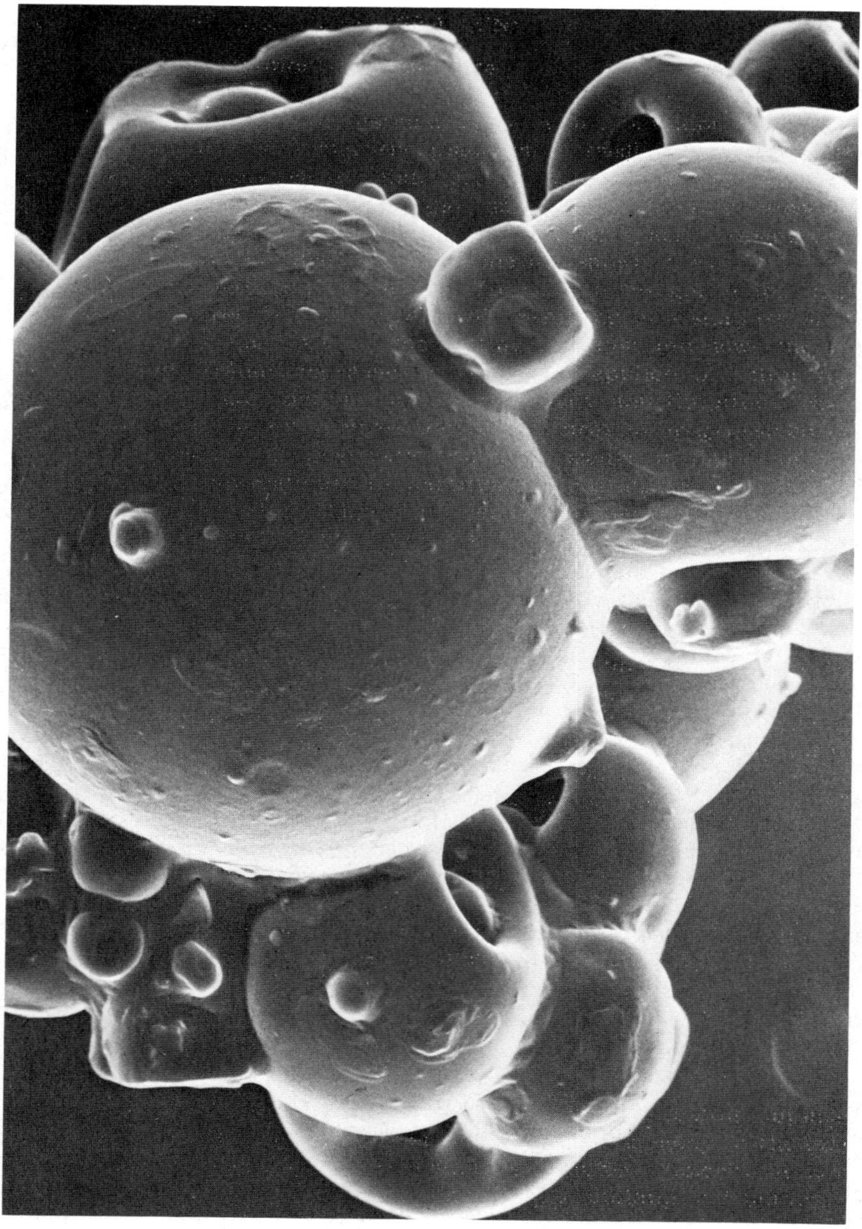

Figure 10.2 Albumin microspheres containing sulfamethizole.

concentration of Span 85 used as emulgent and the intensity of mechanical agitation employed. There was no evidence of thermal degradation of the drug caused by the heating procedure. In vitro release of drug was noted for 1 week. Upon intravenous injection into healthy mice, the nanospheres accumulated mainly in the liver and other elements of the reticuloendothelial system having high phagocytic activity, and showed a sustained-release effect of the entrapped drug. However, when given by intravenous injection into mice bearing Ehrlich solid carcinoma, the drug-loaded spheres were not preferentially accumulated in the tumor cells, being again accumulated mainly in the liver, which would be of obvious advantage in the treatment of hepatoma cells. If the nanospheres were injected interperitoneally into interperitoneally injected Ehrlich ascites carcinoma-bearing mice, phagocytosis of the microspheres by the ascites cells was observed and suppression of tumor growth for approximately 1 week was noted. Upon injection of multiple doses of the microspheres the lifespan of the mice was increased by about 30% compared to a control treated with the free drug. Even better suppression of carcinoma growth was obtained when studying the chemotherapy of directly injected solid tumors in mice.

Ishizaka et al. [42] reported on the preparation of egg albumin "microspheres" containing no drug and "microcapsules" containing hydrophobic particles such as spherical silica gel beads, iron hydrous oxide-modified nylon 12, or phenacetin. The egg albumin was purified somewhat by dissolving in a phosphate buffer at pH 8.0 and then centrifuged and filtered to remove undissolved material. The "microspheres" and "microcapsules" were prepared by a method similar to that of Farhadieh. Increasing concentration of sorbitan trioleate up to 0.5%, increasing stirrer speed, and decreasing concentration of albumin tended to produce a smaller product with a narrower size distribution. There was no observable difference in diameter of the "microspheres" when examined in isooctane or in 2% aqueous polyoxethylene sorbitan monolaurate solution, indicating an absence of swelling when the microspheres were immersed in the latter solution. The phenacetin particles were much more effectively encapsulated than the silica gel beads, presumably because of differences in physicochemical properties.

Another way of preparing human serum albumin microspheres was described by Longo et al. [43]. An aqueous albumin solution was dispersed in a solution of polymethyl methacrylate in chloroform:toluene or in a solution of polyoxyethylene-polyoxypropylene in chloroform. Glutaraldehyde in toulene was added to the continuous phase of the emulsion and diffused into the albumin droplets to crosslink them. After quenching, the microspheres formed were recovered by repeated centrifugation and washing.

A very interesting application of albumin encapsulation was reported by Widder et al. [44, 45]. They prepared, by a method similar to that of Scheffel et al., human albumin microspheres 0.2 to 3 μm in

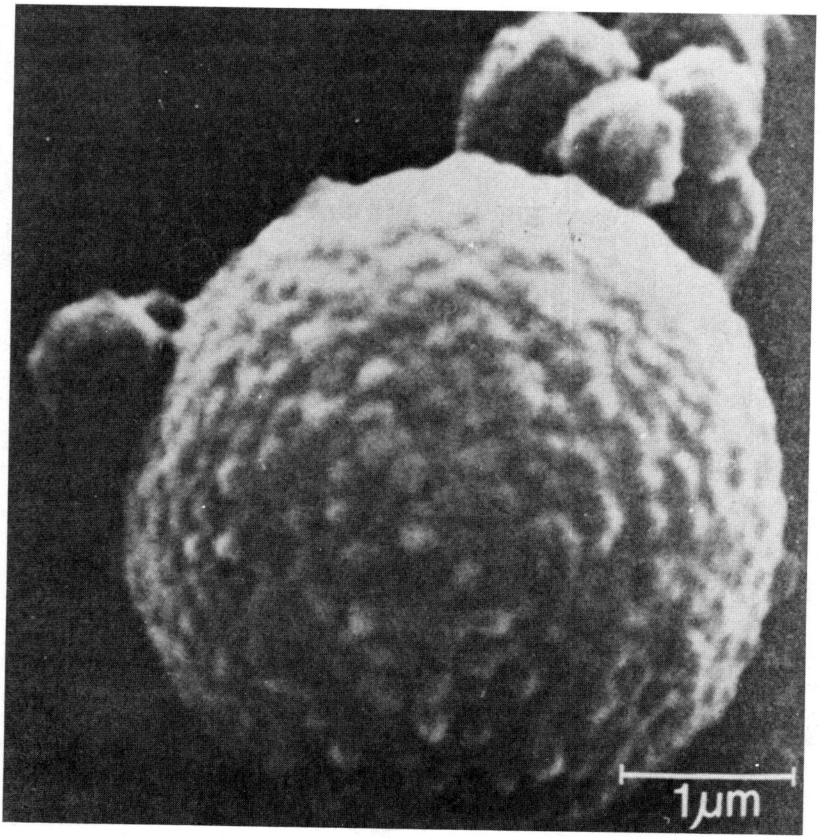

Figure 10.3 Scanning electron micrograph of a rat lymphocyte with adherent microspheres on its surface. The micrograph is representative of cells obtained after coincubation of staphlococcal protein A microspheres bearing rabbit antirat immunoglobulin and splenocytes and subsequent magnetic isolation of microsphere-cell complexes. (Reproduced with permission of the copyright owner, American Pharmaceutical Association, from Ref. 48.)

diameter containing the anticancer agent doxorubicin hydrochloride and also ultrafine magnetite (Fe_3O_4) particles 10 to 20 nm in size, which caused the drug-loaded spheres to congregate at a specific site within the model animal's body (rat tail) under the influence of a magnetic field. In this way effective levels of the drug could be achieved locally in the target tissue while reducing undesirable release of the drug throughout the body. Increasing heat denaturation of the microspheres decreased the release of drug. Upon injection only about 50% of the microspheres were retained at the target site under the influence of a field strength of 8000 Oe, presumably as a result of shunting of blood upstream of the site. After removal of the magnetic field, the microspheres still continued to localize at the target site, presumably because they had lodged in the vascular endothelium, penetrated into the interstitial space, or their conglutination had caused partial thrombosis resulting in their retention. The disadvantages associated with these microspheres include the unknown toxicity of the magnetite particles, the possible unwanted localization of the product in the liver and other regions of the reticuloendothelial system, and the danger of self-flocculation of the magnetic particles causing vascular obstruction to vital organs in the body.

In two other papers by Widder et al. [46] and Senyei et al. [47], similar magnetic microspheres also containing the anticancer agent doxorubicin hydrochloride were prepared and evaluated. Crosslinking of the albumin matrix was effective in delaying drug release in vitro by varying the intensity of either thermal or chemical (formaldehyde or 2,3-butanedione) treatment used. Widder et al. [48] incorporated staphlococcal protein A into such magnetic albumin microspheres, which permitted coupling of cell-specific immunoglobulin G without the use of the normal chemical agents. These modified microspheres were then used to isolate cells having specific antigens from mixed populations of cells using the magnetic property of the microspheres. Figure 10.3 shows a typical cell isolated by the procedure, which may be useful in protein or enzyme purification and in radioimmunoassay.

Studies on magnetic albumin microspheres were also reported by Morimoto et al. [49, 50].

10.3 POLYALKYL CYANOACRYLATE PRODUCTS

Polymers based on alkyl cyanoacrylate monomers are used as biodegradable tissue adhesives in surgery. The various polymers degrade by hydrolysis of the carbon chain, forming formaldehyde and alkyl cyanoacetate as shown below, where R is an alkyl group.

$$\sim CH_2-\underset{\underset{OR}{\overset{\overset{CN}{|}}{C=O}}}{C}-CH_2-\underset{\underset{OR}{\overset{\overset{CN}{|}}{C=O}}}{C}\sim + OH^- \rightarrow \sim CH_2-\underset{\underset{OR}{\overset{\overset{CN}{|}}{C=O}}}{C}-CH_2OH + {}^-\underset{\underset{OR}{\overset{\overset{CN}{|}}{C=O}}}{C}\sim$$

$$-\underset{\underset{OR}{\overset{\overset{CN}{|}}{C=O}}}{C}\sim + HOH \rightarrow H\underset{\underset{OR}{\overset{\overset{CN}{|}}{C=O}}}{C}\sim + OH^-$$

$$\sim CH_2-\underset{\underset{OR}{\overset{\overset{CN}{|}}{C=O}}}{C}-CH_2OH + OH^- \rightarrow \sim CH_2-\underset{\underset{OR}{\overset{\overset{CN}{|}}{C=O}}}{C}^- + CH_2\underset{OH}{\overset{OH}{<}}$$

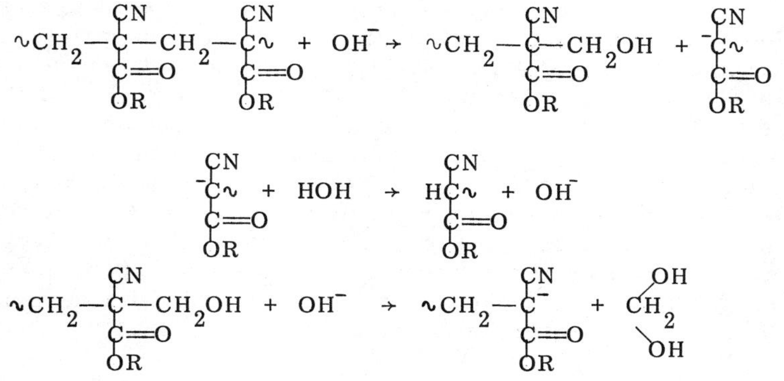

The rate of degradation is a function of the alkyl chain length, being most rapid with shorter chain length. As the methyl ester quickly liberates formaldehyde, it has been associated with histotoxicity. However, the ethyl and particularly the butyl ester and the higher homologs are reported to be well tolerated in vivo by Leonard et al. [51], Pani et al. [52], and Leonard [53].

Florence et al. [54] described an interfacial polymerization procedure for preparing polybutyl cyanoacrylate microcapsules that does not require reactive monomer in the core material. As a typical procedure, an aqueous solution of radiolabeled albumin was emulsified as a w/o emulsion into a continuous phase of chloroform and cyclohexane using sorbitan trioleate as emulgent. Butyl cyanoacrylate monomer in continuous phase was then added and polymerization occurred at the droplet interface, being base-catalyzed in contact with the water phase and being presumably initiated by the protein, which was also extensively incorporated into the membrane. Polymerization was quenched by adding more continuous phase. After decanting the organic phase, the microcapsules were repeatedly washed in a hydroalcoholic solution of polysorbate 20 and finally resuspended in aqueous buffer solution. In a later publication Wood et al. [55] gave more extensive details of the procedure.

Couvreur et al. [56, 57] reported the production of nanoparticles of about 200 nm diameter by the polymerization of mechanically dispersed methyl or ethyl cyanoacrylate in an aqueous acidic medium in the presence of polysorbate 20 as surfactant. No initiator or irradiation was required. The concentration of polysorbate 20 used did not affect the appearance or size of the nanoparticles obtained, except that in its absence larger and more agglomerated particles were obtained. Samples of nanoparticles, spray-frozen and examined by electron microscopy after freeze-fracture, showed the inner structure to be a highly porous matrix, unlike the wall-containing microcapsules prepared by Florence and co-workers. At acid pH values the sorption of nonionized fluorescein by the nanoparticles was appreciable because of the porous nature of the product, particularly by those prepared from ethyl cyanoacrylate. Likewise, the nonionized form of the antimitotic drug daunorubicin was more readily sorbed at alkaline pH. Their potential as lysosomotropic carriers was stressed.

Couvreur et al. [58] subsequently showed that two other cytotoxic drugs, methotrexate and dactinomycin, were bound 15 to 40% and 90%, respectively, by these polyalkyl cyanoacrylate nanoparticles. These binding capacities far exceed those achievable for these drugs entrapped in liposomes. In vitro studies in a calf serum medium showed that the methyl derivative was more readily degraded with release of drug than the ethyl one. In a subsequent paper by Couvreur et al. [59], the tissue distribution in rats following intravenous injection of polymethyl cyanoacrylate and polyethyl cyanoacrylate nanoparticles containing the labeled cytotoxic drugs ^3H-dactinomycin and ^3H-vinblastine was studied. Such drugs require specific targeting to improve anticancer activity and to reduce general toxicity. In comparison to free drug, vinblastine-containing nanoparticles promoted drug accumulation in a broad range of tissues, particularly those higher in endocytic activity, such as the liver, spleen, lungs, and muscle. In contrast, selective accumulation of dactinomycin from the nanoparticles was associated mainly with the small intestine and lungs, indicating that tissue distribution is a function not only of the dosage form but also of the active agent. Polybutyl cyanoacrylate nanoparticles containing adsorbed ^{3}H-dactinomycin were concentrated mainly in rat liver and spleen and to a lesser extent in muscle and lung after injection [60] as indicated in Table 10.1. These results also indicate that the identity of the polymeric carrier affects tissue distribution.

The acute toxicity of free polyalkyl cyanoacrylate nanoparticles was shown to be very low [61], whereas the toxicity of the anticancer drug doxorubicin adsorbed onto such particles was markedly reduced [62].

10.4 EPOXY PRODUCTS

Epoxy compound was condensed with basic or acidic curing agents to prepare resins by bulk or suspension procedures, the bead form being used to entrap drug (Khanna and Speiser [63]). The epoxy compound was prepared by heating 2,2'-bis-(4'-hydroxyphenyl)propane (i.e., bisphenol A) and epichlorhydrin at approximately 110°C for 3.5 hr in 40% aqueous sodium hydroxide solution as shown below.

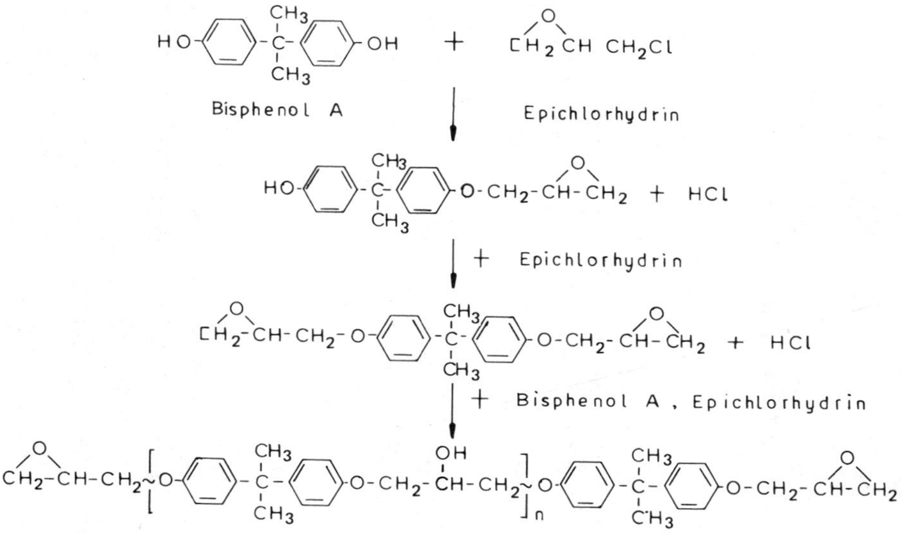

Table 10.1 Tissue Content of ^{3}H-Dactinomycin, 24 hr after Injection (Percent of Injected Dose per Gram of Wet Tissue)[a]

Organs	Free ^{3}H-dactinomycin (F)[b]	^{3}H-dactinomycin loaded nanoparticles (PBN)[b]	Ratio PBN/F	Student's t-test
Blood	0.23 ± 0.40	0.32 ± 0.05	1.39	$0.02 < P \leqslant 0.05$
Spleen	1.10 ± 1.04	48.75 ± 20.90	44.31	$P \leqslant 0.002$
Small intestine	0.99 ± 0.25	1.90 ± 0.34	1.91	$P \leqslant 0.001$
Muscle	0.13 ± 0.03	0.73 ± 0.16	5.61	$P \leqslant 0.001$
Kidneys	4.22 ± 0.46	5.05 ± 2.38	1.19	$P > 0.1$
Liver	0.70 ± 0.21	44.85 ± 8.54	64.07	$P \leqslant 0.001$
Lungs	0.95 ± 0.31	4.46 ± 2.31	4.69	$P \leqslant 0.001$

[a]Drug concentration in microcuries per gram of wet organ weight and per injected millicurie.

[b]Results are the mean ± S.E. from 6 animals.

Source: Ref. 60.

The linear epoxy resins formed had a variable degree of condensation, depending on the conditions of polymerization, and were then cross-linked through epoxy group reaction by addition of the basic curing agent 2-amino-2-ethyl-1,3-propandiol or the acidic curing agent adipic acid, *o*-phosphoric acid, oxalic acid, citric acid, or tartaric acid. Epoxy compounds cured with the primary amine dissolved in buffer solutions below pH 4 and consequently might be expected to liberate drug in the stomach. Acid-cured resins generally dissolved in buffer solutions above pH 3.2 and consequently might release drug in the small intestine. The solubility in buffers was aided by increasing the amount of curing agent used, thus allowing further control over dissolution rate of the polymer. Drug-loaded beads 0.5 to 10 mm in diameter were prepared by dispersing epoxy compound, curing agent, and chloramphenicol, barium sulfate, or dehydroemetine dihydrochloride in a continuous phase of silicone oil using vigorous agitation and heating.

In a subsequent paper, Khanna et al. [64] studied the dissolution behavior of the amine-cured beads in various acidic buffer solutions. The dissolution rate of the beads was affected by the pH and ionic strength of the buffer used. When chloramphenicol was incorporated into the beads, low concentrations of the drug did not affect the dissolution rate of the polymer, whereas higher concentrations decreased the rate by an unknown mechanism.

Unsworth and Helsby [65], in a patent assigned to Magnesium Elektron, Ltd., England, described an epoxy encapsulation process that may have pharmaceutical applications. A water-insoluble epoxy monomer such as Epikote 828 (Shell Chemical Co.) is dissolved in a solvent such as acetone, dioxan, or tetrahydrofuran that has a higher affinity for water than for the monomer. The monomer solution and the material to be encapsulated are then dispersed in water, and a curing agent such as a polyamine or polyamide in the organic solvent is added. Polymerization is affected by stirring at an elevated temperature around 50°C for several hours. A suspending agent such as gelatin, acacia, or methylcellulose may be used to prevent the partially polymerized microcapsules from adhering together.

10.5 MISCELLANEOUS PRODUCTS

Controlled release of naltrexone from biodegradable implants composed of glutamic acid/leucine copolymers has been reported by Sidman et al. [66], which could also have been tested in particle form. Interest in other biodegradable polymers such as polyglutamic acid [67] and hydroxypropylcellulose and related celluloses [3] whereby the active ingredient is covalently bonded to the polymer through ester, amide, or other linkage has been reported. Release of active ingredient is usually controlled by the rate of cleavage of the labile bond and the rate of diffusion of the liberated drug from the polymer device.

REFERENCES

1. D. A. Wood, Biodegradable drug delivery systems, *Int. J. Pharm.* 7:1–18 (1980).
2. D. L. Wise, T. D. Fellman, J. E. Sanderson, and R. L. Wentworth, Lactic/glycolic acid polymers, in *Drug Carriers in Biology and Medicine* (G. Gregoriadis, ed.), Academic, London, 1979, pp. 237–270.
3. S. Yolles, and M. F. Sartori, Erodible matrices, in *Controlled Release Technologies: Methods, Theory and Applications, Vol. II* (A. F. Kydonieus, ed.), CRC Press, Boca Raton, Fla., 1980, pp. 1–6.
4. S. Yolles, T. Leafe, L. Ward, and F. Boettner, Controlled release of biologically active drugs, *Bull. Parent. Drug. Assoc.* *30*:306–312 (1976).
5. R. G. Sinclair, Slow-release pesticide system, polymers of lactic and glycolic acids as ecologically beneficial, cost-effective encapsulating materials, *Environ. Sci. Technol.* *1*:955–956 (1973).
6. H. Jaffe, J. A. Miller, P. A. Gung, and D. K. Hayes, Implantable systems for delivery of insect growth regulators to livestock, in *Controlled Release of Bioactive Materials* (R. Baker, ed.), Academic, New York, 1980, pp. 237–250.
7. T. M. Jackanicz, H. A. Nash, D. L. Wise, and J. B. Gregory, Polylactic acid as a biodegradable carrier for contraceptive steroids, *Contraception* *8*:227–234 (1973).
8. L. C. Anderson, D. L. Wise, and J. F. Howes, An injectable sustained release fertility control system, *Contraception 13*: 375–384 (1976).
9. S. E. Sadek, U.S. Patent 3,976,071 (August 24, 1976).
10. A. Schindler, R. Jeffcoat, G. L. Kimmel, C. G. Pitt, M. E. Wall, and R. Zweidinger, Biodegradable polymers for sustained drug delivery, in *Contempory Topics in Polymer Science, Vol. 2* (E. M. Pearse and J. R. Schaefgen, eds.), Plenum, New York, 1977, pp. 251–289.
11. C. G. Pitt, M. M. Gratzi, A. R. Jeffcoat, R. Zweidinger, and A. Schindler, Sustained drug delivery systems II: factors affecting release rates for poly(ε-caprolactone) and related biodegradable polyesters, *J. Pharm. Sci.* *68*:1534–1538 (1979).
12. C. G. Pitt, T. A. Marks, and A. Schindler, Biodegradable drug delivery systems based on aliphatic polyesters: application to contraceptives and narcotic antagonists, in *Controlled Release of Bioactive Materials* (R. Baker, ed.), Academic, New York, 1980, pp. 19–43.
13. C. G. Pitt and A. E. Schindler, U.S. Patent 4,148,871 (April 10, 1979).

14. J. H. R. Woodland, S. Yolles, D. A. Blake, M. Helrich, and F. J. Meyer, Long-acting delivery systems for narcotic antagonists, I, *J. Med. Chem. 16*:897–901 (1973).
15. S. Yolles, T. D. Leafe, J. H. R. Woodland, and F. J. Meyer, Long-acting delivery systems for narcotic antagonists II: release rates of naltrexone from poly(lactic acid) composites, *J. Pharm. Sci. 64*:348–349 (1975).
16. A. D. Schwope, D. L. Wise, and J. F. Howes, Lactic/glycolic acid polymers as narcotic antagonist delivery systems, *Life Sci. 17*:1877–1886 (1975).
17. A. D. Schwope, D. L. Wise, and J. F. Howes, Development of polylactic/glycolic acid delivery systems for use in treatment of narcotic addiction, *Natl. Inst. Drug Abuse Res. Monogr. Ser. 4*:13–18 (1976).
18. D. L. Wise, A. D. Schwope, S. E. Harrigan, D. A. McCarthy, and J. F. Howes, Sustained delivery of a narcotic antagonist from lactic/glycolic acid copolymer implants, in *Polymeric Delivery Systems* (R. J. Kostelnik, ed.), Gordon & Breach, New York, 1978, pp. 75–89.
19. R. H. Reuning, L. Malspeis, S. Frank, and R. E. Notari, Testing of drug delivery systems for use in the treatment of narcotic addiction, *Natl. Inst. Drug Abuse Res. Monogr. Ser. 4*:43–45 (1976).
20. C. Thies, Development of injectable microcapsules for use in the treatment of narcotic addiction, *Natl. Inst. Drug Abuse Res. Monogr. Ser. 4*:19–20 (1976).
21. N. Mason, C. Thies, and T. J. Cicero, In vivo and in vitro evaluation of a microencapsulated narcotic antagonist, *J. Pharm. Sci. 66*:847–850 (1976).
22. S. E. Harrigan, D. A. McCarthy, R. Reuning, and C. Thies, Release of naltrexone pamoate from injectable microcapsules, in *Polymeric Delivery Systems* (R. J. Kostelnik, ed.), Gordon & Breach, New York, 1978, pp. 91–100.
23. S. Yolles, T. D. Leafe, and F. J. Meyer, Timed-release depot for anticancer agents, *J. Pharm. Sci. 64*:115–116 (1975).
24. S. Yolles, J. F. Morton, and B. Rosenberg, Timed-released depot for anticancer agents, II, *Acta Pharm. Suec. 15*:382–388 (1978).
25. D. L. Wise, G. J. McCormick, G. P. Willet, L. C. Anderson, and J. F. Harris, Sustained release of sulphadiazine, *J. Pharm. Pharmacol. 30*:686–689 (1978).
26. J. Kosar and G. M. Arkins, U.S. Patent 3,406,119 (October 15, 1968).
27. B. Farhadieh, U.S. Patent 3,922,379 (November 25, 1975).
28. H. S. Stein, I. Zolle, and J. G. McAfee, Preparation of technetium (Tc^{99m}) labeled serum albumin (human), *Int. J. Appl. Radiat. 16*:283–288 (1965).

29. B. A. Rhodes, I. Zolle, J. W. Buchanan, and H. N. Wagner, Radioactive albumin microspheres for study of the pulmonary circulation, *Radiology 92*:1453–1460 (1969).
30. I. Zolle, B. A. Rhodes, and H. N. Wagner, Preparation of metabolizable radioactive human serum albumin microspheres for studies of the circulation, *Int. J. Appl. Radiat. 21*:155–167 (1970).
31. R. L. Littenberg, Anaphylactoid reaction to human albumin microspheres, *J. Nucl. Med. 16*:236–237 (1975).
32. H. L. Atkins, W. Hauser, P. Richards, and J. Klopper, Adverse reactions to radiopharmaceuticals, *J. Nucl. Med. 13*: 232–233 (1972).
33. U. Scheffel, B. A. Rhodes, T. K. Natarajan, and H. N. Wagner, Albumin microspheres for study of the reticuloendothelial system, *J. Nucl. Med. 13*:498–503 (1972).
34. R. L. Evans, U.S. Patent 3,663,687 (May 16, 1972).
35. I. Zolle, U.S. Patent 3,937,668 (February 10, 1976).
36. P. A. Kramer, Albumin microspheres as vehicles for achieving specificity in drug delivery, *J. Pharm. Sci. 63*:1646–1647 (1974).
37. P. A. Kramer and T. Burnstein, Phagocytosis of microspheres containing an anticancer agent by tumor cells in vitro, *Life Sci. 19*:515–520 (1976).
38. K. Sugibayashi, Y. Morimoto, T. Nadai, and Y. Kato, Drug-carrier property of albumin microspheres in chemotherapy. I. Tissue distribution of microsphere-entrapped 5-fluorouracil in mice, *Chem. Pharm. Bull. 25*:3433–3434 (1977).
39. K. Sugibayashi, Y. Morimoto, T. Nadai, Y. Kato, A. Hasegawa, and T. Arita, Drug-carrier property of albumin microspheres in chemotherapy. II. Preparation and tissue distribution in mice of microsphere-entrapped 5-fluorouracil, *Chem. Pharm. Bull. 27*: 204–209 (1979).
40. K. Sugibayashi, M. Akimoto, Y. Morimoto, T. Nadai, and Y. Kato, Drug-carrier property of albumin microspheres in chemotherapy. III. Effect of microsphere-entrapped 5-fluorouracil on Ehrlich ascites carcinoma in mice. *J. Pharm. Dyn. 2*: 350–355 (1979).
41. Y. Morimoto, M. Akimoto, K. Sugibayashi, T. Nadai, and Y. Kato, Drug-carrier property of albumin microspheres in chemotherapy. IV. Antitumor effect of single-shot or multiple shot administration of microsphere-entrapped 5-fluorouracil on Ehrlich ascites or solid tumor in mice, *Chem. Pharm. Bull. 28*: 3087–3092 (1980).
42. T. Ishizaka, K. Endo, and M. Koishi, Preparation of egg albumin microcapsules and microspheres, *J. Pharm. Sci. 70*:358–363 (1981).

43. W. E. Longo, H. Iwata, T. A. Lindheimer, and E. P. Goldberg, Preparation of hydrophilic albumin microspheres using polymeric dispersing agents, *J. Pharm. Sci.* *71*:1323–1328 (1982).
44. K. J. Widder, A. E. Senyei, and D. G. Scarpelli, Magnetic microspheres: a model system for site specific drug delivery in vivo, *Proc. Soc. Exp. Biol. Med.* *58*:141–146 (1978).
45. K. J. Widder, A. E. Senyei, and D. F. Ranney, Magnetically responsive microspheres and other carriers for the biophysical targeting of antitumor agents, *Adv. Pharmacol. Chemother.* *16*: 213–271 (1979).
46. K. Widder, G. Flouret, and A. Senyei, Magnetic microspheres: synthesis of a novel parenteral drug carrier, *J. Pharm. Sci.* *68*: 79–82 (1979).
47. A. E. Senyei, S. D. Reich, C. Gonczy, and K. J. Widder, In vivo kinetics of magnetically targeted low-dose doxorubicin, *J. Pharm. Sci.* *70*:389–391 (1981).
48. K. J. Widder, A. E. Senyei, H. Ovadia, and P. Y. Paterson, Specific cell binding using staphlococcal protein A magnetic microspheres, *J. Pharm. Sci.* *70*:387–389 (1981).
49. Y. Morimoto, K. Sugibayashi, M. Okumura, and Y. Kato, Biomedical applications of magnetic fluids. I. Magnetic guidance of ferrocolloid entrapped albumin microsphere for site specific drug delivery in vivo, *J. Pharm. Dyn.* *3*:264–267 (1980).
50. Y. Morimoto, M. Okumura, K. Sugibayashi, and Y. Kato, Biomedical applications of magnetic fluids. II. Preparation and magnetic guidance of magnetic albumin microsphere for site specific drug delivery in vivo, *J. Pharm. Dyn.* *4*:624–631 (1981).
51. F. Leonard, R. K. Kulkarni, G. Brandes, J. Nelson, and J. L. Cameron, Synthesis and degradation of poly(alkyl α-cyanoacrylates), *J. Appl. Polym. Sci.* *10*:259–272 (1966).
52. K. C. Pani, G. Gladieux, G. Brandes, R. K. Kulkarni, and F. Leonard, The degradation of *n*-butyl alpha-cyanoacrylate tissue adhesive, II, *Surgery* *63*:481–489 (1968).
53. F. Leonard, Hemostatic applications of alpha cyanoacrylates: bonding mechanism and physiological degradation of bonds, in *Adhesion in Biological Systems* (R. S. Manly, ed.), Academic, New York, 1970, pp. 185–199.
54. A. T. Florence, T. L. Whateley, and D. A. Wood, Potentially biodegradable microcapsules with poly(alkyl 2-cyanoacrylate) membranes, *J. Pharm. Pharmacol.* *31*:422–424 (1979).
55. D. A. Wood, T. L. Whateley, and A. T. Florence, Formation of poly(butyl 2-cyanoacrylate) microcapsules and the microencapsulation of aqueous solution of (^{125}I) labelled proteins, *Int. J. Pharm.* *8*:35–43 (1981).

56. P. Couvreur, B. Kante, and M. Roland, Les perspectives d'utilisation des formes microdispersées comme vecteurs intracellulaires, *Pharm. Acta Helv.* *53*:341–347 (1978).
57. P. Couvreur, B. Kante, M. Roland, P. Guiot, P. Bauduin, and P. Speiser, Polycyanoacrylate nanocapsules as potential lysosomotropic carriers: preparation, morphology and sorptive properties, *J. Pharm. Pharmacol.* *31*:331–332 (1979).
58. P. Couvreur, B. Kante, M. Roland, and P. Speiser, Adsorption of antineoplastic drugs to polyalkylcyanoacrylate nanoparticles and their release in calf serum, *J. Pharm. Sci.* *68*:1521–1524 (1979).
59. P. Couvreur, B. Kante, V. Lenaerts, V. Scailteur, M. Roland, and P. Speiser, Tissue distribution of antitumor drugs associated with polyalkylcyanoacrylate nanoparticles, *J. Pharm. Sci.* *69*:199–202 (1980).
60. B. Kante, P. Couvreur, V. Lenaerts, P. Guiot, M. Roland, P. Baudhuim, and P. Speiser, Tissue distribution of (^{3}H) actinomycin D adsorbed on polybutylcyanoacrylate nanoparticles, *Int. J. Pharm.* *7*:45–53 (1980).
61. B. Kante, P. Couvreur, G. Dubois-Krack, C. De Meester, P. Guiot, M. Roland, M. Mercier, and P. Speiser, Toxicity of polyalkylcyanoacrylate nanoparticles I: Free nanoparticles, *J. Pharm. Sci.* *71*:786–790 (1982).
62. P. Couvreur, B. Kante, L. Grislain, M. Roland, and P. Speiser, Toxicity of polyalkylcyanoacrylate nanoparticles II: Doxorubicin-loaded nanoparticles, *J. Pharm. Sci.* *71*:790–792 (1982).
63. S. C. Khanna and P. Speiser, Epoxy resin beads as a pharmaceutical dosage form I: method of preparation, *J. Pharm. Sci.* *58*:1114–1117 (1969).
64. S. C. Khanna, M. Soliva, and P. Speiser, Epoxy resin beads as a pharmaceutical dosage form II: dissolution studies of epoxy-amine beads and release of drug, *J. Pharm. Sci.* *58*:1385–1388 (1969).
65. M. Unsworth and G. H. Helsby, U.S. Patent 3,928,230 (December 23, 1975).
66. K. R. Sidman, D. L. Arnold, W. D. Steber, L. Nelsen, F. E. Granchelli, P. Strong, and S. G. Sheth, Use of synthetic polypeptides in the preparation of biodegradable delivery vehicles for narcotic antagonists, *Natl. Inst. Drug Abuse Res. Monogr. Ser.* *4*:33–38 (1976).
67. N. Tani, M. Van Dress, and J. M. Anderson, Hydrophilic/hydrophobic control of steroid release from a cortisol-polyglutamic acid sustained release system, in *Controlled Release of Pesticides and Pharmaceuticals* (D. H. Lewis, ed.), Plenum, New York, 1981, pp. 79–98.

11

Ion-Exchange Resins

11.1 INTRODUCTION

Ion-exchange resins contain ionizable groups attached onto an insoluble crosslinked synthetic polymer and can exchange these groups for ions present in the solution with which they are in contact. The most frequently employed polymeric network used is a copolymer of styrene and divinylbenzene that may be produced by a suspension polymerization process in a spherical bead form. Suitable exchangeable cationic or anionic groups that may be subclassified as either strong or weak are substituted on the styrene portion of the copolymer as shown in Fig. 11.1. These insoluble ion-exchange resins may be supplied in the case of cation exchangers as sodium, potassium, or ammonium salts and of anion exchangers usually as the chloride. The corresponding acid or hydrogen form and base or hydroxyl form may be regenerated by treatment with acid or base, and may be reacted with suitable cationic and anionic drugs for the production of sustained release or other type products. The strength and permeability of the final resin is influenced mainly by the degree of crosslinking produced by the divinylbenzene content, which usually varies between 2 and 12% of the copolymer. Weak cation-exchange resins have a pK_a value of about 6, so that at pH 4 or above their exchange capacity tends to increase. Strong cation-exchange resins have a pK_a value of 1 to 2 and so are normally highly dissociated at all pH's encountered in the gastrointestinal tract. Accordingly, their exchange capacity tends to be independent of pH and to exchange one cation for another depending on concentration and affinity, decreasing usually in the order calcium, potassium, ammonium, sodium, and hydrogen. Carboxylic acid-type cation-exchange resins tend to have approximately twice the exchange capacity of the sulfonic acid type.

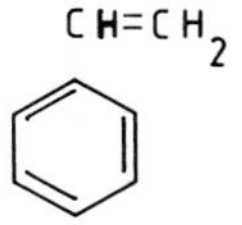

Styrene (vinylbenzene)

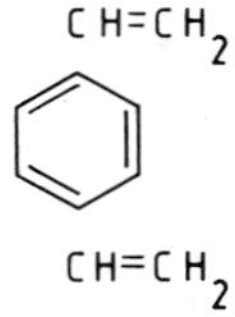

Divinylbenzene

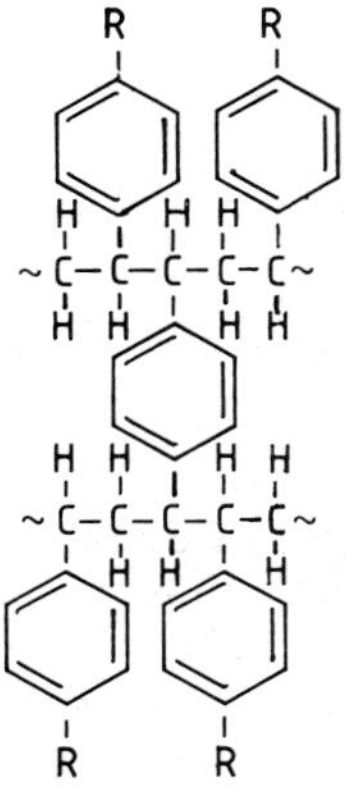

Copolymer

Cation-exchangers

strong: $R = -SO_3^- H^+$ — Sulfonic acid groups

weak: $R = -COO^- H^+$ — Carboxylic acid groups

Anion-exchangers

strong: $R = -CH_2N^+(CH_3)_3\ OH^-$ — Quaternary ammonium groups

weak: $R = -NH_2$ — Primary amino groups

$= \ =NH$ — Secondary amino groups

$= \ =N-$ — Tertiary amino groups

Figure 11.1 Examples of strong and weak cation- and anion-exchange resins based on a copolymer of styrene and divinylbenzene.

Ion-exchange resins will undergo an equilibrium exchange reaction when placed in contact with a solution of anions and cations. The rate of reaction is influenced by the permeability of the solvent and solute through the pores of the resin, whose number and size are influenced by the amount of crosslinking. The diffusion path length is obviously also related to the size of the resin particles. It is the solution that is trapped within the physical structure of the resin that reaches equilibrium and not the external bulk solution, thus complicating the application of normal mass balance equations. Release of ion displaced from the resin will depend on its diffusion rate from the resin.

A strong cation-exchange resin containing substituted sulfonic acid groups could exchange for cations in solution as follows:

$$\text{Resin } SO_3^-A^+ + B^+C^- \leftrightarrow \text{Resin } SO_3^-B^+ + A^+C^-$$

Likewise, a strong anion-exchange resin containing substituted quaternary ammonium groups could exchange for anions in solution as follows:

$$\text{Resin } (N(CH_3)_3^+)X^- + Y^-Z^+ \leftrightarrow \text{Resin } (N(CH_3)_3^+)Y^- + X^-Z^+$$

Equilibrium constants for both reactions could be expressed as

$$k_{\text{cation}} = \frac{[\text{Resin } SO_3^-B^+][A^+C^-]}{[\text{Resin } SO_3^-A^+][B^+C^-]}$$

and

$$k_{\text{anion}} = \frac{[\text{Resin } (N(CH_3)_3^+)Y^-][X^-Z^+]}{[\text{Resin } (N(CH_3)_3^+)X^-][Y^-Z^+]}$$

However, the precise meaning of the activity or concentration of resin [Resin] would be unclear, and so a selectivity coefficient is sometimes used, which for a cation-exchange resin is defined as

$$\underline{k} = \frac{[B^+]_{\text{resin}}[A^+]_{\text{solution}}}{[A^+]_{\text{resin}}[B^+]_{\text{solution}}}$$

A specific ion such as lithium for cation-exchange may be given an arbitrary value of 1 so that other data may be related to its exchange capacity.

Accordingly, it is difficult to model reliably by in vitro experiments appropriate concentrations and types of ions that drug-loaded ion-exchange resins experience in vivo. Incomplete drug release occurs. The nonselective properties of ion-exchange resins mean that they will interact with any food or coadministered drug with an appropriate charge under pH conditions that favor ionization of both species. They may also cause unwanted electrolyte disturbances.

Apart from the styrene/divinylbenzene resins, phenol/formaldehyde copolymers are also used in the production of ion-exchangers with sulfonic acid, phosphoric acid, hydroxyl, or amine as functional groups. Likewise, other polymers, such as those of acrylic and methacrylic acid crosslinked with divinylbenzene and containing appropriate functional groups, have been used as ion-exchange drug carriers.

Charged drugs are normally loaded onto ion-exchange resins by either of two methods. A highly concentrated drug solution, buffered if necessary to enhance dissociation, is passed through a column of resin particles (column procedure). Gyselinck et al. [1] reported that heating the column increased the loading of propranolol onto a strong cation-exchange resin and decreased its release rate. Alternatively, the drug solution is agitated with a quantity of resin particles until equilibrium is established (batch procedure). The reactions involved may be indicated as follows:

Basic drug + acidic ion-exchange resin $\rightleftarrows$ drug resinate

Acidic drug + basic ion-exchange resin $\rightleftarrows$ resin salt

The resin is then washed with deionized water to free it from unreacted drug and other ions prior to drying. Obviously, the process is unsuitable for nonionizable drugs.

Upon ingestion, drugs are most likely eluted from cation-exchange resins by H^+, Na^+, or K^+ ions and from anion-exchange resins by Cl^-, as these ions are most plentiful in gastrointestinal secretions. Typical reactions involved may be envisaged as follows:

In the stomach:

(1) Drug resinate + HCl $\rightleftarrows$ acid resin + drug hydrochloride
(2) Resin salt + HCl $\rightleftarrows$ resin chloride + acidic drug

In the intestine:

(1) Drug resinate + NaCl $\rightleftarrows$ sodium resinate + drug hydrochloride
(2) Resin salt + NaCl $\rightleftarrows$ resin chloride + sodium salt of drug

The drug-depleted resin is not absorbed but is excreted in the feces. Because of the nonbiodegradable nature of most of the resins employed, their use in parenteral dosage forms is unlikely to be medically acceptable.

Ion-exchange resins alone are used clinically in the treatment of various disorders. Cholestyramine is a strong anion-exchange resin supplied as the chloride and containing quaternary ammonium functional groups attached to a styrene 98%:divinylbenzene 2% (approx.) copolymer. It is used orally to bind bile salts, thus reducing cholesterol and lipid absorption. Colestipol HCl, which is the hydrochloride of a copolymer of diethylenetriamine and 1-chloro-2,3-epoxypropane (partly in the acid form), and polidexide, which is a poly[2(diethylamino)ethyl] polyglycerylene dextran polymer, are both anion-exchange resins used medically similarly to cholestyramine. Sodium, calcium, or aluminum polystyrene sulfonate are used in the treatment of hyperkalemia. The ammonium form of the resin has been used to reduce sodium absorption. Weak anion-exchange resins have been used as antacids, but their unpalatability detracts from their usefulness. Ion-exchange resins are also widely used, for example, in the purification of water, the separation of drugs by chromatography, and in their analysis. References 2–6 contain reviews on ion-exchange resins in medicine and pharmacy.

11.2 DRUG RELEASE FROM ION-EXCHANGE RESINS

This section is concerned mainly with the use of ion-exchange resins to form chemical complexes or resinates with drugs to delay absorption from the gastrointestinal tract and so prolong duration of action. These resinates have also been investigated for taste-masking purposes, for reducing nausea associated with the ingestion of certain irritant drugs, and for the improvement of drug stability. Their use in the formulation of controlled-release topical preparations has also been studied [7,8].

Initial interest in ion-exchange resins as drug carriers for sustained release was reported by Chaudhry and Saunders [9]. Ephedrine was found during in vitro studies to be rapidly released in 0.1 *N*-hydrochloric acid solution from carboxylic acid-type cation-exchange resins but not from the sulfonic acid type, which exhibited comparable sustained release over 6 hr when tested also in 0.1 *N* sodium chloride or 0.1 *N* sodium bicarbonate solutions. Similar results were obtained using dexamphetamine sulfate, it being noted that the use of infinite sink conditions gave greater drug release, as might be expected from the equilibrium nature of the reaction. The release rate could be made more zero-order-like by using resins only partly converted to the ephedrine form or by using a mixture of the alkaloid and hydrogen forms of the resin. Subsequently, Bajpai et al. [10] showed that tablets (Asmapax, Nicholas Laboratories, Ltd., England) containing ephedrine hydrochloride combined with a cation-exchange resin and also containing theophylline and phenobarbitone caused a prolonged relief of bronchospasm in 57 patients. Also, Abrahams and Linnell [11] reported that a clinical trial of tablets containing dexamphetamine resinate indicated that a dose of 1 tablet was usually effective in suppressing appetite for 12 hr, producing an average weight loss in 53 patients of 6 lb in 4 weeks. In comparison to unbound drug, the resinate produced fewer side-effects and of lower severity.

Brudney [12] reviewed the use of ion-exchange resins to form drug complexes with a number of drugs including vitamin B_2, vitamin B_{12}, and phenoxymethylpenicillin. Vitamin B_2 was effectively taste-masked by forming a resin complex. Vitamin B_{12} was reported to be considerably more stable as the resinate than as the free drug. Phenoxymethylpenicillin in a highly crosslinked resin was found to release the drug more rapidly in the initial stages of dissolution studies using dilute hydrochloric acid and more slowly in the latter stages compared with a resin of lower crosslinkage.

Smith et al. [13] investigated the use of various sulfonic acid and carboxylic acid cation-exchange resins for the preparation of a sustained-release liquid preparation containing the antihistamine methapyrilene. The sulfonic acid-type resin gave a much greater prolonged release of drug than the carboxylic acid type in dilute hydrochloric acid, and was studied further. Sorption of drug and its subsequent in vitro release decreased with an increase in the degree of crosslinking in the resin. Particle size had little effect on the ultimate amount of drug bound by the resin but tended to affect drug release inversely. However, in vivo studies in guinea pigs failed to show a significant extended pharmacological response of the drug in its resinate form. Hirscher and Miller [14] obtained similar in vitro results when investigating a range of cation-exchange resins with the drugs amphetamine sulfate, atropine sulfate, and thiopropazate hydrochloride. Likewise various cation-exchange resins based on polystyrene, polyacrylic acid, and polymethacrylic acid polymers were investigated by Schlichting [15] to produce a sustained-release form of the antihistamine drug carbinoxamine for oral use. Carboxylic acid resins having a pK_a of 5.2 or greater tended to have too rapid a release in simulated gastric fluid to be of use as a sustained-release product. Resins having a pK_a less than 5.2 gave sustained release of the drug during dissolution studies in simulated gastric fluid followed by simulated intestinal fluid. Increase in crosslinkage of the polymer network decreased both drug binding and its subsequent release. Decreasing the size of the resin bead employed tended to increase in vitro drug release only slightly. The sulfonic acid resins of polystyrene/divinylbenzene were considered to be very effective for sustained-release purposes, though an in vivo study in a human subject showed evidence of incomplete drug absorption from this type of product.

More recently Hinsvark et al. [16] determined the plasma levels of amphetamine and phentermine in humans in a crossover study involving oral administration of salt and resin-bound formulations (Biphetamine® and Ionamin®, respectively). The bioavailability of either drug was not affected by formulation as resinate, though the rate and variability of absorption was lower.

11.3 DRUG RELEASE FROM COATED ION-EXCHANGE RESINS

A lot of interest in ion-exchange resins has been centered around their use as drug carriers to form cores that are then coated. Though the process would be more expensive than coating the drug particles directly, it has the advantage that the controlled-release properties of the core can be combined with those of the coating. Also, it would provide spherical core material of narrow particle size range, which would aid application of coating and reproducibility of drug release. However, as ion-exchange resins tend to absorb water from the atmosphere or upon immersion into aqueous media, which produces swelling, undesirable premature rupture of coating by the expanding core can be a troublesome problem. In fact, ion-exchange resins have been investigated by Van Abbé and Rees [17] and by others as tablet disintegrants.

A modification of the coating approach was first reported by Nash and Crabtree [18], who enteric-coated macrocapsule shells with cellulose acetate phthalate, having previously filled them with tritium-labeled methylamphetamine bound onto either carboxylic acid or sulfonic acid types of cation-exchange resins. Upon administration to dogs, the enteric-coated shells delayed drug absoprtion by 3 to 4 hr. Greater reduction in absorption was observed with the sulfonic acid type of drug resinate.

Phares and Sperandio [19] successfully encapsulated a nondrug-loaded micronized ion-exchange resin in gelatin using a simple coacervation procedure as previously described in Chap. 3. The schistosomicide, lucanthone hydrochloride, has a high incidence of producing adverse side effects. Hussein et al. [20] combined this drug with the ion-exchange resin Nalcite-HCR in micronized form and then coated it with Pharmagel A or cellulose acetate phthalate-gelatin coacervation systems to further reduce release in simulated gastric fluid over 2 hr and so decrease side effects. The encapsulated resins were found to tablet better. No in vivo studies were presented.

Borodkin and Yunker [21] showed that maximum interaction of amine drugs such as methapyrilene hydrochloride, pseudoephedrine hydrochloride, ephedrine base, and dextromethorphan hydrobromide with a weak cation-exchange resin having a polymethacrylic/divinylbenzene matrix occurred at pH 4.5 to 5.5. The selectivity coefficient of drugs with tertiary amine groups was much higher than those with primary, secondary, or quaternary amine groups. In a subsequent paper by Borodkin and Sundberg [22], the above-mentioned four bitter drugs were combined with a polycarboxylic acid ion-exchange resin and

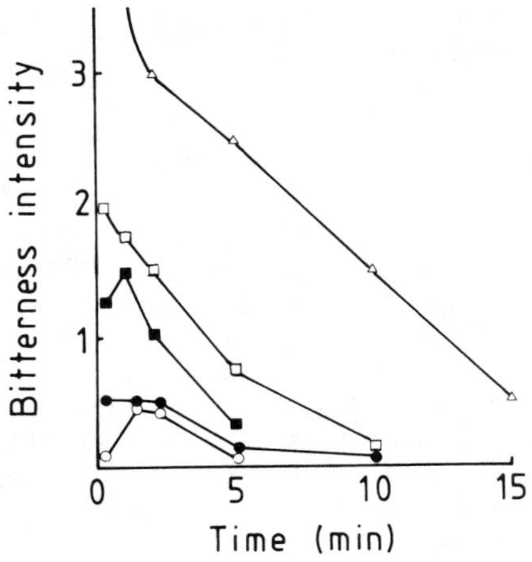

Figure 11.2 Bitterness intensity curves for various ephedrine dosages. (Reproduced with permission of the copyright owner, American Pharmaceutical Association, from Ref. 22.)

coated for taste-masking purposes with a 4:1 mixture of ethylcellulose: hydroxypropylmethylcellulose using an air suspension coater (see Chap. 7). Products with about 25% coating exhibited marked reduction in bitterness compared to the drug resinates, as shown in Fig. 11.2. Upon compression of coated powder into chewable tablets, bitterness was sometimes increased, presumably due to rupture of coating during compression or chewing. Complete availability and sustained release of pseudoephedrine from coated resin particles in tablet form was demonstrated by means of a urinary excretion study in 8 human subjects.

A number of patents involving coating of either ion-exchange resins or drug-loaded resins have been reported. For example, Macek et al. [23], in a patent assigned to Merck & Co., Inc., coated anion-exchange resin particles with an acrylic polymer crosslinked with allyl sucrose to reduce the astringent and gritty taste. Koff [24], in a patent assigned to Hoffmann-La Roche, Inc., described the use of castor wax to improve the palatability of cation-exchange resins loaded with amprotropine. Clarke [25], in a patent assigned to N. V. Philips, applied cellulose acetate phthalate as an enteric coating onto 0.1–1 mm diameter drug-loaded cation-exchange resin beads using a coating pan.

Geneidi and Hamacher [26] reported how in vitro release of chlorpheniramine ions varied from a strong sulfonic acid-type cation-exchanger prepared containing different percentages of crosslinking agent from 4 to 30% in the polystyrene/divinylbenzene matrix and different solvent concentrations (60 and 100%). The highly expanded 4/60 macroporous beads exhibited no sustained release, unlike the less expanded 10/60, 15/100, and 30/100 beads. They could, however, be made to exhibit sustained-release properties by wax coating.

Motycka and Nairn [27] reported the influence of wax coating on the release rate of the benzoate anion from a strongly basic ion-exchange resin. The resin beads were coated with beeswax, polyethylene glycol 6000, carnauba wax, paraffin wax, stearic acid, or stearyl alcohol by agitation in molten material at elevated temperature, the excess being removed by filtration prior to hardening the coating by cooling. The product was gently separated into bead form by mechanical fragmentation. Large differences in the wax-to-resin ratios for different waxes were obtained, even though the same procedure was used. Increasing hydrophobicity of the waxy material used tended to progressively delay drug release, carnauba wax and paraffin wax being most effective. Later the same workers [28] used several other microencapsulation techniques to coat the benzoate-loaded resin beads. Gelatin-acacia complex coacervation with crosslinking by formaldehyde (see Chap. 3), variable-ethylcellulose-grade encapsulation by phase separation induced by temperature change with or without low-molecular-weight polyethylene, wax treatment or plasticizer (see Chap. 4), and cellulose acetate butyrate encapsulation by phase separation using a nonsolvent addition process (see Chap. 4) were tried. In vitro dissolution studies in a phosphate buffer pH 9.2 showed that all coatings decreased the rate of the anion release from the resin complex, the greatest reduction being observed with the ethylcellulose standard 100 cP plus polyethylene and paraffin coating. Suitable choice of coating content allowed benzoate release rate and extent from the resinate to be varied over a wide range. However, recently Raghunathan et al. [29] reported that direct application of ethylcellulose-vegetable oil coating by an air suspension technique on drug resinate particles, usually composed of phenylpropanolamine bound to sulfonic acid cation-exchange resin, was unsuccessful in further prolonging drug release, as the coating tended to rupture in the 0.1 *N* hydrochloric acid medium due to the swell of the core. Pretreatment of the core particles with an agent such as polyethylene glycol 4000 overcame this problem by filling the resin-drug complex matrix, so retarding water uptake and maintaining coat integrity during release studies.

11.4 MISCELLANEOUS

A concept of forming acid-base complexes with other macromolecules for controlled drug release and other purposes must be mentioned. For example, Spross et al. [30] formed drug complexes with the sulfoethyl derivative and sulfate ester of Sephadex (crosslinked dextran) to improve the palatability of a number of drugs without altering bioavailability. O'Brien and Roe [31] in a patent assigned to the National Cash Registrar Co., combined dextromethorphan or spiramycin with cellulose acetate phthalate for taste-masking purposes. Okano et al. [32] combined quinidine sulfate with acid polysaccharides such as the sodium salt of dextran sulfate, sodium carboxymethylcellulose, or sodium alginate to reduce intestinal drug absorption in rats. Also, Kojima et al. [33] found that, compared to the parent drug, a mitomycin C-dextran conjugate was more effective in reducing the toxicity and extending the survival of mice bearing experimental tumors. DEAE cellulose microspheres, which are a basic ion-exchange material available commercially in the size range 40 to 160 μm have been investigated for parenteral sustained delivery of drugs by Illum and Davis [34].

REFERENCES

1. P. Gyselinck, R. Vanseveren, P. Braeckman, and E. Schacht, Drug-polymer combinations I. The preparation of sustained release drugs by combination with ion exchanging resins, *Pharmazie 36*:769–772 (1981).
2. L. Saunders, Sustained release of drugs from ion exchange resins, *J. Mond. Pharm. 4*:36–43 (1961).
3. D. Jack, Prolongation of drug action, *Pharm. J. 134*:581–586 (1962).
4. S. Eriksen, Sustained action dosage forms, in *The Theory and Practice of Industrial Pharmacy* (L. Lachman, H. A. Lieberman, and J. L. Kanig, eds.), Lea & Febiger, Philadelphia, 1970, pp. 408–436.
5. Martindale, *The Extra Pharmacopoeia*, 28th ed., Pharmaceutical Press, London, 1982, pp. 411–414.
6. A. T. Florence and D. Attwood, *Physicochemical Principles of Pharmacy*, Macmillan, London, 1981, pp. 297–300, 414–415.
7. W. C. Fiedler and G. J. Sperandio, The formulation of ointments containing medication adsorbed on ion exchange resins, *J. Amer. Pharm. Assoc. Sci. Ed. 46*:44–47 (1957).
8. J. B. Ward and G. J. Sperandio, Formulation of ion exchange adsorbates in topical products, *Amer. Perfum. Cosmet. 81*:23–27 (1966).

9. N. C. Chaudhry and L. Saunders, Sustained release of drugs from ion exchange resins, *J. Pharm. Pharmacol.* *8*:975–986 (1956).
10. H. S. Bajpai, J. P. Gupta, and R. C. Gupta, Bronchial asthma: therapeutic evaluation of ephedrine resinate, *Clin. Med.* *76*:29–31 (1969).
11. A. Abrahams, and W. H. Linnell, Oral depot therapy with a new long-acting dexamphetamine salt, *Lancet* *273*(2):1317–1318 (1957).
12. N. Brudney, Ion-exchange resin complexes in oral therapy, *Can. Pharm. J.* *45*:245–248 (1959).
13. H. A. Smith, R. V. Evanson, and G. J. Sperandio, The development of a liquid antihistaminic preparation with sustained release properties, *J. Amer. Pharm. Assoc. Sci. Ed.* *49*:94–97 (1960).
14. D. A. Hirscher and O. H. Miller, Drug release from cation exchange resins, *J. Amer. Pharm. Assoc.* *NS2*:105–108 (1962).
15. D. A. Schlichting, Ion exchange resin salts for oral therapy I. carbinoxamine, *J. Pharm. Sci.* *51*:134–136 (1962).
16. O. N. Hinsvark, A. P. Truant, D. J. Jenden, and J. A. Steinborn, The oral bioavailability and pharmacokinetics of soluble and resin-bound forms of amphetamine and phentermine in man, *J. Pharmacokinet. Biopharm.* *1*:319–328 (1973).
17. N. J. Van Abbé and J. T. Rees, Amberlite resin XE-88 as a tablet disintegrant, *J. Amer. Pharm. Assoc. Sci. Ed.* *47*:487–489 (1958).
18. J. F. Nash and R. E. Crabtree, Absorption of tritiated d-desoxyephedrine in sustained-release dosage forms, *J. Pharm. Sci.* *50*:134–137 (1961).
19. R. E. Phares and G. J. Sperandio, Coating pharmaceuticals by coacervation, *J. Pharm. Sci.* *53*:515–518 (1964).
20. A. M. Hussein, A. A. Kassem, A. Sina, and A. A. Badawy, Release of lucanthone from its encapsulated nalcite-HCR resinate, *U.A.R. J. Pharm. Sci.* *11*:1–8 (1970).
21. S. Borodkin and M. H. Yunker, Interaction of amine drugs with a polycarboxylic acid ion-exchange resin, *J. Pharm. Sci.* *59*:481–486 (1970).
22. S. Borodkin and D. P. Sundberg, Polycarboxylic acid ion-exchange resin adsorbates for taste coverage in chewable tablets, *J. Pharm. Sci.* *60*:1523–1527 (1971).
23. T. J. Macek, C. E. Shoop, and D. R. Stauffer, U.S. Patent 3,499,960 (March 10, 1070).
24. A. Koff, U.S. Patent 3,138,525 (June 23, 1964).
25. C. A. Clarke, British Patent 1,218,102 (January 6, 1971).
26. A. S. Geneidi and H. Hamacher, Sustained release of chlorphenamine from macroporous KY-23 ion exchange resin beads, in

F.I.P. Abstracts, 39th International Congress of Pharmaceutical Sciences, Brighton, September, 1979, p. 72.

27. S. Motycka and J. G. Nairn, Influence of wax coatings on release rate of anions from ion-exchange resins beads, *J. Pharm. Sci. 67*:500–503 (1978).
28. S. Motycka and J. G. Nairn, Preparation and evaluation of microencapsulated ion-exchange resin beads, *J. Pharm. Sci. 68*:211–215 (1979).
29. Y. Raghunathan, L. Amsel, O. Hinsvark, and W. Bryant, Sustained release drug delivery system I: coated ion-exchange resin system for phenylpropanolamine and other drugs, *J. Pharm. Sci. 70*:379–384 (1981).
30. B. Spross, M. Ryde, and B. Nystrom, Sephadex cation exchangers as carriers of drug components for improvement of their palatability, *Acta Pharm.Suec. 2*:1–12 (1965).
31. P. O'Brien and H. L. Roe, U.S. Patent 3,242,049 (March 22, 1966).
32. T. Okano, Y. Watanabe, and K. Yoshimura, Effect of acid polysaccharides on the intestinal absorption and blood level patterns of quinidine in rats and its mechanism from the point of intermolecular interaction, *Yakugaku Zasshi 97*:1359–1365 (1977).
33. T. Kojima, M. Hashida, S. Muranishi, and H. Sezaki, Mitomycin C-dextran conjugate: a novel high molecular weight pro-drug of mitomycin C, *J. Pharm. Pharmacol. 32*:30–34 (1980).
34. L. Illum and S. S. Davis, Cellulose microspheres as a sustained release system for parenteral administration, *Int. J. Pharm. 11*: 323–327 (1982).

12

Congealable Disperse-Phase Encapsulation Procedures

12.1 INTRODUCTION

A simple type of encapsulation procedure involves dispersing fine particles of drug at high temperature in a hydrophilic or hydrophobic liquid vehicle that will solidify when cooled to normal ambient temperature. Alternatively, the drug may be dissolved in the liquid vehicle. Suitable hydrophilic vehicles include gelatin, agar, and starch. Suitable hydrophobic vehicles include various waxes such as Japanese, Glycowax, and beeswax, and hardened oils and fats such as hydrogenated castor oil and hydrogenated beef tallow. Obviously, candidate drugs for the process must be adequately heat-stable under the conditions of manufacture.

It may be necessary to employ a wetting agent to promote coverage of dispersed core by the coating material. The microcapsules formed often tend to have an irregular thickness of coating, which may affect their release properties. When solutions of drugs in the coating material are used, the microparticles formed act as hydrophilic or hydrophobic matrices.

12.2 EXTRUSION DEVICES

In order to form these spherical particles, extrusion through a capillary device or emulsification procedures may be employed. Spray congealing may also be used, applications of which have been previously discussed in Chap. 8.

Various capillary extrusion devices have been described in the literature. For example, Fisher and Wilson [1] described an apparatus for forming spherical drug containing beads from the hot feed material by pneumatic compression through a series of needles. The droplets formed were hardened by cooling in a transversely moving stream of immiscible liquid. The variables that altered particle size included orifice diameter, viscosity and delivery rate of feed material, and velocity of the coolant liquid. Madan et al. [2] formed spherical monodisperse stearyl alcohol particles up to 200 μm in diameter using a vibrating reed dipped into a receiving liquid similar in design to the apparatus described by Wolf [3]. To form larger particles a vibrating capillary apparatus was used based on the design of Mason et al. [4]. Figure 12.1 shows such a type of apparatus constructed in the author's own laboratory. It consists of a hypodermic needle (A), which is attached to the diaphragm of a speaker (B). Vibration of the diaphragm and needle is caused by a function generator/amplifier unit (C). The heated feed material is pneumatically delivered to the vibrating needle from a reservoir (D). The vibrating needle causes the filament issuing from the needle to divide into very uniform droplets, which may be hardened by passage through cold air or by immersion into a coolant liquid. Again, control of droplet size may be achieved by variation in orifice diameter, vibration frequency and amplitude, and viscosity and delivery rate of the feed material. The apparatus was capable of producing very uniformly sized spherical particles. A more elaborate but similar type of apparatus for producing radioisotope microspheres and employing an oscillating needle driven by an electromagnetic transducer has been described by Lysher [5] in a patent assigned to McDonnell Douglas Corp. Another apparatus was shown in a paper by Nolen and Kool [6].

A simple apparatus whereby the feed material drips out of a reservoir through a fine capillary using positive pressure if necessary is suitable for the formation of larger microcapsules or microparticles. Shovers and Sandine [7], in a patent assigned to Pfizer, Inc., used such a procedure for the encapsulation of the enzyme diacetyl reductase and its cofactor reduced nicotinamide adenine dinucleotide in either gelatin or calcium alginate. Likewise, Madan and Shanbhag [8] described such equipment for the encapsulation of sodium salicylate in cellulose acetate phthalate.

The various capillary extrusion devices may be used to form droplets containing drug and an aqueous solution of sodium alginate at room temperature. Coagulation is accomplished chemically by allowing the droplets to fall into an agitated dilute aqueous solution of calcium chloride, which causes the immediate formation of water-insoluble calcium alginate.

Details of various emulsification procedures used will be mentioned in the following sections of the chapter. Where a product is formed in a manufacturing liquid, it usually must be separated by filtration or centrifugation, washed free of the residual liquid if necessary, and dried by conventional means before use.

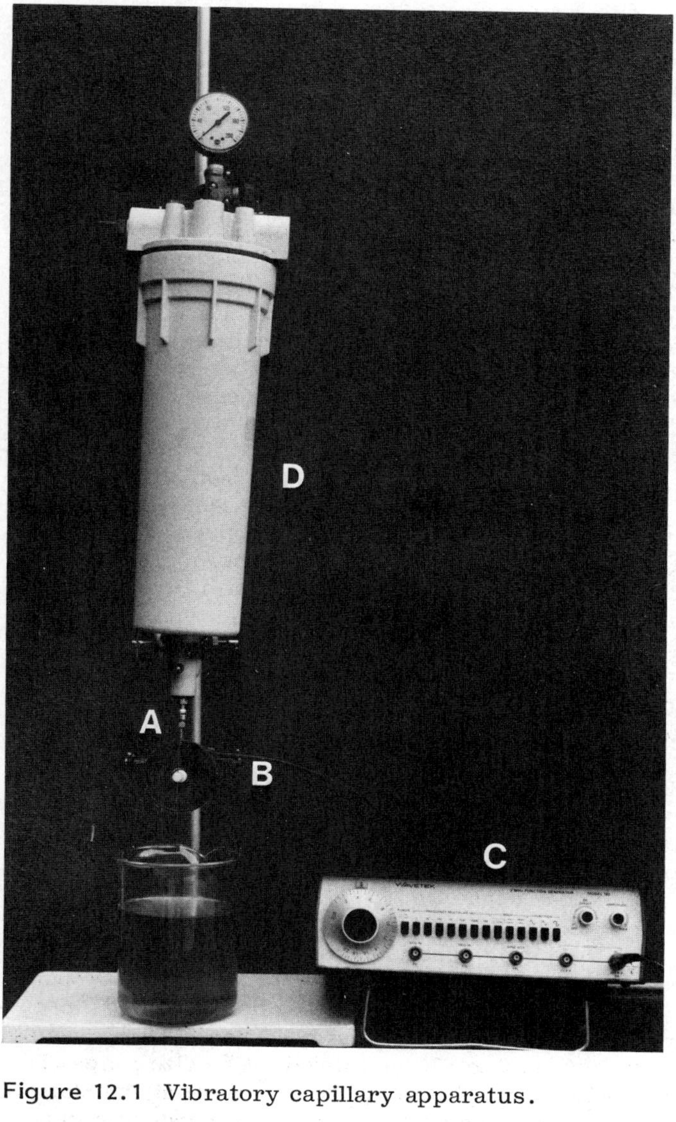

Figure 12.1 Vibratory capillary apparatus.

12.3 HYDROPHILIC CONGEALABLE SYSTEMS

12.3.1 Gelatin

Probably the most frequently employed hydrophilic colloid used for this type of encapsulation is gelatin, which in adequate concentration in water undergoes sol-gel transformation at the convenient temperature range of 30 to 40°C. Thus Milosovich [9], in a patent assigned to Parke Davis & Co., used a capillary extrusion procedure into a coolant nonsolvent liquid to encapsulate a heated (70°C) aqueous solution of the antihistamine drug, diphenhydramine hydrochloride, in gelatin and glycerin. The 590 to 840 μm beads or microparticles obtained were then coated with ethylcellulose films of varying thickness applied from a chloroform and ethanol solvent using a Wurster coater so as to reduce the surface tackiness and to further control their release properties. Fisher and Wilson [1] encapsulated the same drug in glycogelatin base using capillary extrusion and petroleum ether at 10°C as the coolant liquid. Likewise, Madan et al. [10] used a capillary method to encapsulate an aqueous solution of sodium salicylate in 30% gelatin at 50°C using mineral oil at 5°C as the liquid coolant. The hardened droplets were removed and washed with acetone to remove adhering oil. Some of the microcapsules were crosslinked by treatment with formaldehyde solution prior to air drying. The product was free-flowing and showed no tendency to agglomerate. When low concentrations of gelatin were used, the product tended to become tacky. In vitro dissolution studies showed that the formalized product took about 2 hr for complete drug release, the nonformalized about 20 min, and an unencapsulated control less than 2 min.

Various emulsification procedures have also been used to form discrete particles of gelatin-encapsulated product. Thus Tanaka et al. [11] dissolved vitamin B_2 or suspended fine particles of sulfanilamide in 23% w/w (approx.) heated aqueous gelatin solution and emulsified it using a stirrer into heated mineral oil at 50 to 60°C. In view of the comments in Chap. 1 on its selective absorption, vitamin B_2 was a poor choice as a model drug requiring sustained-release formulation. The emulsion formed was cooled to 5°C to harden the droplets. Isopropanol was added to dehydrate the droplets prior to isolation, washing, and drying. The micropellets obtained were then hardened by immersion in 10% formalin-isopropanol at 2 to 5°C and dried. The size range of the product was relatively large (0.3 to 0.5 mm) because of the slow stirrer speed employed. Lengthening the duration of hardening gave the product increasing resistance to hydrolysis by protease in simulated gastric juice and gave increased sustained-release effect when tested in dogs and humans. Paridissis and Parrott [12] used a similar procedure to encapsulate dispersed particles of aspirin, carbon, D&C yellow 11 and sulfadiazine, and solutions of amaranth, sulfadiazine sodium, and mepyramine maleate. Based on urinary recovery data in humans, the aspirin- and sulfadiazine-formalized products were shown to have sustained-release properties. In a series of papers by Hashida and co-workers [13–16] various anticancer agents such as 5-fluorouracil and bleomycin or tracer compounds such as ^{131}I-labeled sodium *o*-iodohippurate were usually dissolved in 20% w/v gelatin solution and emulsified into sesame oil containing a nonionic emulgent at about 50°C. An ultrasonic vibrator was used to promote the formation of the w/o emulsion. The average diameter of the gelatin microspheres formed after cooling was 1 to 2 μm. No attempt was made to separate the spherical microparticles from the oil phase. These emulsions were shown to be particularly effective upon injection at delivering the various anticancer agents into regions of the lymphatic system of animals to reduce lymphatic dissemination of tumor cells causing metastasis. Also, the drug containing gelatin microspheres in the emulsion had reduced cytotoxicity to normal tissue.

Spherical gelatin nanoparticles by an emulsification-congealable procedure have also been prepared by Yoshioka et al. [17]. The anticancer agent mitomycin C was rapidly liberated from the nanospheres during in vitro release studies, though not from larger microspheres. Release was markedly slowed if the drug was initially conjugated with dextran. Upon intravenous injection, nanospheres containing the drug-dextran complex were selectively accumulated in the liver and spleen, whereas larger particles tended to lodge in the lungs. Upon intraperitoneal injection of the encapsulated forms, the survival time of BDF_1 mice bearing P388 leukemia was not extended over that achieveable with an aqueous solution of the drug, though such dosage forms should be advantageous for treating tumors localized in the liver, spleen, or lungs.

12.3.2 Agar and Agarose

Agar is prepared from various species of *Gelidium* and other red algae that belong to the Rhodophyceae. It may be separated into neutral agarose and acidic agaropectin. Both agar and agarose have been used for encapsulation purposes, the latter having a very low sulfate content, superior optical clarity, increased gel strength, but is considerably more expensive. Both materials dissolve as colloidal sols in water if heated to greater than 90°C and form gels when used in adequate concentration upon cooling to about 40°C that remelt on being heated to about 85°C.

Nakano et al. [18] prepared spherical agar or agarose beads containing a dispersion of sulfamethizole particles by adding the drug to the polymer sol at about 70°C. This system was extruded as droplets using a plastic syringe into various cooled organic solvents, of which ethyl acetate was considered to be the most suitable for promoting solidification of the product as spheres. The apparatus is shown in

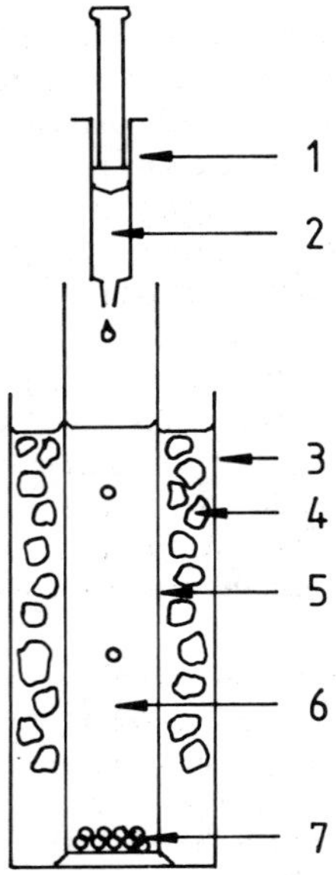

Figure 12.2 Schematic drawing of the apparatus used for preparation of agar beads. 1, syringe; 2, hot agar sol with suspended drug particles; 3, wide-mouthed bottle; 4, ice water; 5, graduated cylinder; 6, cold ethyl acetate; 7, agar beads. (From Ref. 18.)

Fig. 12.2. The beads formed were separated from the coolant liquid by filtration and dried. They had a diameter of about 3mm. In vitro dissolution testing showed that encapsulation delayed drug release, but there was little difference between the effect of agar and agarose. In a subsequent paper, Nakano et al. [19] showed that the sulfamethizole-containing agar beads had a sustained-release effect when administered to humans. However, larger beads showed a reduced bioavailability, presumably due to incomplete absorption of contained drug.

In two papers by Hashida et al. [20] and Kojima et al. [21], the anticancer agent mitomycin C was covalently linked through its amino groups to preprepared agarose spheres using the cyanogen bromide method. In vitro dissolution studies showed that about 50% of the drug was liberated from the agarose bead conjugate in about 6 days. In vivo studies using urinary excretion data following subcutaneous injection in rats and intraperitoneal injection in mice showed that the conjugated form of the drug had a marked sustained-release effect in comparison to the free form. The conjugate exhibited almost identical suppression of the growth of locally implanted Ehrlich ascites carcinoma cells in mice as free mitomycin C, but its toxicity was considerably reduced. However, the activity of the conjugate was less against BDF_1 mouse-transplanted L1210 leukemia, presumably due to more complete hepatic inactivation of the locally and slowly liberated drug, prior to its entry into the systemic circulation to affect the growth of this more widely disseminated tumor.

12.3.3 Starch

Various starches are composed of different ratios of amylose, which is a linear polysaccharide, and amylopectin, which has a branched polysaccharide structure. Wurzburg et al. [22], in a patent assigned to National Starch and Chemical Corp., used either separated amylose or amylose-rich starch that could be partially hydrolyzed or derivatized if necessary to reduce the temperature needed for encapsulation. As an example of the process, a solubilized vitamin A palmitate solution was dispersed in a 30% aqueous gelatinized amylose-rich corn starch and emulsified into corn oil at 25°C using suitable agitation. Upon cooling to 15°C, the amylose-containing disperse phase hardened around its encapsulated vitamin A-containing droplets to allow its isolation as spherical beadlets about 1 mm in diameter by flitration, washing with hexane if required, and drying. The procedure improved the thermal and oxygen stability of the encapsulated vitamin.

12.3.4 Alginates

The chemical reaction between water-soluble sodium alginate and calcium chloride to form water-insoluble alginate can be utilized for encapsulation. Thus Shovers and Sandine [7] mixed brewers' yeast and diacetyl reductase at 5°C in a dilute 0.66% w/v aqueous solution of sodium alginate. The mixture was then extruded dropwise with agitation into an excess of 2.5% w/v aqueous calcium chloride solution. The spherical microparticles formed were filtered and dried.

Recently Lim and Moss [23] microencapsulated viable mammalian cells or tissues by suspending them in 0.6% sodium alginate-saline solution that was gelled by extrusion as droplets into 1.5% calcium chloride solution. A permanent semipermeable coating was then formed on the surface of these harvested temporary microcapsules by suspending them in a solution of 0.02% polylysine for 3 to 5 min. The calcium alginate gel inside the resultant microcapsules was then "liquefied" by immersing them in isotonic sodium citrate solution at pH 7.2 for 5 min, wherein Ca^{2+} and Na^+ ion exchange occurred. The encapsulation procedure successfully maintained the viability of hepatoma cells or pancreatic islets, allowing superior proliferation within the microcapsules in tissue culture experiments (see Fig. 12.3).

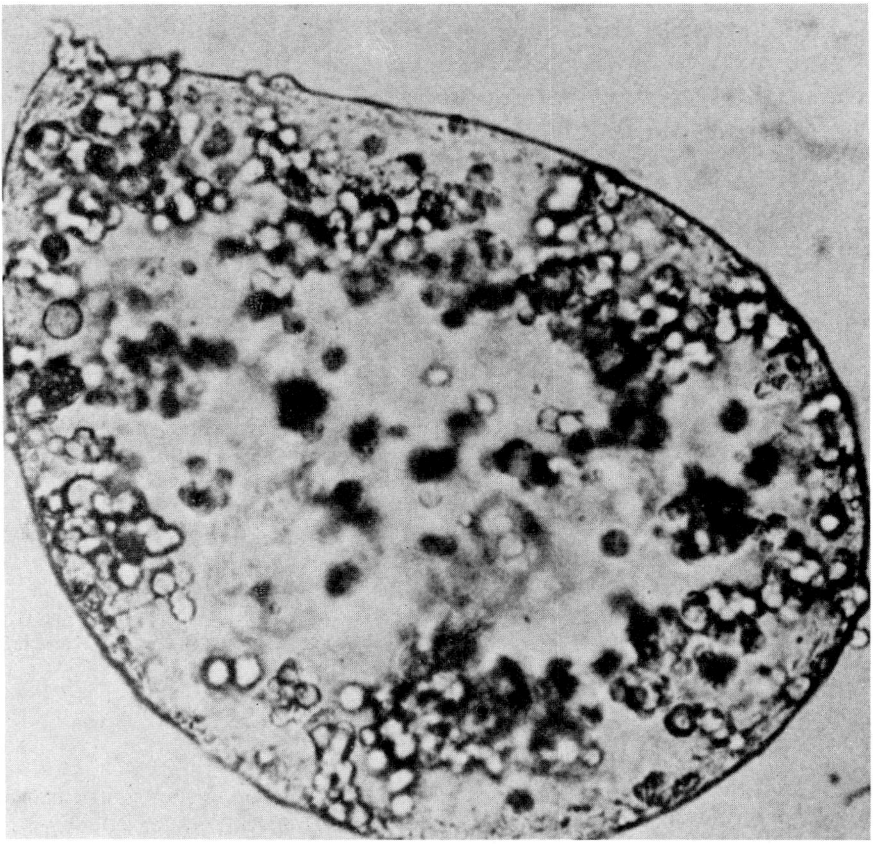

Figure 12.3 Hepatoma cells growing inside a microcapsule. (Reproduced with permission of the copyright owner, American Pharmaceutical Association, from Ref. 23 and courtesy of Dr. F. Lim.)

12.3.5. Polyethylene Glycol

High-molecular-weight polyethylene glycols are solid at room temperature. Thus El-Egakey et al. [24] added methylene blue, phenazone, or vitamin C to molten polyethylene glycol 6000 (melting point 60 to 61°C) and emulsified it into liquid paraffin at 70°C using mechanical agitation and either Span 40 or Span 60 as w/o emulgent. Cold petroleum ether was added to solidify the droplets, which were then filtered off, washed with further petroleum ether, and air-dried. Particle size of the product was influenced by the stirring speed, the concentration of emulgent, and the core:coat ratio used.

Fisher and Wilson [1] also recommended the use of polyethylene glycol for the encapsulation of pharmaceuticals using a capillary extrusion procedure.

12.4 HYDROPHOBIC CONGEALABLE SYSTEMS—WAXES, FATS, AND OILS

Various waxes have been used to encapsulate drugs using a meltable disperse-phase procedure. Many of these systems have been described under the process of spray congealing in Chap. 8. It may be necessary to add a suitable surfactant to cause the hydrophobic coating to wet the core material effectively.

Kowarski et al. [25] dispersed 1 part of fine particles (8 to 10 μm) of sulfamethazine in 2 parts of molten Japanese synthetic wax at elevated temperature. The mixture was slowly poured into cold water at 5°C, the water being continuously agitated. The microcapsules formed were filtered off and dried at room temperature to form a free-flowing powder whose particles passed through a 40-mesh sieve. The coated drug was shown to have a sustained-release effect when tested orally in rabbits. In retrospect, it may be stated that this animal model selected for the in vivo studies was unwise, as Chiou et al. [26] have shown that the rabbit stomach was always full of solid, potentially interactant material and that gastric clearance time is very prolonged, particularly if fasted. Perhaps not suprisingly, incomplete drug absorption was reported by Kowarski et al., which absorption was improved by adding a wetting agent to the oral dose given.

Draper and Becker [27] dispersed 1 part of sulfaethylthiadiazole in 3 parts of molten bleached beeswax or Glycowax S-932 and emulsified it into hot water containing suitable emulgents. While maintaining agitation the temperature of the o/w emulsion was gradually lowered to room temperature and the drug-wax particles obtained were filtered off the aqueous phase, washed with water, and air dried. If polyvinyl alcohol was used to increase the viscosity of the continuous phase of the emulsion, finer microcapsules of wider size distribution tended to be produced. Most of the larger particles were aggregates of smaller ones. A selected product was shown to have sustained-release properties when tested in humans.

Sullivan and Kalkwarf [28] dispersed the narcotic antagonist, naltrexone, in molten combinations of α-glyceryl monopalmitate, 1,2- and 1,3-glyceryl dipalmitate, and glyceryl tripalmitate at 80°C, and rapidly cooled the systems in cylindrical molds until they solidified. The solid dispersions were then ground and sieved to obtain particles in two size ranges, 125 to 250 μm and <125 μm, prior to suspension in 1.5% methylcellulose for injection. In vivo testing in mice showed that the larger particles were more effective in prolonging the duration of action of the drug, and there was no evidence of tissue damage at the site of injection.

The rate of glyceride biodegradation and drug release could be increased by raising the proportion of lower-molecular-weight glyceride used.

A large number of encapsulation procedures involving a meltable dispersion followed by a cooling process were listed by Kondo [29]. Suitable wall materials for drugs included paraffin wax, glyceryl tristearate, and Japanese wax.

Yazawa et al. [30], in a patent assigned to Fuji Photo Film Co., Ltd., described how amino acids or polypeptides could be dispersed in a molten mixture of a fat and/or oil and hardened as microcapsules by being extruded as droplets into cold water. The oil was used to reduce the mechanical strength of the microcapsules. As an example of the procedure, DL-methionine was encapsulated in a mixture of equal parts of hydrogenated beef tallow and soybean oil using a surfactant or thickening agent in the cold water to reduce aggregation of the product. The microcapsules were approximately 3 mm in diameter and exhibited a sustained release of drug during dissolution studies.

REFERENCES

1. C. E. Fisher and C. H. Wilson, U.S. Patent 3,092,553 (June 4, 1963).
2. P. L. Madan, L. A. Luzzi, and J. C. Price, Factors influencing microencapsulation of a waxy solid by complex coacervation, *J. Pharm. Sci. 61*:1586–1588 (1972).
3. W. R. Wolf, Study of the vibrating reed in the production of small droplets and solid particles of uniform size, *Rev. Sci. Instrum. 32*:1124–1129 (1961).
4. B. J. Mason, O. W. Jayaratne, and J. D. Woods, An improved vibrating capillary device for producing uniform water droplets of 15 to 500 μm radius, *J. Sci. Instrum. 40*:247–249 (1963).
5. W. M. Lysher, U.S. Patent 3,933,955 (January 20, 1976).
6. R. L. Nolen and L. B. Kool, Microencapsulation and fabrication of fuel pellets for inertial confinement fusion, *J. Pharm. Sci. 70*:364–367 (1981).
7. J. Shovers and W. E. Sandine, U.S. Patent 3,733,205 (May 15, 1973).
8. P. L. Madan and S. R. Shanbhag, Cellulose acetate phthalate microcapsules: method of preparation, *J. Pharm. Pharmacol. 30*:65–67 (1978).
9. G. Milosovich, U.S. Patent 3,247,066 (April 19, 1966).
10. P. L. Madan, R. K. Jani, and A. J. Bartilucci, New method for preparing gelatin microcapsules of soluble pharmaceuticals, *J. Pharm. Sci. 67*:409–411 (1978).
11. N. Tanaka, S. Takino, and I. Utsumi, A new oral gelatinized sustained-release dosage form, *J. Pharm. Sci. 52*:664–667 (1963).

12. G. N. Paridissis and E. L. Parrott, Gelatin encapsulation of pharmaceuticals, *J. Clin. Pharmacol.* *8*:54–59 (1968).
13. M. Hashida, M. Egawa, S. Muranishi, and H. Sezaki, Role of intramuscular administration of water-in-oil emulsions as a method for increasing the delivery of anticancer agents to regional lymphatics, *J. Pharmacokin. Biopharm.* *5*:225–239 (1977).
14. M. Hashida, Y. Takahashi, S. Muranishi, and H. Sezaki, An application of water-in-oil and gelatin-microsphere-in-oil emulsions to specific delivery of anticancer agent into stomach lymphatics, *J. Pharmacokin. Biopharm.* *52*:241–255 (1977).
15. M. Hashida, S. Muranishi, and H. Sezaki, Evaluation of water in oil and microsphere in oil emulsions as a specific delivery system of 5-fluorouracil into lymphatics, *Chem. Pharm. Bull.* *25*:2410–2418 (1977).
16. M. Hashida, S. Muranishi, H. Sezaki, N. Tanigawa, K. Satomura, and Y. Hikasa, Increased lymphatic delivery of bleomycin by microsphere in oil emulsion and its effect on lymph node metastasis, *Int. J. Pharm.* *2*:245–256 (1979).
17. T. Yoshioka, M. Hashida, S. Muranishi, and H. Sezaki, Specific delivery of mitomycin C to the liver, spleen and lungs: nano- and microspherical carriers of gelatin, *Int. J. Pharm.* *81*:131–141 (1981).
18. M. Nakano, Y. Nakamura, K. Takikawa, M. Kouketsu, and T. Arita, Sustained release of sulphamethizole from agar beads, *J. Pharm. Pharmacol.* *31*:869–872 (1979).
19. M. Nakano, M. Kouketsu, Y. Nakamura, and K. Juni, Sustained release of sulfamethizole from agar beads after oral administration in humans, *Chem. Pharm. Bull.* *28*:2905–2908 (1980).
20. M. Hashida, T. Kojima, Y. Takahashi, S. Muranishi, and H. Sezaki, Timed-release of mitomycin C from its agarose bead conjugate, *Chem. Pharm. Bull.* *25*:2456–2458 (1977).
21. T. Kojima, M. Hashida, S. Muranishi, and H. Sezaki, Antitumor activity of timed-release derivative of mitomycin C, agarose bead conjugate, *Chem. Pharm. Bull.* *26*:1818–1824 (1978).
22. O. B. Wurzburg, P. C. Trubiano, and W. Herbst, U.S. Patent 3,499,962 (March 10, 1970).
23. F. Lim and R. D. Moss, Microencapsulation of living cells and tissues, *J. Pharm. Sci.* *70*:351–354 (1981).
24. M. A. El-Egakey, F. El-Khawas, A. G. Iskandar, and M. M Abdel-Khalek, Micro-encapsulation by polyethyleneglycols. Part I: Development of the technique, *Pharmazie* *29*:466–468 (1974).
25. C. R. Kowarski, B. Volberger, J. Versanno, and A. Kowarski, A method of preparing sustained release sulfamethazine in small size batches, *Amer. J. Hosp. Pharm.* *21*:409–410 (1964).
26. W. L. Chiou, S. Riegelman, and J. R. Amberg, Complications in using rabbits for the study of oral drug absorption, *Chem. Pharm. Bull.* *17*:2170–2173 (1969).

27. E. B. Draper and C. H. Becker, Some wax formulations of sulfaethylthiadiazole produced by aqueous dispersion for prolonged-release medication, *J. Pharm. Sci.* 55:376–380 (1966).
28. M. F. Sullivan and D. R. Kalkwarf, Sustained release of naltrexone from glyceride implants, *Natl. Inst. Drug Abuse Res. Monogr. Ser.* 4:27–32 (1976).
29. A. Kondo, Encapsulation utilizing meltable dispersion and cooling process, in *Microcapsule Processing and Technology* (J. Wade Van Valkenburg, ed.), Dekker, New York, 1979, pp. 121–130.
30. K. Yazawa, F. Arai, M. Kitajima, and A. Kondo, U.S. Patent 3,804,776 (April 16, 1974).

13

Miscellaneous Other Methods of Encapsulation and Entrapment

In this chapter various other methods of encapsulation and entrapment of drugs and other biologically active materials are discussed. These are briefly reviewed to provide a comprehensive coverage of other microencapsulation and related procedures that do not warrant treatment in a separate chapter. This may be due to their infrequent use in the pharmaceutical industry, lack of published literature, or because they are considered outside the primary scope of this book.

13.1 PHYSICAL METHODS

A number of physical methods have been investigated for the application of coatings to drugs. Apart from those mentioned in previous chapters, others are discussed in the following sections.

13.1.1 Centrifugal and Other Extrusion Devices

Various types of centrifugal coating apparatus have been developed at the Southwest Research Institute (SwRI), San Antonio, Texas, and have been described by Goodwin and Somerville [1]. The design of a multiorifice centrifugal head is shown in Fig. 13.1. Coating solution is fed into grooves located above and below a series of orifices located around the periphery of the cylindrical head. It overflows the internal weirs to form a membrane across the individual orifices. The liquid core material is fed onto a disk rotating concentrically inside the head at the level of the orifices, where it is spun off as droplets that repeatedly cause the membrane to distend and break to form a series of encapsulated droplets that emerge from the orifices. A more advanced type of multiorifice centrifugal extrusion head, is shown in Fig. 13.2. Here the encapsulation head rotates around a central axis. A concentric feed tube for liquid core and coating material enters through a

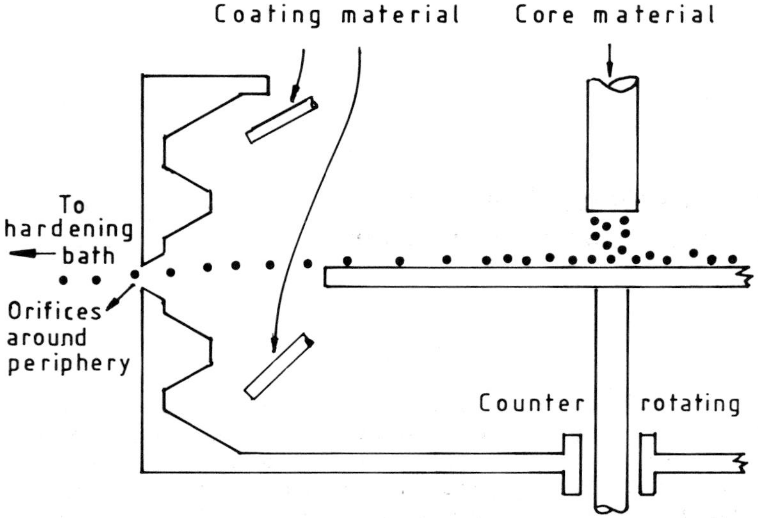

Figure 13.1 Multiorifice centrifugal head.

seal arrangement into an inner chamber. The core material is pumped through the inner central feed tube and flows through a series of radially dispersed tubes that penetrate orifices around the periphery of the rotating head. The heated coating material is pumped through the outer central feed tube and flows through the annuli formed by the sets of radial tubes and orifices. As the head rotates, the bi-liquid columns of immiscible inner core and outer coating liquids formed at the nozzles break up at nodes to form individual micro-capsules. Their coatings congeal by air cooling and the microcapsules are caught by a plastic lined cone, as shown in Fig. 13.3, to be fed downward to a vibrating conveyor for sieving and packing. A typical production rate of 500 lb hr^{-1} can be achieved with a 16-nozzle head. Capsule size may be altered by controlling feed rate, orifice size, rotation speed, and the surface tension of the coating material. The process is also adaptable for use with noncongealable types of coating materials employing solvent extraction, solvent evaporation, or chemical hardening.

Somerville et al. [2], in work financially supported by Beecham Laboratories, reported the use of the SwRI multiorifice centrifugal extrusion process for the microencapsulation of quinidine sulfate. A slurry of drug particles in molten hydrogenated triglycerides, mono-diglycerides, and several types of natural and synthetic waxes was

Figure 13.2 Multiorifice centrifugal extrusion head. (Reprinted with permission from Ref. 1. Copyright 1974 Americal Chemical Society and courtesy of W. W. Harlowe, Southwest Research Institute, San Antonio, Texas.)

encapsulated in similar molten materials except that no mono-diglycerides were used. Upon air cooling a product of size range 420 to 841 μm was recovered, which consisted of a dispersion of drug particles in a solid waxy matrix. Figure 13.4 shows blood levels of drug obtained in humans following oral administration of a relatively slow-release formulation and a relatively fast-release formulation in comparison to uncoated drug. Release patterns of formulations containing beeswax and hydrogenated triglycerides changed significantly upon aging. Mangold et al. [3] also reported preliminary results using the SwRI centrifugal extrusion device for the microencapsulation of ethynylestradiol, norethindrone, and norethindrone acetate in mixtures

Figure 13.3 Pilot plant for encapsulation with molten waxy coating. A multiorifice centrifugal extrusion head, mounted at the end of the elevated platform, throws out streams of microcapsules that are hardened by air cooling before reaching the plastic collection cone. (Reprinted with permission from Ref. 1. Copyright 1974 American Chemical Society and courtesy of W. W. Harlowe, Southwest Research Institute, San Antonio, Texas.)

of various glycerides and fatty acids or alcohols in order to reduce undesirable side effects associated with the daily peaks produced by conventional orally administered contraceptive therapy.

A number of other centrifugal encapsulation processes have also been patented. For example, Timreck [4], in a patent assigned to

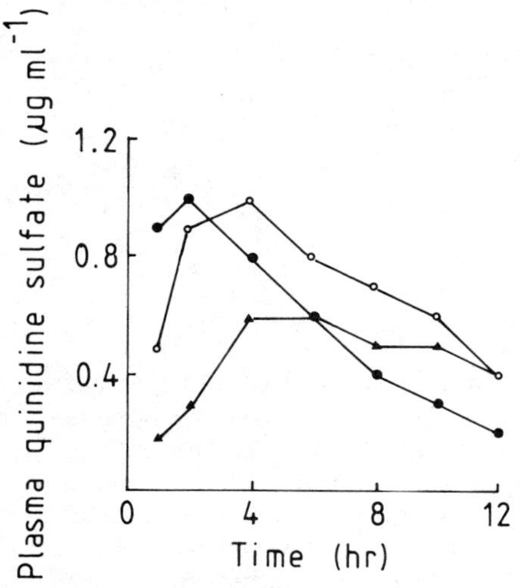

Key

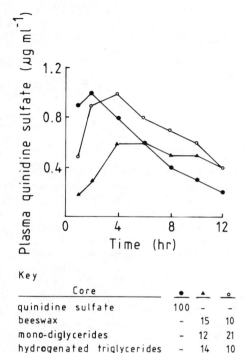

Core	●	▲	○
quinidine sulfate	100	-	-
beeswax	-	15	10
mono-diglycerides	-	12	21
hydrogenated triglycerides	-	14	10

Figure 13.4 In vivo blood levels obtained with various formulations. (Reprinted with permission from Ref. 2. Copyright 1976 American Chemical Society.)

Chas. Pfizer & Co., described how a rotating orifice device could be used to coat aqueous dispersions of fat-soluble vitamins in molten gelatin solution. The coating was rapidly gelled by air cooling to lessen diffusion occurring between core and coating. Also Dannelly [5], in a patent assigned to Eastman Kodak Co., described the use of a centrifugal encapsulation apparatus employing a rotating nozzle.

For microcapsules with delicate coatings that tended to be damaged during collection in the centrifugal process, SwRI developed a submerged extrusion nozzle device. Figure 13.5 shows a diagram of the nozzle. It consists of a concentric tube located in a circular jacket that tapers just beyond the nozzle tip. Core material is pumped through the inner concentric tube and the coating through the annulus. An inner carrier fluid flows through the jacket, and its velocity increases in the tapered region to carry the biliquid column emerging from the nozzle, which fragments into microcapsules. The coating used is often a molten wax, and the carrier is water, which is cooled

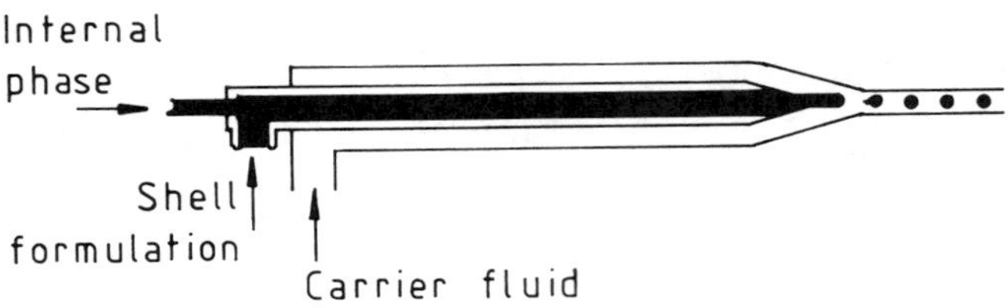

Figure 13.5 Submerged extrusion nozzle. (From Ref. 6.)

downstream of the nozzle to promote congealing of the coating. Vibratory devices can be fitted to both rotary and fixed nozzle designs to improve the uniformity of size distribution of the product obtained.

A process that is suitable for the encapsulation of both polar and nonpolar liquids in insoluble coatings to produce spherical macrocapsules with a usual size of 0.5 to 3 mm has been described by Arens and Sweeney [7] in a patent assigned to the Minnesota Mining and Manufacturing Co. Figure 13.6 shows a diagram of the process. A concentric biliquid column is formed by forcing a jet of core liquid to be encapsulated through the bulk of a hardenable liquid coating material, which is often a molten wax. The stream travels in the desired tragectory usually through cold air for a time sufficient to allow the column to fragment and contract into a series of microcapsules whose coating

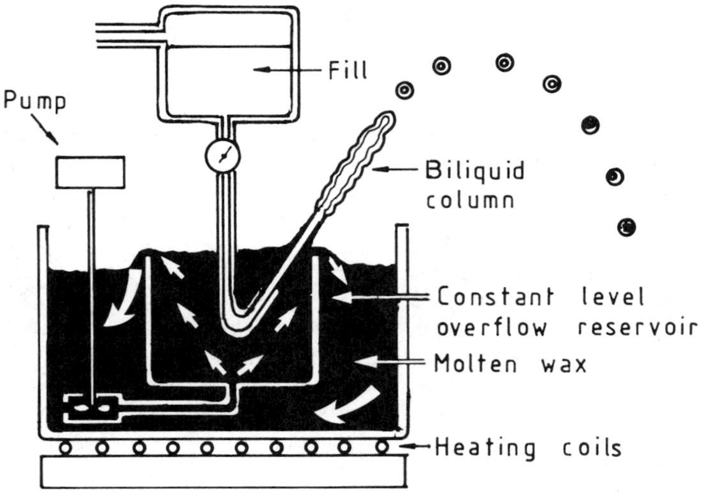

Figure 13.6 Diagrammatic representation of the 3M microencapsulation process. (From Ref. 8.)

is hard enough to withstand the impact of falling. In a subsequent patent assigned to the same company, Bolles [9] described the use of surfactant in the coating formulation to lower its surface tension and so promote its more uniform deposition around the core material. The process has been used to encapsulate flavored oils in aqueous gelatin or sugar-based coatings.

13.1.2 Electrostatic Encapsulation

Microencapsulation utilizing electrostatic deposition was developed at the Illinois Institute of Technology Research Institute, Chicago (see Fig. 13.7). In this method of encapsulation a liquid or suspension of core material is atomized into the coating chamber. The core droplets formed are electrically charged as they leave the atomizer and electrostatically interact with similar size droplets of coating material, which should bear an opposite charge of similar magnitude. Provided that

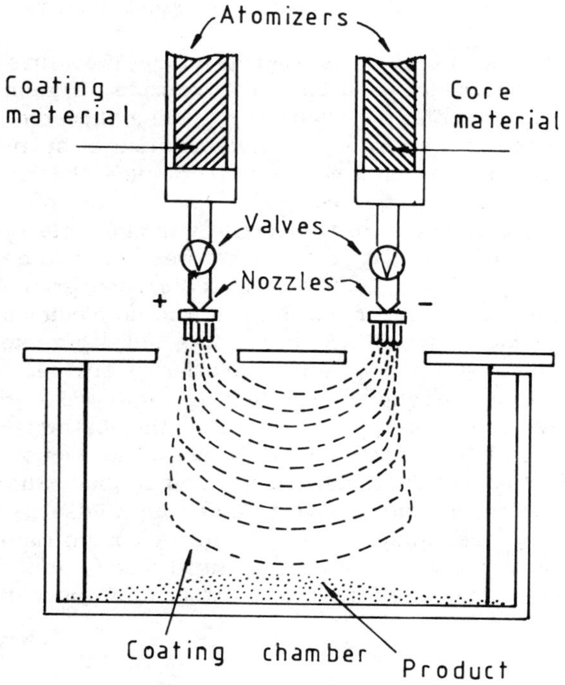

Figure 13.7 Schematic diagram of a typical electrostatic microencapsulation apparatus.

the coating material has a lower surface tension than the core material , it spreads over the core droplets to encapsulate them. Neither the viscosity or temperature of the coating material is critical, provided that it remains fluid during the encapsulation process. In the only pharmaceutical example reported in the literature, Langer and Yamate [10] successfully encapsulated glycerin with carnauba wax using similar equipment. The majority of the microcapsules were below 1 μm, as the formation of bigger microcapsules was difficult due to the massive applied voltages required to charge larger aerosol droplets adequately. These larger droplets also tended to lack the desirable property of Brownian motion.

13.2 DIP COATING

An apparently simple means of encapsulation involves single or repeated dipping of cores into a coating solution, allowing the excess to drain off and drying. However, because of the static nature of the process, the major difficulty encountered is damage to the coating as cores separate from contact with each other and the mechanical support used to hold them while being coated.

Accordingly, there have been only a few reports in the literature of the use of dip coating for microencapsulation. For example, Gagnon et al. [11] dip-coated aspirin-containing granules by placing them on a sieve prior to immersion once into a heated alcoholic solution of stearic acid. The granules were then spread on a wire screen to allow the alcohol to evaporate and the stearic acid to congeal. The problem of sticking of coated granules together and to the screen was encountered. Disintegration of the granules in artificial gastric fluid was noted to be delayed. Also, Andrade et al. [12] dip-coated activated carbon granules by placing them in a mesh sack and immersing them in an alcoholic solution of hydroxyethyl methacrylate (HEMA) containing butyl peroctoate as initiator. Alternatively, an aqueous buffer solution of bovine serum albumin containing glutaraldehyde as crosslinking agent was used as the coating system. The sack was removed from the coating, placed on absorbent paper, and the coating was polymerized in vacuo at 80°C for 2 hr. Provided that not too much coating was applied, the poly-HEMA or albumin-coated activated charcoal retained high adsorptive capacity for uric acid and creatinine. The poly-HEMA-coated material caused a negligible drop in platelet count of perfused blood, unlike coated charcoal. However, for this application completely uniform coating is probably unnecessary.

13.3 IN SITU POLYMERIZATION

The process of in situ polymerization differs from that of interfacial polycondensation described in Chap. 5 in that monomer or precondensate

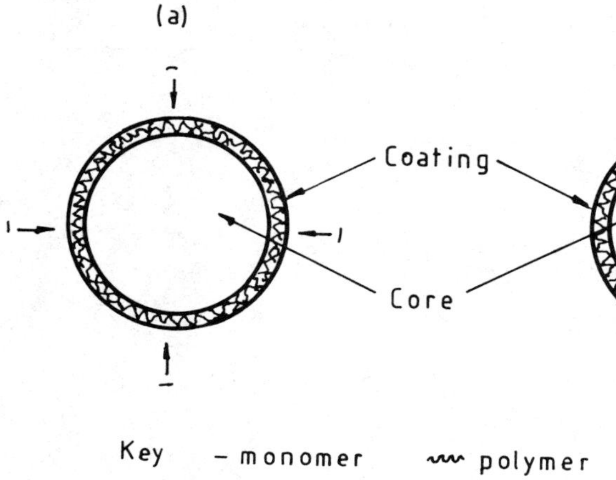

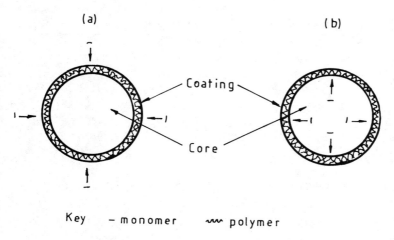

Figure 13.8 Schematic representation of in situ polymerization.

is dissolved only in either the core material or the surrounding manufacturing vehicle. In the presence of a catalyst, an insoluble polymer coating forms around the core material as illustrated in Fig. 13.8. The core material, which may be solid or liquid, must be insoluble in the manufacturing vehicle, which is usually either a hydrophilic or hydrophobic liquid. To achieve a uniform dispersion of the core material it may be necessary to add a surface-active agent or suspending agent. Deposition of the polymer at the interface is often favored by its hydrophilic-lipophilic balance or by the preferential location of the catalyst at the interface.

The in situ polymerization procedure of Florence and co-workers for the production of polyalkyl cyanoacrylate microcapsules was discussed in Chap. 10. Kondo [13] outlined a number of patents for other in situ polymerization processes suitable for the production of polystyrene, polyurethane, crosslinked formaldehyde-urea, and other polymer-coated microcapsules, mainly for nonpharmaceutical applications. However, problems associated with unfavorable conditions of polymerization, such as use of high temperatures for prolonged periods and toxicity associated with residual monomers, catalysts, and other manufacturing components, have tended to limit the use of in situ polymerization procedures for the encapsulation of drugs and similar biological materials.

13.4 LIPOSOMES

A dosage form that has attracted enormous interest in recent years is liposomes. These are obtained by dispersing phospholipids in water

Table 13.1 Some Typical Components of Liposomes

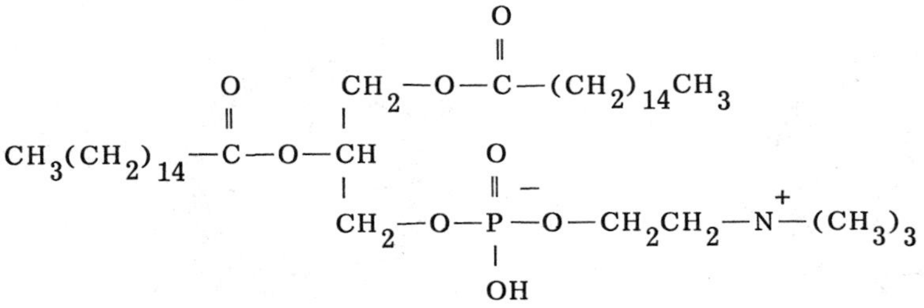

Dipalmitoyl phosphatidyl choline (lecithin)

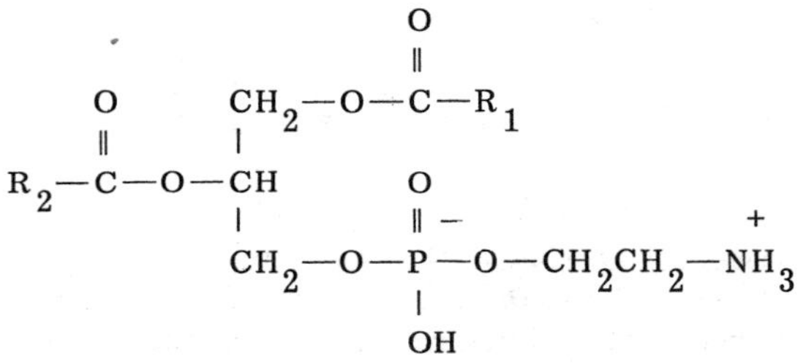

R_1 and R_2 are long-chain acyl groups

Phosphatidyl ethanolamine

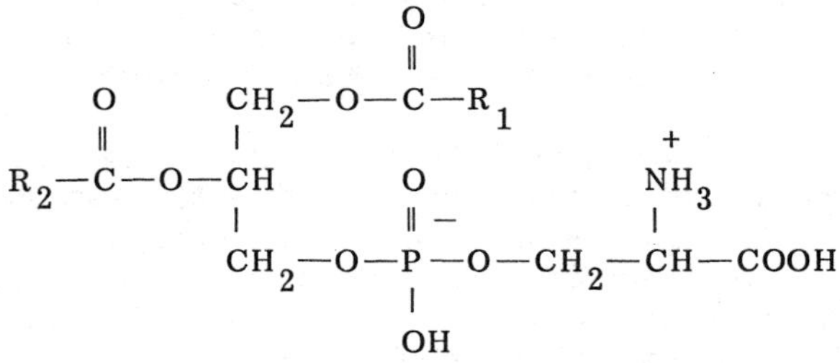

Phosphatidyl serine

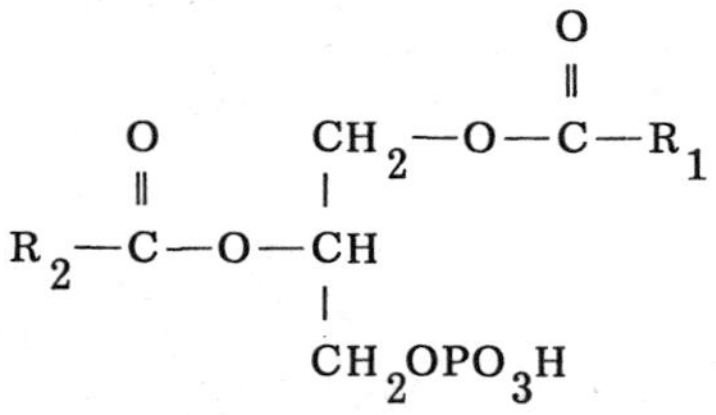

Phosphatidic acid (to impart - charge)

Table 13.1 (Continued)

$CH_3(CH_2)_{17}NH_2$

Stearylamine (to impart + charge)

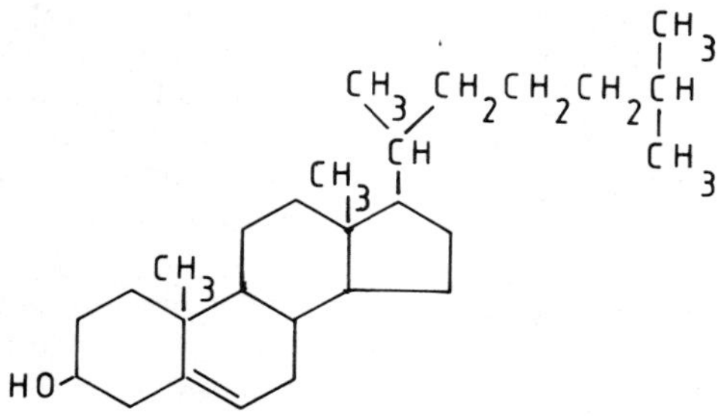

Cholesterol (to modify thermotropic phase transition)

so that they orientate their molecules because of unfavorable entropy to form one or more concentric membranes around an aqueous compartment(s). Other possible additions include sterols, organic acids, or bases, and membrane protein. Some typical phospholipids and other substances used, shown in Table 13.1, can impart positive, negative, or neutral charge to the liposomes depending on the components used and on the pH, ionic composition, and temperature of the dispersion medium. Having a hydrophilic-lipophilic balance, the phospholipid molecules will orientate in solution to form a bilayer that will tend to adsorb plasma protein and other available substances as illustrated in Fig. 13.9a. At a particular temperature range the fatty acyl side

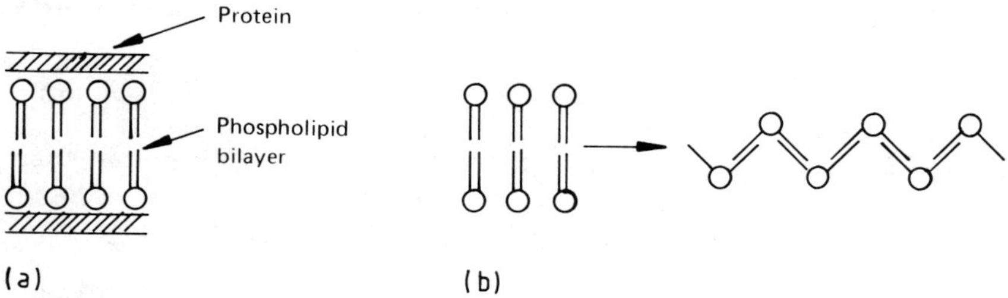

Figure 13.9 (a) Phospholipid bilayer with adsorbed protein. (b) Idealized phase transition.

chains of the phospholipid become more loosely packed, as illustrated in Fig. 13.9b. This thermotropic phase transition occurs over a narrow temperature range with homogeneous phospholipids and causes the bilayer(s) to become much more permeable. Cholesterol is often incorporated into such liposome preparations to alter chain fluidity and so diminish the effect of phase transition on permeability. Depending on the means used to prepare the liposomes, multilamellar vesicles (MLV), small unilamellar vesicles (SUV), or large unilamellar vesicles (LUV) can be formed as illustrated in Fig. 13.10. Polar drugs and biologically active materials can dissolve in the aqueous internal compartment(s) of liposomes. Likewise, nonpolar substances can become entrapped in liposomes, being associated with the lipophilic regions as illustrated in Fig. 13.10.

Being colloidal in dimensions and composed of biocompatible materials, liposomes can be injected intravenously, where they are mainly concentrated in the liver and spleen by endocytosis. However, if injected locally, MLV tend to remain at the site of injection, where they slowly disintegrate to release their bioactive material, whereas smaller vesicles rapidly enter the circulation to be removed by the reticuloendothelial system. There have been various attempts made to improve the targeting of liposomes to other tissues in the body by the use of selective determinants and antibodies.

Apart from their use as model biological membranes, liposomes have been extensively investigated for use as drug carriers. Drug classes examined include antibiotics, anticancer agents, antiparasitic agents, corticosteroids, heavy metal chelating agents, hormonal replacement therapy, and vitamins. Liposomes have also been used in enzyme replacement therapy, immunopotentiation, gene transfer, and radiopharmaceuticals. Most of the studies reported relate to in vitro, in tissue culture or in vivo animal studies, and have shown some degree of promise. Studies in humans are infrequently reported, and as yet no liposome preparation appears to have progressed beyond the clinical trial stage. Except perhaps for certain selective drugs and enzymes that are required to act in the liver or for depot therapy in the lungs, it seems at present unlikely that liposome preparations will be widely used as controlled delivery systems until the following difficulties are overcome.

1. It is difficult to prepare liposomes of reproducible type and hence biological response.
2. The instability of liposome preparations does not permit their storage for longer than a few weeks—it is not possible to isolate liposomes in dry powder form.
3. Drug loadings achievable are usually low.
4. The purified phospholipids and other materials used in their preparation are very costly.

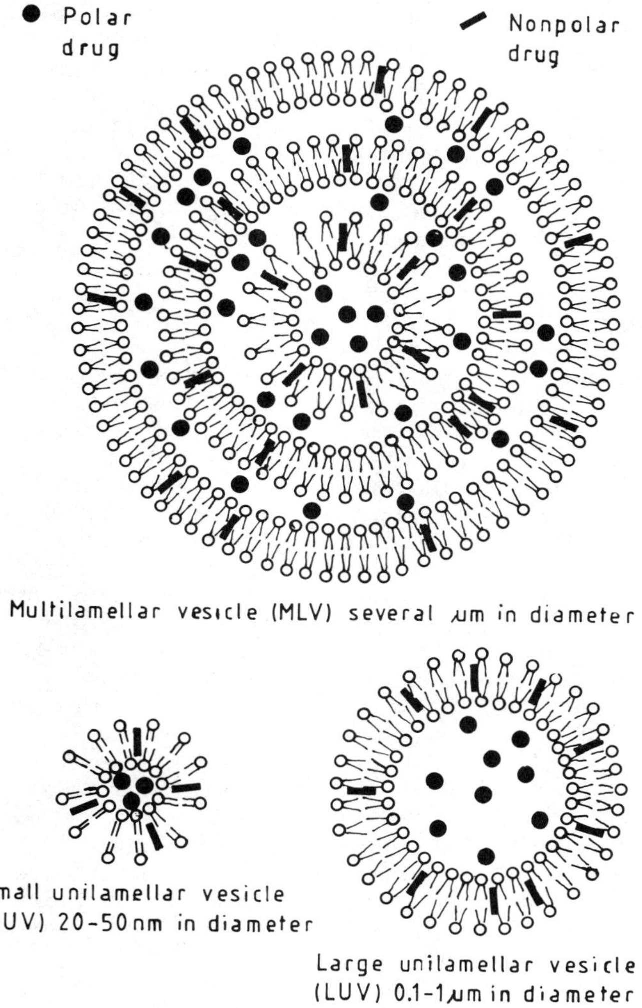

Figure 13.10 Schematic representation of different types of liposomes formed (not to scale).

5. It is difficult to prevent rapid uptake of liposome preparations by the liver and other elements of the reticuloendothelial system.
6. There is a possibility of unwanted vascular obstruction caused by larger liposomes.

The treatment of this important and interesting topic has been kept at the introductory level, as liposomes have been considered not to fall within the central theme of this book. There is an enormous literature on the subject, and the interested reader should initially consult a textbook [14, 15] or a recent review [16–18] on the subject. Perusal of this literature should be of particular interest to those concerned with the uses of nanocapsules and nanoparticles (see Chaps. 3, 9, and 10), where obvious similarities between the classes of drugs investigated and methodologies employed are apparent.

13.5 SPHERICAL MATRICES FROM LIQUID SUSPENSION

As an alternative to the use of spray congealing (see Chap. 8), Kawashima et al. [19] described a simple procedure for embedding poorly water-soluble drug particles in small, spherical, waxy or other matrices. Initially fine sulfamethoxazole particles were rendered strongly hydrophobic to increase their affinity for the matrix material. This was done by dispersing the drug particles in a dilute solution of palmitic acid in alcohol, filtering, and drying. The surface-treated drug particles were then dispersed with vigorous agitation in distilled water at 90°C, and powdered beeswax was added. The molten beeswax collected the drug particles and formed spherical aggregates, which were recovered upon cooling by filtration and drying. Alternatively, the surface-treated drug particles were agglomerated by addition of a small amount of a solution of white beeswax or ethylcellulose in benzene added to the aqueous drug suspension at elevated temperature.

The size range of the product obtained varied from 50 μm to 4 mm. Apart from variation in stirrer speed, surface treatment gave larger aggregates, as these particles could be more easily wetted by the molten wax. Dissolution studies showed that the matrices, containing in particular surface-treated drug, released their active ingredient more slowly than the untreated drug. The wax was more effective than ethylcellulose at delaying drug release, and it also exhibited enteric properties.

13.6 GRANULATION PROCESSES

Various granulation procedures are widely used in the pharmaceutical industry for the production of solid dosage forms. Whereas these do not produce microcapsules with distinct inner core and outer coating, they can be used to form small, irregular-shaped masses that contain drug particles embedded in a matrix that often gives the product controlled-release properties. These granules have structural similarities to various microparticles previously discussed in this book. They are normally tableted as, for example, in the patent of Kornblum and Stoopak [20], assigned to Sandoz Inc., where different colored granules containing the drugs phenobarbitone and bellafoline, together with variable composition of release-retardant materials, were compressed. The different wet and dry granulation processes and equipment used in their production are covered in various standard texts on tableting. However, a small number of references on granulation will be considered where the product was not intended primarily for tableting but for purposes similar to those of microcapsules and microparticles.

Jacobsen [21], in a patent assigned to Novo Terapeutisk Laboratorium A/S, Denmark, dispersed fine particles of sterile procaine penicillin in a heated sterile solution of aluminum monostearate in trichloroethylene. The organic solvent was evaporated and the mass obtained was granulated in a sterile disintegrator. Microparticles obtained were suspended in sterile sesame oil to produce a sustained-release injectable form of the antibiotic.

Grass and Robinson [22], in a patent assigned to Smith, Kline & French Laboratories, described a process for embedding drug particles in waxy coating material for subsequent use in the formulation of sustained-release suspensions. As an example of the process, finely divided particles of sulfaethylthiadiazole were dispersed in a solution of hydrogenated castor oil in chloroform at 55°C. After cooling and solvent evaporation, the mass obtained was granulated to fine particles. These were then dispersed in molten glyceryl distearate at a temperature of 65°C, which was insufficient to melt the hydrogenated castor oil. This compound oil phase was then emulsified into water at 58°C. The emulsion was allowed to cool below 45°C, which caused hardening of glyceryl distearate around the primary drug-wax particles. The product was then collected by filtration, dried at 45°C, and sieved. Any other suitable combination of a high-melting-point lipid for the primary particles and a low-melting-point lipid for the outer coating could be used.

13.7 SPHERONIZATION

Spheronization is a process whereby solid spherical particles are formed using a special spheronizing machine such as designed by Fugi Denki Kogyo Co., Ltd., Japan. Their marumerizer forms spheres by breaking up extruded lengths of a plastic wet mass of drugs and excipients, which are then rolled into small spherical balls on a rotating grooved plate mounted horizontally in a smooth-walled cylindrical apparatus. An available extruder having a radical or cylindrical screen is suitable for the formation of spherical particles below 3 mm, whereas if supplied with an axial die a larger size product can be obtained. The size of the spheres produced is also affected by the composition of the feed material, the roughness and spacings of the grooved plate, and by its speed of rotation. Figure 13.11a shows a working extruder and marumerizer, and Fig. 13.11b shows a view of the grooved plate. Exudate and

Figure 13.11 (a) Extruder and marumerizer. (b) Grooved plate of marumerizer. (c) Exudate and spherical granules. (Courtesy of U.K. agent, Russel Finex, Ltd., London.)

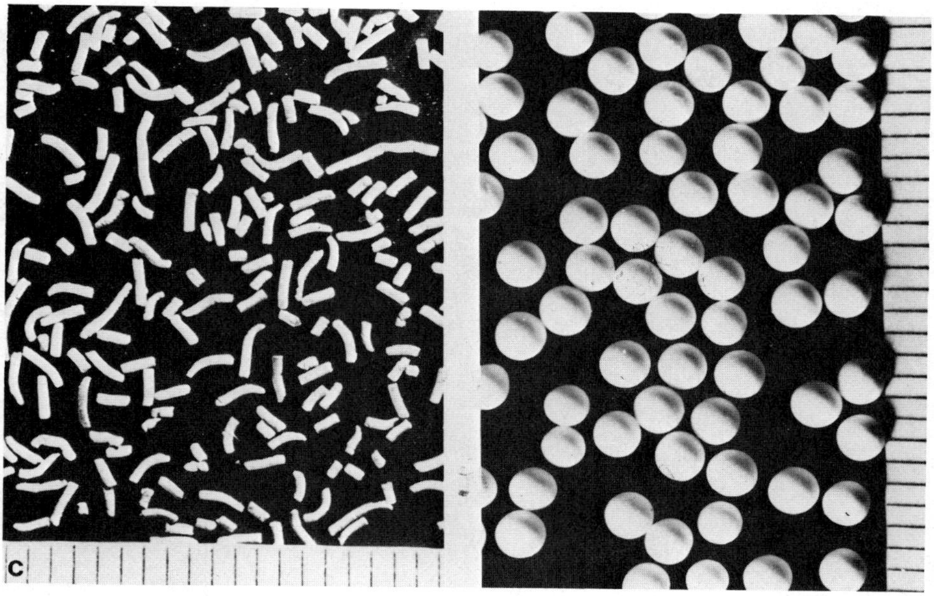

Figure 13.11 (Continued)

spherical granules are shown in Fig. 13.11c. Similar spheronization equipment is available from G. B. Caleva, Ltd., Sunninghill, U.K. The exudate is usually spheronized in less than 5 min of treatment, and after the spheres have been emptied from the marumerizer through an opening in the cylindrical wall, they must be dried using a fluidized bed dryer or other means. Greater detail of the technique and the effect of processing variables are given by Reynolds [23] and by Woodruff and Nuessle [24]. Because of their smooth, rounded profile and uniformity of size, spheres produced by the marumerization process have been used as micro- and macroparticulate drug carriers and for the production of core material with improved physical properties for subsequent coating. Table 13.2 summarizes some examples of these applications.

13.8 MOLECULAR-SCALE ENTRAPMENT

Banker and colleagues have investigated the process of molecular-scale entrapment of drugs in a variety of different polymers. The usual technique involves the addition of a solution of drug to a commercially available polymer latex. Under suitable conditions the drug molecules interact

Table 13.2 Some Pharmaceutical Applications of Spherical Particles Produced by Use of Spheronization

Drug	Main carrier materials	Primary use of spheres	Comments	Reference
Furosemide	Lactose/microcrystalline cellulose	As controlled-release dosage form	Effect of different disintegrants studied	Lovgren and Bogentoft [25]
Salicylic Acid	Microcrystalline cellulose	As cores for air suspension coating with polymethacrylate	Product evaluated for controlled drug release	Gurny et al. [26]
Phenazone	Microcrystalline cellulose/potato starch	As cores for air suspension coating with polymethacrylate	Product compressed into tablets evaluated for controlled drug release	Juslin and Puumalainen [27]

with the colloidal polymeric dispersion to cause flocculation or gelation. This material is then filtered if necessary, dried, and size-reduced. Figure 13.12 shows the principal steps involved in the process. Banker [28] and Yang and Banker [29] envisaged the interaction between the drug and the polymer and/or surfactant of the latex to be probably due to chemisorption or physisorption and to be due to bridge formation when an entrapment facilitator such as a dicarboxylic acid (e.g., succinic acid) was used.

As an example of the process, Goodman and Banker [30] entrapped the cationic antihistamine drug methapyrilene hydrochloride in an anionic acrylic copolymer emulsion. Flocculation was promoted by charge neutralization. The particulate product obtained was characterized by excellent uniformity of drug distribution, high drug loading, and sustained drug release upon in vitro and in vivo testing. A range of 10

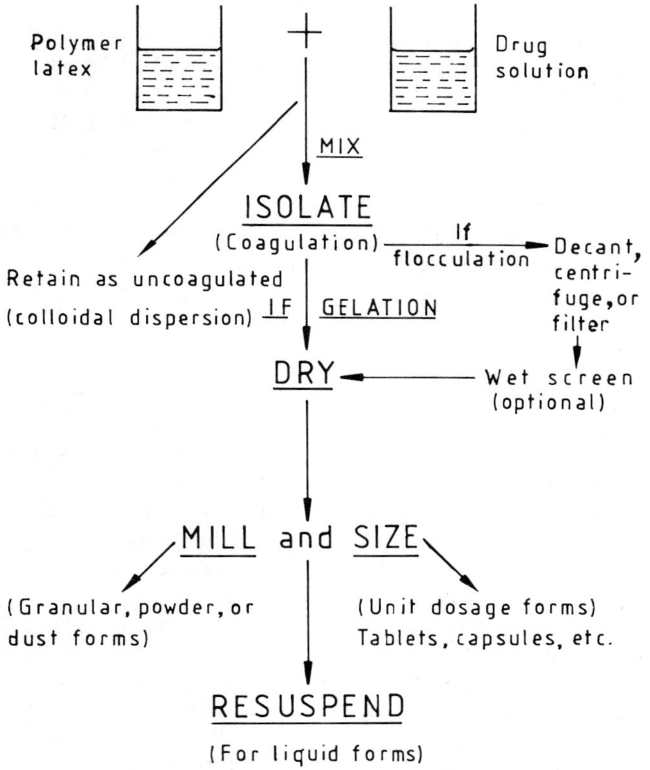

Figure 13.12 Processing steps in the molecular-scale entrapment of drugs employing polymeric latices. (From Ref. 28.)

other widely used cationic nitrogen-containing drugs were also found to flocculate the acrylic copolymer emulsion. This type of simple drug-polymer flocculation procedure has been further investigated [31, 32].

Boylan and Banker [33] entrapped the anionic drug phenobarbitone in a styrene-acrylic copolymer latex using the divalent cation Mg^{2+} in magnesium sulfate as the gelling agent. The gel formed was dried and mechanically ground to produce particles that showed sustained drug-release properties. A wide variety of other polymeric emulsions and drugs were also screened.

Gurny et al. [34] molecularly entrapped the model drug noreleagnine hydrochloride in ethylcellulose. This was done by stirring an aqueous solution of the drug into an ethylcellulose latex and adding methanol to form a coagulum. The supernatant was discarded and the coagulum was dried at 37°C for 24 hr. The in vitro release rate of the drug from the particles of the product was significantly decreased by the entrapment procedure. When implanted into rat brain there was no evidence of adverse cytological damage or depression of growth.

Recently Gurny et al. [35] developed a parenteral drug delivery system containing a molecular dispersion of the model drug testosterone in poly-DL-lactic acid pseudolatex. The polymer particles obtained had an initial diameter of 0.45 μm and showed some coalescence over 6 months storage at room temperature, causing a slight rise in the apparent viscosity of the system. Unfortunately, in vivo tests in rats showed an initial major drug release over 2 days, which was probably due to poorly entrapped drug on the surface of the particles. This was followed by an almost zero-order release for a further 12 days. PLA/PGA copolymer latices were also investigated.

13.9 OTHER APPROACHES AND CONSIDERATIONS

There have been a number of other approaches investigated for the production of controlled-release dosage forms that have similarities in design and end use to those already discussed in this book. For example, various biological and other macromolecules, such as albumin, antibodies, dextran, DNA, fibrinogen, lectin, and polylysine, have been used as carrier molecules for anticancer agents and enzymes. These complexes have been formulated into injections and used for improving the targeting or stability of these bioactive materials [36]. Of course, chemical modification of drugs to achieve controlled delivery has been extensively investigated [37]. Another potential approach involves injecting a drug solution and biodegradable starch microspheres upstream of the target region in animal models. The spheres become temporarily lodged in the vascular supply, causing partial obstruction in blood flow, with drug accumulation and transient ischemia in the organ. The residence time of the spheres was controlled by their size distribution and degree of crosslinking [38]. Also, erythrocytes [39] or yeast cells [40] may be useful as carriers for drugs or enzymes. These areas of research are often worth examining, as they may provide many useful ideas on methodology and applications.

This book is concerned mainly with the preparation and use of microcapsules in human medicine. However, microencapsulation is widely employed in many other areas, such as foods and food additives, cosmetics and fragrances, laundry aids, veterinary products, pesticides and herbicides, pigments and paints, paper making, printing, photographic agents, adhesives, catalysts, and many more. A review of the published and patent literature relating to these types of products can be helpful for discovering approaches not constrained by health hazards associated with materials or processes employed and that may be exploited in the design of products intended for the treatment of human disorders. Likewise, it is hoped that the many approaches outlined in this book may be of use to those working on microencapsulation problems outside the pharmaceutical area.

Many of the microencapsulation and related procedures outlined in this book are still at the research and development stage and have not been commercially exploited. Other microencapsulation procedures, such as vacuum deposition, have not been discussed because they have no pharmaceutical applications. It should be obvious from the variety of encapsulation procedures discussed that it is possible to encapsulate bioactive materials by many different methods. It is hoped that this book will suggest suitable methods and opportunities for technological transfer between existing processes. The final method chosen for a particular material will be influenced mainly by the intended end use of the product, the physicochemical properties of the core and coating, and by the availability of required production facilities. Its selection will require pilot-scale studies on a number of potential methods. Problems will frequently arise later during scaling-up operations, such as containment of organic solvents as discussed by Gardner [41].

It should be appreciated that regulatory agencies for medicines worldwide will normally seek additional evidence of safety and efficacy before approving microencapsulated and related products for clinical trials or subsequent marketing. Apart from the usual submissions required for a new or existing drug, the therapeutic or other claims, such as taste-masking, associated with these products over conventional dosage forms will have to be demonstrated. The normal dosage of a microencapsulated product must not contain enough drug to cause a fatal overdosage should the encapsulation mechanism fail. Adequate controlled-release properties must be demonstrated in vivo. Coating materials should be *generally regarded as safe* (GRAS). Novel polymers and other adjuvants are unlikely to be approved. Limits on solvents and polymer intermediates remaining in the formulation must be acceptable. The additional evidence sought will also include stability testing to determine the expiration date of the product. During this time the content of active ingredient and unique release properties should not significantly alter under expected environmental conditions of storage in the finished dosage form and packaging.

REFERENCES

1. J. T. Goodwin and G. R. Somerville, Microencapsulation by physical methods, *Chemtech 4*:623–626 (1974).
2. G. R. Somerville, J. T. Goodwin, and D. E. Johnson, Controlled release of quinidine sulfate microcapsules, in *Controlled Release Polymeric Formulations* (D. R. Paul and F. W. Harris, eds.), American Chemical Society, Washington, D.C., 1976, pp. 182–189.
3. D. J. Mangold, H. W. Schlameus, J. W. Goldzieher, and J. T. Doluisio, Development of orally-active sustained release dosage forms for steroids, in *Proceedings of the 8th International Symposium on Controlled Release of Bioactive Materials, Ft. Lauderdale, July 26–29*, 1981, pp. 173–174.
4. A. E. Timreck, U.S. Patent 3,526,682 (September 1, 1970).
5. C. C. Dannelly, U.S. Patent 4,123,206 (October 31, 1978).
6. J. T. Goodwin and G. R. Somerville, Physical methods for preparing microcapsules, in *Microencapsulation Processes and Applications* (J. E. Vandegaer, ed.), Plenum, New York, 1974, pp. 155–163.
7. R. P. Arens and N. P. Sweeney, U.S. Patent 3,423,489 (January 21, 1969).
8. R. H. Sudekum, Microcapsules for topical and other applications, in *Microencapsulation* (J. R. Nixon, ed.), Dekker, New York, 1976, pp. 119–128.
9. T. F. Bolles, U.S. Patent 3,779,942 (December 18, 1973).
10. G. Langer and G. Yamate, Encapsulation of liquid and solid aerosol particles to form dry powders, *J. Colloid Interface Sci. 29*:450–455 (1969).
11. L. P. Gagnon, G. DeKay, and C. O. Lee, Coating of granules, *Drug Standards 23*:47–52 (1955).
12. J. D. Andrade, K. Kunitomo, R. Van Wagenon, B. Kastigir, D. Gough, and W. J. Kolff, Coated adsorbents for direct blood perfusion: HEMA/activated carbon, *Trans. Amer. Soc. Artif. Int. Organs 17*:222–228 (1971).
13. A. Kondo, Microencapsulation by in situ polymerization, in *Microcapsule Processing and Technology* (J. Wade Van Valkenburg, ed.), Dekker, New York, 1979, pp. 46–58.
14. A. C. Allison and G. Gregoriadis, eds., *Liposomes in Biological Systems*, Wiley, New York, 1979.

15. C. G. Knight, ed., *Liposomes from Physical Structure to Therapeutic Applications*, Elsevier, Amsterdam, 1981.
16. G. Gregoriadis, Liposomes, in *Drug Carriers in Biology and Medicine* (G. Gregoriadis, ed.), Academic, London, 1979, pp. 287–341.
17. R. L. Juliano and D. Layton, Liposomes as a drug delivery system, in *Drug Delivery Systems: Characteristics and Biomedical Applications* (R. L. Juliano, ed.), Oxford University Press, New York, 1980, pp. 189–236.
18. I. W. Kellaway, J. Hadgraft, M. Ahmed, M. Arrowsmith, R. S. Chawla, F. H. Farah, M. J. James, and M. J. Taylor, Liposomes—model membranes for modern medicines, *Manuf. Chem. Aerosol News 51*(8):43–44 (1980).
19. Y. Kawashima, H. Ohno, and H. Takenaka, Preparation of spherical matrixes of prolonged-release drugs from liquid suspension, *J. Pharm. Sci. 70*:913–916 (1981).
20. S. S. Kornblum and S. B. Stoopak, U.S. Patent 4,012,498 (March 15, 1977).
21. H. Jacobsen, U.S. Patent 3,016,330 (January 9, 1962).
22. G. M. Grass and M. J. Robinson, U.S. Patent 2,875,130 (February 24, 1959).
23. A. D. Reynolds, A new technique for the production of spherical particles, *Manuf. Chem. Aerosol News 41*(6):40–43 (1970).
24. C. W. Woodruff and N. O. Nuessle, Effect of processing variables on particles obtained by extrusion-spheronization processing, *J. Pharm. Sci. 61*:787–790 (1972).
25. K. Lovgren and C. Bogentoft, Influence of different disintegrants on dissolution rate and hardness of furosemide granules prepared by spheronization technique, *Acta Pharm. Suec. 18*: 108-109 (1981).
26. R. Gurny, P. Guitard, P. Buri, and H. Sucker, Réalisation et développement théorique de formes médicamenteuses à libération contrôlée par des films méthacryliques, *Pharm. Acta Helv. 52*: 182–187 (1977).
27. M. Juslin and P. Puumalainen, Pellets coated with acrylic plastic in tabletting, in *Abstracts of the 37th International Congress of Pharmaceutical Sciences, The Hague, September 5–9*, 1977, p. 48.
28. G. S. Banker, Controlled release of effectors from polymers and microcapsules, in *Polymeric Delivery Systems* (R. J. Kostelnik, ed.), Gordon & Breach, New York, 1978, pp. 25–58.
29. W. Yang and G. S. Banker, Mechanism of molecular-scale drug entrapment using colloidal polymeric latices, *Drug Develop. Indust. Pharm. 8*:27–40 (1982).
30. H. Goodman and G. S. Banker, Molecular-scale entrapment as a precise method of controlled drug release I: Entrapment of

cationic drugs by polymeric flocculation, *J. Pharm. Sci. 59*: 1131–1137 (1970).

31. C. T. Rhodes, K. Wai, and G. S. Banker, Molecular scale drug entrapment as a precise method of controlled drug release II: Facilitated drug entrapment in polymeric colloid dispersions, *J. Pharm. Sci. 59*:1578–1581 (1970).
32. C. T. Rhodes, K. Wai, and G. S. Banker, Molecular scale drug entrapment as a precise method of controlled drug release III: In vitro and in vivo studies of drug release, *J. Pharm. Sci. 59*: 1581–1584 (1970).
33. J. C. Boylan and G. S. Banker, Molecular scale drug entrapment as a precise method of controlled drug release IV; Entrapment of anionic drugs by polymer gelation, *J. Pharm. Sci. 62*: 1177–1184 (1973).
34. R. Gurny, S. P. Simmons, G. S. Banker, R. Meeker, and R. D. Myers, A new biocompatible drug delivery system for chronic implantation in animal brain, *Pharm. Acta Helv. 54*: 349–352 (1979).
35. R. Gurny, N. A. Peppas, D. D. Harrington, and G. S. Banker, Development of biodegradable and injectable latices for controlled release of potent drugs, *Drug Develop. Indust. Pharm.* 7:1–25 (1981).
36. M. J. Poznansky and L. G. Cleland, Biological micromolecules as carriers of drugs and enzymes, in *Drug Delivery Systems, Characteristics and Biomedical Applications* (R. L. Juliano, ed.), Oxford University Press, New York, 1980, pp. 253–315.
37. A. A. Sinkula, Methods to achieve sustained drug delivery. The chemical approach, in *Sustained and Controlled Release Drug Delivery Systems* (J. R. Robinson, ed.), Dekker, New York, 1978, pp. 411–555.
38. B. Lindberg, A. Nygren, G. Malson, T. Malson, and R. Lindblom, Injectable starch microspheres for targeting of drugs, in *Abstracts of the 4th International Conference on Surface and Colloid Science, Jerusalem, July 5–10*, 1981, p. 212.
39. G. M. Ihler, Potential use of erythrocytes as carriers for enzymes and drugs, in *Drug Carriers in Biology and Medicine* (G. Gregoriadis, ed.), Academic, London, 1979, pp. 129–153.
40. J. L. Shank, U.S. Patent 4,001,480 (January 4, 1977).
41. G. L. Gardner, Manufacturing encapsulated products, *Chem. Eng. Prog. 62*(4):87–91 (1966).

14

Release of Drug from Microcapsules and Microparticles

14.1 INTRODUCTION

Release of drug from microcapsules and microparticles is a mass transport phenomenon involving diffusion of drug molecules from a region of high concentration in the dosage form to a region of low concentration in the surrounding environment. Attempts to model drug release from such preparations have been reported by several researchers [1, 2] to be less than satisfactory. This is due to the great diversity in the physical form of microcapsules and microparticles with regard to size, shape, and arrangement of core and coating materials. Also, physicochemical properties of core material, such as solubility, diffusivity, and partition coefficient, and of coating material, such as variable thickness, porosity, and inertness, make modeling of drug release difficult. No single approach to quantifying drug release from such products will be universally suitable, and in this chapter a number of physical models will be reviewed. These approaches have generally been developed in relation to macro dosage forms such as capsular (reservoir) and matrix (monolithic) devices whose configuration is known and whose drug release attains a pseudo-steady-state condition, which may not apply to microcapsules and microparticles. However, a review of suitable models, stressing their limitations in relation to micro dosage forms, is very valuable in understanding those parameters of importance that may be varied to control drug release.

14.2 PERMEATION CONSIDERATIONS

Permeation theory, whereby drug molecules are transported through one or more polymeric membranes comprising the coating material, is important to an understanding of how core substances are released from microcapsules and microparticles. The coating normally acts as a barrier, the resistance of which is influenced by factors such as the identity of the film former, its degree of crystallinity, the inclusion of plasticizers and fillers, its thickness, the occurrence of pores, and the presence of a stagnant diffusion layer in contact with the outer coating surface. Drug transport through such coating material is a very complex subject and will be considered primarily in terms of passive diffusion, where factors that influence transport, such as electrochemical and thermal gradients, and active transport, are considered to be negligible. The coating material may be considered as consisting of one or more homogeneous barriers that offer resistance to the transport of a drug substance in the direction of the flux vector. Figure 14.1 shows a typical microcapsule arrangement whereby a homogeneous coating is interposed between a core composed of a solid or liquid drug and an aqueous receiving phase that contains a negligible bulk drug concentration. The first barrier to diffusion of the drug would be the coating and the second barrier would be the unstirred diffusion layer between the coating and the aqueous receiving phase. Such a microcapsule arrangement would be initially similar to that reported by Si-Nang et al. [3], whereby the liquid mucolytic drug eprazinone was microencapsulated in gelatin-acacia by a complex coacervation procedure, the coating being hardened by crosslinking with glutaraldehyde prior to immersion of the microcapsules into an agitated aqueous medium. These workers described how to determine coating thickness either directly after microtome sectioning or by computational techniques from density and drug composition data. Mean values of 7 and 6 μm,

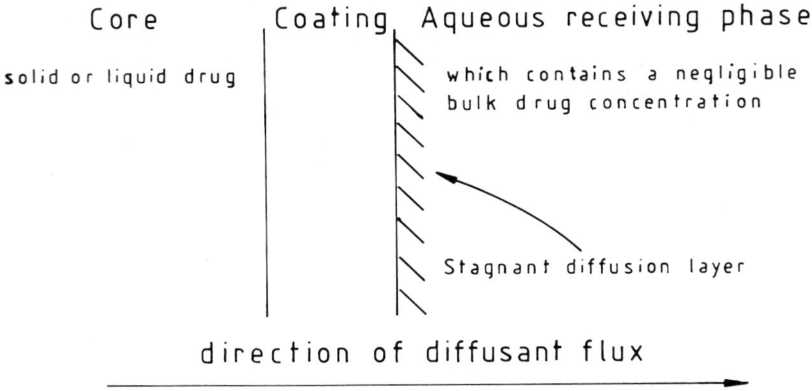

Figure 14.1 Schematic representation of drug passage through an idealized homogeneous coating of a microcapsule.

respectively, were obtained, indicating the comparative thinness of the microcapsule walls. The coating thickness also tended to be quite variable over the surface of the microcapsules. Those produced by interfacial polycondensation have coating thicknesses measureable in angstrom units. Obviously, if a gelatin-acacia coating were applied to cores precoated with a cellulose derivative as reported by Harris [4], then permeation through two coating barriers in series would have to be considered. The thickness of the stagnant diffusion layer is a function of stirring rate in the bulk aqueous phase, its composition and viscosity, and the molecular volume of the diffusant at a particular temperature. Its width is usually estimated indirectly from permeation experiments and is usually of the order of 100 μm for well-stirred systems. Because of the comparative thickness of the boundary layer relative to the coating, it is often an important factor controlling drug release.

Unlike the fluid phase, which is usually of uniform diffusional property, the coating material is normally less uniform, being composed of continuum, pores, and frequently, as in pan coating, dispersed solid phase. The continuum is generally composed of a high-molecular-weight polymer that gives the coating structure and whose composition, including degree of crosslinking, crystallinity, and plasticization, may provide an approximately uniform diffusional pathway for the penetrant drug molecules or alternatively may act as a heterogeneous barrier. Pores are frequently observed in the surface of microcapsules and microparticles and may penetrate directly to the core material or alternatively form a system of interconnecting channels that may extend partially or completely through the coating. Because of the comparative thinness of their coatings, such pores and other coating defects, such as minute cracks, are much more likely to be important in providing parallel or sole diffusional pathways in these dosage forms than in coated or matrix-type tablets. Vidmar et al. [5] have shown that parallel pathways for drug diffusion through pores and coating material exist for ethylcellulose microcapsules. Permeation through such pores can be reduced by use of a waxy sealant, which occludes them as shown in Fig. 14.2. Dispersed solids such as pigments and fillers may be embedded in the continuum or in pores and generally provide a discontinuous barrier that hinders diffusion of the drug. Diffusion of drug molecules through the coating material may also be affected by the simultaneous diffusion of water and other molecules through the coating from the external aqueous environment, which may affect the continuum. For example, enteric polymers such as cellulose acetate phthalate become soluble at pH values in excess of about 5.5, progressively reducing the magnitude of the continuum. Lippold [6] reported that water dissolved the polyethylene glycol content of microcapsule coating composed of quaternary polymethacrylic acid esters, increasing their porosity and release of encapsulated chloramphenicol. Polylactic

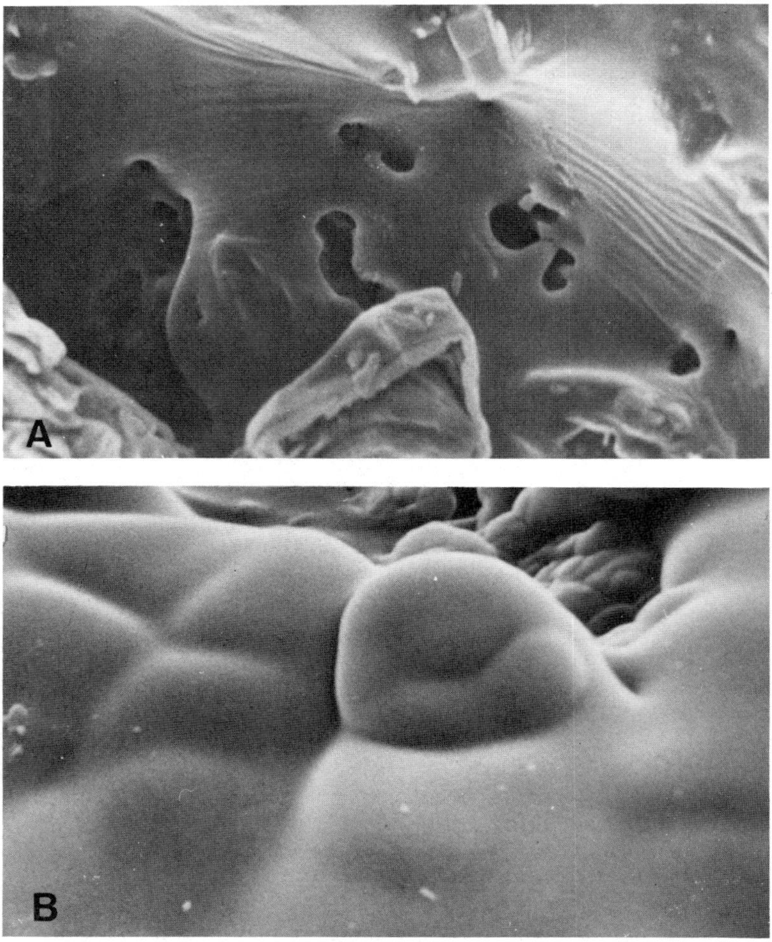

Figure 14.2 Ethylcellulose microcapsules without (A) and with (B) waxy sealant (paraffin wax), showing effect on the pores and surface.

acid and other biodegradable polymers slowly hydrolyze upon immersion in water with progressive loss in molecular weight and barrier resistance. Other polymers, such as methylcellulose and albumin, also swell upon immersion into aqueous media, causing altered porosity and hinderence to drug diffusion.

It is necessary to consider the interaction of the diffusing drug with the various phases it encounters while traversing the coating.

Hence partitioning of drug between the coating and aqueous phases and drug adsorption onto solid inclusions dispersed in the film are of importance.

14.3 DIFFUSION—SOME INITIAL MATHEMATICAL AND OTHER CONSIDERATIONS

Diffusion is the process whereby concentration differences are reduced by the spontaneous movement of matter. The drug flux J or drug mass dM in time dt passing through a reference plane perpendicular to the direction of diffusive flow is given by Fick's first-law equation,

$$J = \frac{dM}{dt} = -DA\frac{dC}{dx} \qquad (14.1)$$

where D is the diffusion coefficient (or diffusivity), A is the area of the reference plane and dC/dx is the concentration gradient. The negative sign implies that the flux is in the direction of decreasing concentration. D is really a proportionality factor, which was originally believed to be a constant but which, as is now known, may be concentration- and/or time-dependent. However, for the purposes of modeling drug release, D is usually assumed to be a constant and has the units $m^2 s^{-1}$. However, in many pharmaceutical applications of diffusion, drug concentration varies with both distance and time, and this relationship is expressed by Fick's second-law equation, or the differential equation of diffusion,

$$\left(\frac{\partial C}{\partial t}\right)_x = D\left(\frac{\partial^2 C}{\partial x^2}\right)_t \qquad (14.2)$$

The derivation of this equation and mathematical aspects of diffusion are presented in a number of standard textbooks [7–9]. It may be generalized to

$$\frac{\partial C}{\partial t} = D\left[\frac{\partial^2 C}{\partial x^2} + \frac{\partial^2 C}{\partial y^2} + \frac{\partial^2 C}{\partial z^2}\right] = D\nabla^2 C \qquad (14.3)$$

Fick's second law implies that the rate of change in concentration in a volume element of the diffusional field is proportional to the rate of change of the concentration gradient at that region of the field.

Over a wide range of temperatures, experimentally measured diffusion coefficients exhibit a temperature variation and may be represented by a simple Arrhenius equation:

$$D = D_0 e^{-E_a/RT} \qquad (14.4)$$

where R is the gas constant and T is the absolute temperature. Both the activation energy E_a and the preexponential temperature-independent factor D_0 may be obtained from a plot of ln D versus T^{-1}. Activation energies tend to increase for larger molecules and to be much higher for the diffusion of the same size molecules through polymers than through fluids. The diffusion coefficient of small, electrically neutral, spherical molecules are frequently related to their radius r by the Stokes-Einstein equation,

$$D = \frac{RT}{N_a} (6\P \eta r)^{-1} \qquad (14.5)$$

where N_a is the Avogadro number and η is the viscosity of the medium.

However, diffusion in microcapsules and microparticles may involve transport not only through an isotropic medium such as a drug in solution, but also through a homogeneous polymeric membrane. Transport of drug through such a membrane involves dissolution of the permeating drug in the polymer at the high concentration side of the membrane interface and diffusion across the membrane in the direction of decreasing concentration as shown in Fig. 14.3, where the effect of boundary diffusion layers has been neglected.

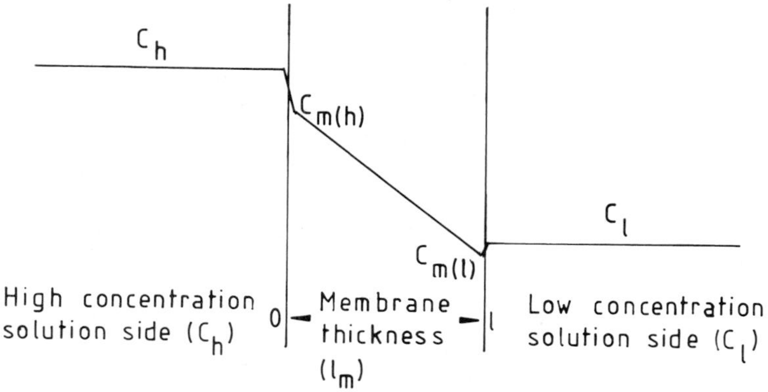

Figure 14.3 Concentration gradient across an ideal isotropic polymeric membrane (K < 1).

Obviously, the drug concentrations just inside the membrane interfaces are

$$C_{m(h)} = KC_h$$

at the high concentration side and

$$C_{m(l)} = KC_l$$

at the low concentration side, where K is the partition or distribution coefficient of the drug toward the polymer and is assumed to be constant. Assuming a steady state, Fick's first-law equation (14.1) may be integrated from x = 0 to x = l and from $C = C_{m(h)}$ to $C = C_{m(l)}$ to give

$$J = \frac{DA}{l_m}\left[C_{m(h)} - C_{m(l)}\right] = \frac{DA\,\Delta C_m}{l_m} \qquad (14.6)$$

where ΔC_m is the difference in drug concentration across the two faces of the membrane and l_m is the membrane thickness. As the concentration of drug dissolved in the membrane is usually unknown, Eq. (14.6) is usually written as

$$J = \frac{DKA\,\Delta C}{l_m} \qquad (14.7)$$

where ΔC is the difference in concentration of drug in solution on either side of the membrane. Membrane permeation is often characterized by the composite term permeability coefficient P, where P = DK. Sometimes the term permeability is used for DK/l_m where the membrane thickness is unknown, which is frequently the case with microcapsules. The term permeability constant is also sometimes used for D/l_m, and Takamura et al. [10] have described how this value for electrolytes may be determined with polyphthalamide microcapsules. From this value the apparent diffusion coefficient can be calculated to study the effect of membrane materials on the permeability of microcapsules as described in a subsequent paper by Takamura et al. [11].

14.4 DIFFUSION COEFFICIENT

Diffusion is always considered to take place through void spaces, that is, in the free volume of the matter under consideration. As the free volume of liquids is small, diffusivities through such phases tend to be low, $2 \times 10^{-10}\ m^2\ s^{-1}$ being a reasonable value for drugs in aqueous media at body temperature. The free volume of solid polymeric films tends to be much smaller, typical values for D being 1×10^{-14} to 1×10^{-16} m^2 s^{-1}.

The diffusion coefficient, which is a measure of the rate of drug movement, will be influenced by some factors shown in Table 14.1, which will be briefly considered as certain aspects thereof have already been covered in Chap. 2.

Diffusivity is very sensitive to the size of the penetrant, decreasing rapidly for large, bulky molecules with molecular weight over 500, which may require the fortuitous alignment of several polymer segments to create adequate voidage for their movement. As a generalization, the diffusion coefficient of molecules vary approximately as the cube root of their molecular weight, but shape factors are also very important. Though nominally assumed to be a constant, D_m can also vary with the concentration of the penetrant.

Polymeric membranes are not homogeneous. Their molecules often show some degree of preferential orientation, depending on their composition and method of casting. A typical linear polymer shows both ordered (crystalline) and discontinuous (amorphous) regions. Most drug molecules are poorly soluble in polymers and are unable to penetrate the crystalline regions of the polymer, so permeation takes place through the amorphous regions. Also, polymers with stiff backbones will decrease D_m, as they will find it more difficult to move than polymers with flexible backbones, such as various elastomers. Obviously, the diffusion of drug molecules will be slowed if large, bulky functional groups are attached to the polymer chain. Of course, as the temperature of a polymer is lowered below its glass transition temperature (T_g), the number of crystallites and related structures formed rapidly increases. As these structures are impermeable to all but very small molecules such as certain low-molecular-weight gases, they hinder diffusion of the penetrant through the polymer. Polymers form

Table 14.1 Some Factors that Influence the Diffusion Coefficient in Polymeric Membranes (D_m)

Factor	Net influence on D_m
Increase in penetrant size	Decrease
Increase in polymer crystallinity	Decrease
Increase in crosslinking	Decrease
Plasticizers	Increase
Dispersed solids	Decrease

crystallites with difficulty because of the low probability of orientating their chains in a regular manner. Thus diffusivity may become more marked with increase in polymer chain length and degree of branching. However, as the molecular weight of a polymer is increased, the glass transition temperature is increased because there are fewer chain ends that have more free volume than segments. This gives rise to a decrease in diffusivity. The net effect is that increase in molecular weight of the polymer may produce only a small reduction in diffusivity, as reported by Deasy et al. [12] for sodium salicylate release from ethylcellulose microcapsules prepared using a 10- or 100-cP grade.

Increasing the degree of crosslinking in a polymer reduces polymer chain mobility, increases its glass transition temperature, and decreases the diffusivity of penetrant molecules, particularly large ones. Plasticizers reduce polymer chain interactions and so increase diffusivity. Dispersed solids such as pigments and fillers decrease diffusivity if present. Assuming that these particles are randomly distributed and orientated in the membrane material, they will slow the rate of diffusion by causing the penetrant molecules to stream around them. They will occupy volume and may obstruct pores, thus reducing the available diffusional pathways. These particles often interact by an adsorption phenomenon with the penetrant molecules, which interaction may be quantified by the Langmuir equation at a particular temperature. Depending on whether the concentration of diffusant is adequate to saturate the adsorption sites and whether physical or chemical adsorption occurs, these active dispersed solids will affect membrane transport primarily in the nonstationary state.

Many microcapsules and microparticles have coating material composed of hydrophilic linear colloid such as gelatin, which forms a swollen, three-dimensional network upon immersion into water as shown in Fig. 14.4. As the water penetrates the polymer, it tends slowly to form a gel layer that is separated from the unaffected polymer by a partially wetted layer. Depending on the volume fraction of polymer, such gels tend to hinder the outward transport of core diffusant, either by mechanically blocking it and forcing it by a more tortuitous pathway through the aqueous phase or by directly interacting with it. Chain length of polymer and degree of crosslinking affect only bulk viscosity and not microscopic viscosity, and so have little effect on the diffusivity of small drug molecules. Larger drug molecules will have increased likelihood of collision with polymer molecules slowing their rate of diffusion. Gel molecules also frequently interact with diffusant molecules, depending on the number of binding sites on each polymer molecule, and this effect will tend to tail off when the gel becomes saturated with diffusant. Obviously this effect could be reduced by the addition of a competitive adsorbant to the core material. As such hydrophilic polymers continue to take up water, they progressively swell. Takenaka et al. [13] attributed retardation in drug release

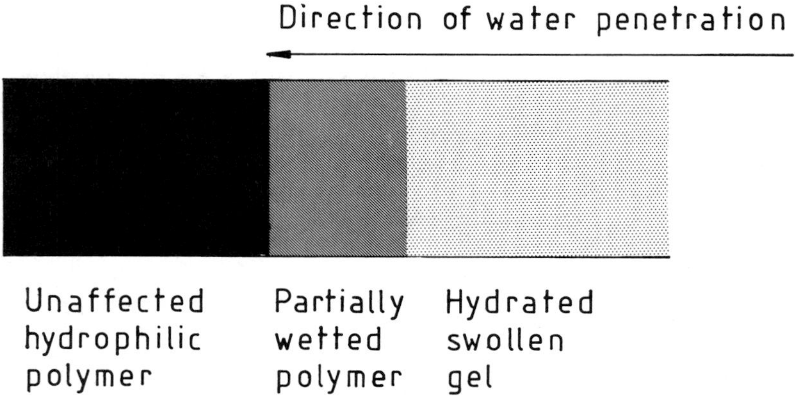

Figure 14.4 Progressive penetration of water into a hydrophilic polymer.

from gelatin-acacia microcapsules in the terminal stages of dissolution experiments to swelling of the coating, which lengthens the diffusional path length of the solute molecules. With progressive imbibition of water such hydrogels tend eventually to exhibit diffusion coefficients that approach those obtainable in water.

The diffusivities of drugs in various polymeric materials that are used for encapsulation can vary a thousand or more fold. Takenaka et al. [13] reported apparent diffusion coefficients of sulfamethoxazole through a gelatin-acacia coacervated microcapsule wall of 1.63×10^{-13} to 2.83×10^{-14} m^2 s^{-1}, the value tending to decrease with increasing coacervation pH and increasing amount of formaldehyde used as a cross-linking agent. The diffusion coefficient of the drug at 37°C was calculated to be 6.87×10^{-10} m^2 s^{-1}. Senjkovic and Jalsenjak [14] reported that the apparent diffusion coefficient of isoniazid in ethylcellulose membranes of microcapsules varied from 0.53×10^{-11} to 2.79×10^{-11} m^2 s^{-1}, depending on size fraction, which indicates that the microcapsules were highly porous.

An apparent permeability coefficient (defined as $P_{appar} = D_{appar}/l$ of gelatin-acacia microcapsules to entry of sodium chloride was determined by Jalsenjak and Kondo [15]. Smaller microcapsules were surprisingly shown to have lower apparent permeability coefficients, possibly because of greater structural water around such microcapsules, which retarded movement of hydrated ions. Similar results were also reported by Senjkovic and Jalsenjak [16] for the apparent diffusion coefficients of phenobarbitone sodium from variable-size ethylcellulose microcapsules. Apart from the influence of structural water, the effect was possibly due to the lower porosity of the membranes of smaller microcapsules as confirmed by density determinations. Differential scanning calorimetry was later used by Vidmar et al. [5] to show that there was little or no structural water associated with ethylcellulose membranes.

14.5 PARTITION COEFFICIENT

The partition coefficient in the usual convention is taken as

$$K = \frac{C_{o\ or\ m}}{C_{aq}} \tag{14.8}$$

where the subscripts o or m and aq stand for the organic or membrane and aqueous phase, respectively. Hence the higher the value for K, the greater is the nonpolar solvent, lipid, or hydrophobic polymer solubility. The penetration of aqueous solutions of drugs through polymeric coating depends on the membrane-water partition coefficient, as indicated in Eq. (14.7). Partition coefficients are often determined for octanol-water systems, which may be related by regression equations to other organic solvent or membrane-water partition coefficients. Generally, by studying the partitioning of a reference derivative in a particular system, the log partition coefficient of an analog or homolog can be determined by summing the group contributions or ¶'-values of the structural modifications in relation to the reference compound and adding these to its log partition coefficient. ¶' may be defined as

$$¶' = \log K_x - \log K_a \tag{14.9}$$

where K_a is the partition coefficient of a parent molecule between two solvent phases and K_x is that for a derivative [17]. Hansch and Dunn [18] have reviewed the relationship between partitioning and biological activity in drugs, as there have been numerous quantitative relationships between activity and log K or log K_{appar} values reported in the literature. Such values have also been widely used to quantify interaction of drugs in dosage forms with packaging materials. As there tends to be a good relationship between structural modification in a class of compounds and their partitioning in varied solvents, it is often possible to make reliable inference about partitioning in membranes, which might be difficult to determine experimentally from simple organic solvent-water systems.

The partition coefficient may be varied by altering the solubility of the drug in the elution medium. It should also be appreciated that as the partition coefficient becomes large, the flux through a polymeric membrane tends to become diffusion layer-controlled and hence to become insensitive to further increase in partition coefficient.

Fick's law may fail to quantify drug release accurately when the partition coefficient is not constant but is dependent on drug concentration. This would occur particularly if the drug is a weak acid or a weak base were the value of K would be pH-dependent. It could also occur if the drug binds in a nonlinear fashion to some component of the system, as with Langmuir-type binding.

14.6 DRUG SOLUBILITY AND SOLUBILITY GRADIENT

The flux of drug released is proportional to the concentration difference across the membrane and will be maximized when the penetrant is presented to the membrane at its saturated concentration and when the receiving phase contains negligible concentration, i.e., tends to act as a perfect sink. Drugs must usually be in a molecularly dispersed form before diffusion across membranes can occur, and consequently the process of solution is very important. Obviously, the magnitude of the concentration difference, which is taken as the driving force for drug transport across the membrane, will tend to decrease as the solubility of the drug on the upstream side of the membrane decreases. For this reason the dissolution rate of poorly soluble drugs can be a very important factor in limiting drug release from microcapsules and microparticles.

Noyes and Whitney [19] described the dissolution of a solid into a liquid when the process is diffusion-controlled by the equation

$$\frac{dC}{dt} = k(C_s - C) \tag{14.10}$$

where k is the dissolution rate constant, C_s is the saturation solubility of the solute in the given solvent, and C is the concentration of solute in the solvent after time t. Brunner [20] defined k to be

$$k = \frac{D_s A}{V l_b} \tag{14.11}$$

where D_s is the diffusion coefficient in the solvent, V is the volume of the solution, and l_b is the boundary diffusion layer thickness. Combining both equations gives

$$\frac{dC}{dt} = \frac{D_s A}{V l_b}(C_s - C) \tag{14.12}$$

Si-Nang and Carlier [21] presented a modification of this equation to quantify drug release from microcapsules as follows:

$$\left(\frac{dC}{dt}\right)_{coating} = \frac{D_s A'k'}{Vl_m} \tag{14.13}$$

where A' is the internal surface area of the coating and k' is a coefficient expressing the porosity and tortuosity of the coating. This equation was then modified to calculate an apparent wall diffusion coefficient.

Khanna et al. [22] presented the following equation derived from the Noyes-Whitney expression:

$$W_0^{1/3} - W_t^{1/3} = kat \tag{14.14}$$

where W_0 is the initial weight of the particle, W_t is the weight of the particle at time t, and a is the surface weight fraction at time t. Hence if k and a are constants, the cube root of the residual weight plotted against time should be linear. Such a plot was used to characterize the dissolution behavior of epoxy resin beads containing chloramphenicol.

The majority of drugs are weak organic electrolytes and their aqueous solubility is affected mainly by increasing ionization. Acidic drugs are less soluble at low pH because they are predominantly undissociated and cannot interact with water to the same extent as ionized forms. The solubility of such drugs at any pH is given by

$$pH - pK_a = \log\left(\frac{S - S_0}{S_0}\right) \tag{14.15}$$

where S is the total saturation solubility of the drug and S_0 is the solubility of the undissociated species. Likewise, basic drugs are less soluble at high pH, and the solubility of such drugs at any pH is given by

$$pH - pK_a = \log\left(\frac{S_0}{S - S_0}\right) \tag{14.16}$$

where S_0 is obtainable at very low pH for an acidic drug and at very high pH for a basic drug. Amphoteric drugs such as sulfonamides and tetracyclines, which display both basic and acidic characteristics, have lowest solubility at the pH of zwitterion formation, with solubility increasing above the isoelectric point in accordance with Eq. (14.15) and below in accordance with Eq. (14.16). However, many coating materials employed are hydrophobic and nonpolar, being more readily penetrated by the nonionized form of the drug unless the membrane is penetrated by aqueous-filled pores of adequate dimension to allow passage of solute molecules. Likewise, the nonionized form of drugs is better absorbed in vivo through the lipoidal membranes surrounding cells. High-molecular-weight compounds often have low aqueous and lipid solubility and hence are poorly transported and absorbed.

An extensive review of other factors that influence solubility of drugs, such as the interaction of substituent groups, molecular size, and solid-state properties such as polymorphism, coprecipitate formation, and particle size, are outside the scope of this book. The interested reader should consult Refs. 23 and 24. However, to obtain microcapsules that release drug at a constant or zero-order rate, it is important to maintain the thermodynamic activity of the drug within the product at a fixed value. For readily soluble drugs this is usually done by using a saturated solution with excess solid phase or more usually the solid or liquid drug alone. Constant release will continue as long as excess solid or liquid phase is present, provided that other factors remain constant.

14.7 COATING AREA AND THICKNESS

The flux of drug passing through microcapsule coating will increase as the core coating interfacial area increases, as indicated in Eq. (14.7). Thus decrease in particle size has been reported by numerous research workers to enhance drug release from such products as it increases the area of contact between a given weight of core and coating. The accompanying decrease in coating thickness in such a situation would also increase the flux of drug released.

14.8 KINETICS OF DRUG RELEASE FROM MICROCAPSULES AND MICROPARTICLES

Mathematical models used to describe the kinetics of drug release from microcapsules or microparticles are usually based on drug release from macrocapsular or macromatrix devices, respectively. Because of the size difference, drug release from the microdosage forms tends more quickly to attain a steady state that is of shorter duration. These dosage forms are often irregular in shape, so that conventional models based on spherical, cylindrical, or other regular geometries often show a poor fit for release data. Also, many microcapsules, particularly those produced by various coacervation procedures, are multinuclear or are composed of aggregates of smaller microcapsules so that their release kinetics do not follow that expected of a reservoir-type device but rather that of a monolithic device. Because of the diversity of inclusion and lack of homogeneity of many polymeric coatings, the following review of mathematical approaches that have been used to quantify drug release in vitro from microcapsules and microparticles is of interest primarily for indicating those factors that influence drug release. By identifying these factors it is possible to vary them to achieve greater control over core release from these products.

14.9 MICROCAPSULES CONFORMING TO RELEASE FROM RESERVOIR-TYPE DEVICES

In this section a number of mathematical expressions derived for reservoir-type devices are considered. Assuming that the thermodynamic activity of the core material is maintained constant within the microcapsule, which is spherical and has inert homogeneous coating as shown in Fig. 14.5, then the steady-state release rate derived from Fick's first law is given by

$$\frac{dM_t}{dt} = 4\pi DK\Delta C \frac{r_o r_i}{r_o - r_i} \qquad (14.17)$$

where r_o and r_i are the outside and inside radii, respectively [25]. Assuming that all the parameters on the right-hand side of Eq. (14.17) remain constant, integration of the equation over a finite period of the steady state would indicate that drug release is constant or zero-order. Obviously, the rate of drug release will be decreased by increase in coating thickness (r_o - r_i). However, as r_o becomes very much larger than r_i, the steady-state release rate will tend to become dependent only on r_i rather than on the outer dimensions of the microcapsule, because $dM_t/dt \rightarrow 4\pi DK\,\Delta C r_i$. Baker and Lonsdale [25] have stated that when r_o/r_i exceeds a value of about 4, further increase in device size for a fixed-core radius will not significantly change the release rate. Thus increasing the thickness of coating on microcapsules already conforming to these disparities of core:coating radii will have little further

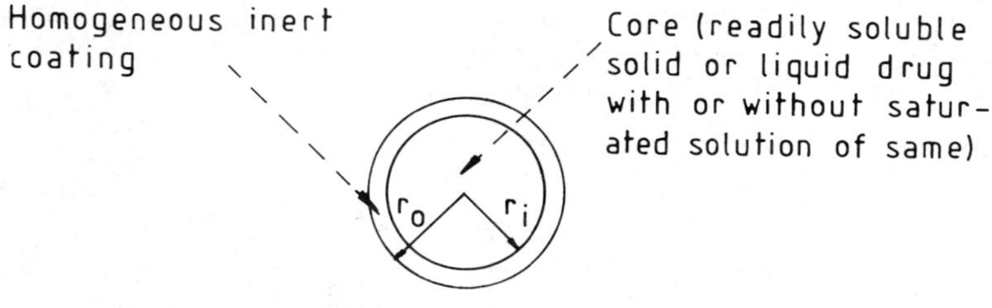

Figure 14.5 Idealized spherical microcapsule.

influence on drug release rate, though it will delay or even abolish the onset of a steady-state release.

If the thermodynamic activity of the core material does not remain constant because initially a saturated or unsaturated drug solution was encapsulated or because similar conditions apply during the terminal release phase, then the rate of release will become exponential or first-order. As an example of the mathematical calculations involved, other relevant equations will now be developed, but subsequently—because of space requirements—only final equations describing aspects of drug release will be presented with appropriate references where applicable. Assume that the volume of the internal reservoir is V_i and the volume of the outside receiving fluid or sink is V_o. The mass of drug remaining at any time in the reservoir is M_{it} and in the sink is M_{ot} such that the total mass of drug M_{tot} is equivalent to

$$M_{tot} = M_{it} + M_{ot} \tag{14.18}$$

such that at $t = 0$, $M_{tot} = M_{it}$. Also, $M_{it} = C_{it}V_i$ and $M_{ot} = C_{ot}V_o$, where C_{it} and C_{ot} are the drug concentrations in the inner and outer compartments, respectively, after time t. Assuming the area of the coating to be

$$\frac{1}{2}(4\pi r_i^2 + 4\pi r_o^2) = 2\pi(r_i^2 + r_o^2)$$

then, by analogy with Eq. (14.7) based on Fick's first law,

$$\frac{dM_{it}}{dt} = \frac{-2\pi(r_i^2 + r_o^2)DK}{r_o - r_i}(C_{it} - C_{ot}) \tag{14.19}$$

$$= \frac{-2\pi(r_i^2 + r_o^2)DK}{r_o - r_i}\left(\frac{M_{it}V_o - M_{ot}V_i}{V_iV_o}\right) \tag{14.20}$$

and rearranging,

$$\frac{dM_{it}}{M_{it}V_o - M_{ot}V_i} = \frac{-2\pi(r_i^2 + r_o^2)DK\ dt}{(r_o - r_i)V_iV_o} \tag{14.21}$$

Substituting $M_{ot} = M_{tot} - M_{it}$,

$$\frac{dM_{it}}{M_{it}(V_o + V_i) - M_{tot}(V_i)} = \frac{-2\pi(r_i^2 + r_o^2)DK\ dt}{(r_o - r_i)V_iV_o} \tag{14.22}$$

Upon integrating using $u = M_{it}(V_o + V_i) - M_{tot}(V_i)$ such that $du = (V_o + V_i)dM_{it}$,

$$\left(\frac{1}{V_o + V_i}\right)\ln[M_{it}(V_o + V_i) - M_{tot}(V_i)] = \frac{-2\P(r_i^2 + r_o^2)DKt}{(r_o - r_i)V_iV_o} + B \tag{14.23}$$

To find a value for the integration constant B, let $M_{it} = M_{tot}$ at $t = 0$; therefore,

$$B = \frac{1}{(V_o + V_i)} \ln(M_{tot}V_o) \tag{14.24}$$

which upon substitution into Eq. (14.23) gives

$$\frac{1}{V_o + V_i} \ln[M_{it}(V_o + V_i) - M_{tot}V_i] = \frac{-2\P(r_i^2 + r_o^2)DKt}{(r_o - r_i)V_iV_o} + \frac{1}{V_o + V_i} \ln(M_{tot}V_o) \tag{14.25}$$

Multiply across by $(V_o + V_i)$ to get

$$M_{it}(V_o + V_i) - M_{tot}V_i = M_{tot}V_o \exp\left[\frac{-2\P(r_i^2 + r_o^2)DK(V_o + V_i)t}{(r_o - r_i)V_iV_o}\right] \tag{14.26}$$

or

$$M_{it} = \frac{M_{tot}}{V_o + V_i} V_o \exp\left[\frac{-2\P(r_i^2 + r_o^2)DK(V_o + V_i)t}{(r_o - r_i)V_iV_o}\right] \tag{14.27}$$

The release rate may be obtained by differentiating the equation to get

$$\frac{dM_{it}}{dt} = \frac{-M_{tot}2\P(r_i^2 + r_o^2)DK}{V_i(r_o - r_i)} \exp\left[\frac{-2\P(r_i^2 + r_o^2)DK(V_o + V_i)t}{(r_o - r_i)V_iV_o}\right] \tag{14.28}$$

Equation (14.28) indicates that the release rate declines exponentially with time and also with coating thickness. Obviously the time to release half the initial drug loading in the microcapsule (i.e., the half-life, $t^{1/2}$ or $t_{50\%}$) may be calculated by setting $M_{it} = M_{tot}/2$ in Eq. (14.27) and rearranging:

$$t^{1/2} = \frac{-(r_o - r_i)V_iV_o}{2\P(r_i^2 + r_o^2)DK(V_o + V_i)} \ln\left(\frac{V_o - V_i}{2V_o}\right) \tag{14.29}$$

As the sink V_o is usually very much larger than the internal volume of the microcapsules V_i, Eqs. (14.27), (14.28), and (14.29), respectively, become

$$M_{it} = M_{tot} \exp\left[\frac{-2\P(r_i^2 + r_o^2)DKt}{(r_o - r_i)V_i}\right] \tag{14.30}$$

$$\frac{dM_{it}}{dt} = \frac{-M_{tot}2\P(r_i^2 + r_o^2)DK}{V_i(r_o - r_i)} \exp\left[\frac{-2\P(r_i^2 + r_o^2)DKt}{(r_o - r_i)V_i}\right] \tag{14.31}$$

and

$$t^{1/2} = \frac{-(r_o - r_i)V_i \ln(0.5)}{2\P(r_i^2 + r_o^2)DK} \tag{14.32}$$

$$= \frac{0.693\,(r_o - r_i)V_i}{2\P(r_i^2 + r_o^2)DK} \tag{14.33}$$

Exponential decline in chlorpheniramine maleate release from waxy-coated microcapsules into aqueous medium was reported by Chambliss et al. [26]. This is to be expected, as the drug is very water-soluble and would quickly decline below its saturated solution concentration within the product. The in vitro release rate was enhanced by addition of the surfactant docusate sodium to the dissolution medium, though the in vivo availability of the drug in humans was not significantly altered by coadministration of the surfactant, as presumably drug release from the microcapsules was not the rate-limiting step in drug absorption.

Recently Christensen et al. [27] have presented complex equations describing in vitro drug release from coated pellets based on non-quasistationary diffusion from nonswelling spheres of equal or different size. Good agreement between predicted and actual drug release was reported for propoxyphene hydrochloride pellets.

14.10 LAG TIME AND BURST EFFECTS

Microcapsules with constant activity cores usually exhibit an initial delay or lag time, t_l (see Fig. 14.6) before achieving the steady-state value if tested very shortly after preparation, because the concentration gradient within the coating will not yet have been fully achieved. Alternatively, if the microcapsules have been stored for a considerable period of time before testing, which is more commonly the case, they will exhibit an initial release rate higher than the steady-state value. This so-called burst effect can be troublesome in practice, as it leads to initial overdosage. It arises from saturation of the coating by drug during storage with rapid release of drug from the outer regions of the coating. A similar effect is observed when drug particles are embedded in the outer surface of the coating. Lag times for microcapsules have been reported by Francois and De Neve [28] and burst effects for microparticles by Rhine et al. [29] or microcapsules by Madan [2].

For flow through a spherical wall, the intercepts t_l of the steady-state section of plot (b) in Fig. 14.6 is given by Crank [7] as

$$t_1 = \frac{(r_o - r_i)^2}{6D} \tag{14.34}$$

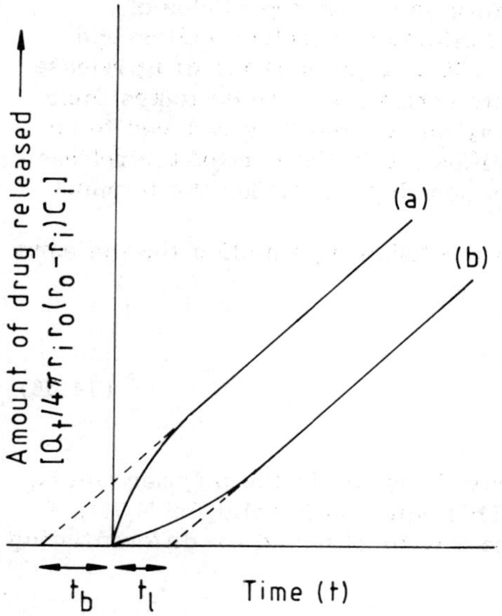

Figure 14.6 Plot of amount of drug released versus time for microcapsules showing a burst effect (a) or a lag time effect (b) before reaching steady state.

In theory the value for t_l could then be used to calculate D if the dimensions of the microcapsule where known, but in practice the variable contribution by prior diffusion of the drug through the coating before testing would be difficult to account for. Likewise, the intercepts on the time axis of plot (a), which is the burst time t_b, is given by analogy with the observations of Baker and Lonsdale [25] as

$$t_b = \frac{-(r_o - r_i)^2}{3D} \tag{14.35}$$

This value may again be used with caution to determine D. From Eqs. (14.34) and (14.35) it can be seen that the time necessary to reach the steady state depends only on the square of the coating thickness and the reciprocal of the diffusion coefficient. Obviously, more varied initial release rates will be observed if the drug is only partially equilibrated with the coating during storage.

14.11 MICROCAPSULES AND MICROPARTICLES CONFORMING TO RELEASE FROM MONOLITHIC DEVICES

14.11.1 Drug Dissolved in Coating Polymer

An example of this type of dosage form is the microparticles of polylactic/glycolic acid copolymers containing dissolved naltrexone base as reported by Schwope et al. [30]. Equations for drug release from such systems contain an infinite series term, which makes their use troublesome. However, they may be conveniently reduced to an early time approximation that quantifies the initial part of the release curve and into a late time approximation that quantifies the terminal portion.

Baker and Lonsdale [25] gave the following equation for the early time approximation for a sphere:

$$\frac{M_t}{M_\infty} = 6\left(\frac{Dt}{r^2\pi}\right)^{1/2} - \frac{3Dt}{r^2} \tag{14.36}$$

where M_∞ is the total amount of drug dissolved in the polymer and M_t is the amount released at time t. This equation is valid for $M_t/M_\infty <$ 0.4. Total drug release at any time may be obtained by differentiating the above equation to give

$$\frac{d(M_t/M_\infty)}{dt} = 3\left(\frac{D}{r^2\pi t}\right)^{1/2} - \frac{3D}{r^2} \tag{14.37}$$

The corresponding late time approximation equations are

$$\frac{M_t}{M_\infty} = 1 - \frac{6}{\P^2}\exp\left[\frac{-\P^2 Dt}{r^2}\right] \tag{14.38}$$

which is valid for $M_t/M_\infty > 0.6$, and

$$\frac{d(M_t/M_\infty)}{dt} = \frac{6Dt}{r^2}\exp\left[\frac{-\P^2 Dt}{r^2}\right] \tag{14.39}$$

These equations indicate that the release rate for the first 40% of the drug is linearly related to the square root of the reciprocal of time, i.e., $t^{-1/2}$ and for the final 40% of drug in an exponential manner. Between 40 and 60% release there is a progressive change from $t^{-1/2}$ to exponential dependence.

Guy et al. [31] considered diffusion from a sphere where a phase boundary existed at the interface such that slow interfacial transfer is the rate-limiting step for drug release. The early time approximation showed that the amount of drug released was proportional to the interfacial rate constant and to the bulk concentration of drug in the sphere. The corresponding late time approximation showed the release to be exponential.

14.11.2 Drug Dispersed in Coating Polymer

Many microcapsules form clusters during preparation, particularly those produced by various coacervation procedures such as gelatin and ethylcellulose microcapsules. However, Luzzi, et al. [32] also reported the formation of clusters in nylon coated phenobarbitone sodium microcapsules recovered by a spray-drying process. The approximately spherical units formed are usually composed of solid drug particles dispersed in the coating material, and because of the relatively small number of irregularly joined nuclei they contain, their uniformity of drug dispersion is usually lower than that produced by microparticle production procedures. Obviously a portion of the drug (C_m) may dissolve in the membrane material, but if this is assumed to be very small compared to the total drug loading (C_{tot}), then the expected drug release has been mathematically derived from Higuchi [33] by Baker and Lonsdale [25] for a homogeneous spherical matrix as follows:

$$\frac{d(M_t/M_\infty)}{dt} = \frac{3C_m D}{r_o^2 C_o}\left[\frac{(1 - M_t/M)^{1/3}}{1 - (1 - M_t/M_\infty)^{1/3}}\right] \quad [14.40]$$

The assumption that $C_{tot} \gg C_m$ is usually valid for most polymeric drug dispersions containing greater than 1% drug. This model (see Fig. 14.7a) assumes that the solid drug dissolves from the surface layer first, and when this layer has become exhausted the next layer begins to dissolve. Obviously, from Eq. (14.40) it is not possible to express the release rate as a single function of time. Accordingly, the release rate from a slab geometry is often used to approximate drug release from such irregularly shaped microcapsules and microparticles. The relevant equation is

$$\frac{dM}{dt} = \frac{A}{2}\left[\frac{DC_m}{t}(2C_{tot} - C_s)\right]^{1/2} \tag{14.41}$$

which upon integration gives

$$Q = \frac{M}{A} = \left[DC_m(2C_{tot} - C_s)t\right]^{1/2} \tag{14.42}$$

where Q is the mass of drug released per unit area of surface at time t. The value for C_{tot} should preferably be at least 10 times that of C_s. Assuming that the diffusion coefficient and other parameters of Eq. (14.42) remain constant during release, this equation may be expressed as

$$Q = k_1 t^{1/2} \tag{14.43}$$

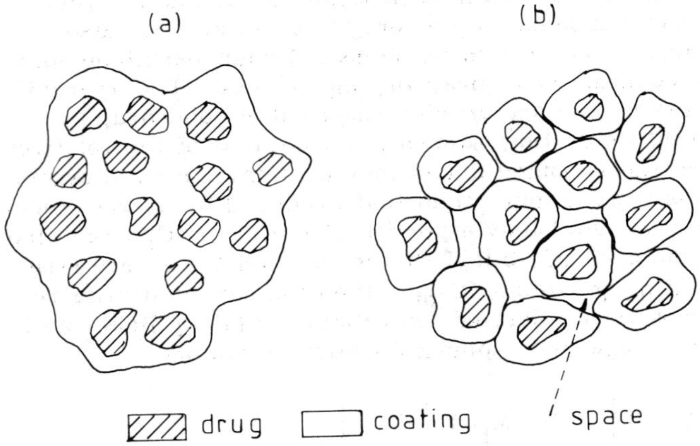

Figure 14.7 Drug release from homogeneous (a) and granular (b) matrices.

when k_1 is a constant. Jalsenjak et al. [34], Madan [35], and many others have reported that drug release from microcapsules produced by coacervation is proportional to the square root of time. From Eq. (14.42) it is also obvious that the release rate is proportional to the square root of the drug loading, which can be readily varied by altering the drug:polymer ratio.

Higuchi [36] subsequently presented the following equation for the steady-state drug release from the planar surface of a granular-type matrix as might be observed with a cluster of microcapsules as shown in Fig. 14.7b, where geometry is usually very irregular:

$$Q = \left[\frac{D\varepsilon}{\tau}(2C_{tot} - C_s)C_s t\right]^{1/2} \tag{14.44}$$

where ε is the porosity of the leached portion of the matrix and τ is the tortuosity of the matrix or degree of nonlinearity of the formed capillaries. It was assumed that $C_{tot} \gg C_s$, that the noninteractant, uniformity dispersed drug particles were much smaller than those comprising the matrix, and that perfect sink conditions prevailed. As the tortuosity factor is often assumed to be about equal to 3, Eq. (14.44) is valid provided that C_{tot} is greater than C_s by a factor of 3 or 4. In reality, many microcapsules and microparticles contain drug loadings not greatly in excess of the drug solubility, and the application of Eqs. (14.42) and (14.44) to such products is questionable. More recently, Fessi et al. [37] have shown that a similar square-root equation of the form

$$Q = C_{tot}(Dt)^{1/2} \tag{14.45}$$

also applied to matrix formulations where $C_{tot} < \varepsilon C_s$.

Lag time and burst effects may be observed with monolithic devices but these cannot be used to estimate the diffusion coefficient of the drug species. However, the polymeric diffusivity can be conveniently determined from the slope ($D_m{}^{1/2}$) of the linear plot of

$$\frac{Q}{t^{1/2}} \quad \text{versus} \quad \left[(2C_{tot} - C_s)C_s\right]^{1/2}$$

Roseman [38] has discussed in greater detail various equations that have been developed to quantify drug release from monolithic polymer devices of slab, cylindrical, and spherical geometry.

Takenaka et al. [13] cited tortuosity values of 86 to 1140 based on dissolution studies in distilled water for gelatin-acacia coacervated microcapsules containing sulfamethizole, which increased with increasing pH of coacervation and amount of crosslinking agent used.

14.12 PORE EFFECTS

The flux of drug passing through the pores of certain microcapsules can be a major source of drug release. Large pores in adequate number provide a release mechanism that is apparently independent of the coating and is controlled by the rate of dissolution of the core. However, when very fine pores are present in the coating whose diameter is only slightly greater than that of the diffusing species, appreciable resistance is offered by the coating to mass transport. The relationship between the diffusivity in such fine pores D_p and in free solution D_s for spherical solute molecules of radius r_s passing through a cylindrical pore of radius r_p has been given by Flynn et al. [39] as

$$\frac{D_p}{D_s} = \left(1 - \frac{r_s}{r_p}\right)^2 \left[1 - 2.104\frac{r_s}{r_p} + 2.09\left(\frac{r_s}{r_p}\right)^3 - 0.95\left(\frac{r_s}{r_p}\right)^5\right] \qquad (14.46)$$

This equation shows that large molecules are slowed to a much greater extent than smaller ones in their passage through fine pores. In microcapsules where diffusion through the polymeric continuum is negligible, as occurs with many microcapsules produced by interfacial polycondensation, these porous membranes act as semipermeable coatings, allowing the transport in or out of such microcapsules of small, low-molecular-weight solutes but effectively retaining high-molecular-weight molecules. Chang [40] effectively used this approach for the production of his range of nylon-coated "artificial cells."

Tateno et al [41] estimated from electrophoretic data that 1 to 3% of space in the walls of ethylcellulose-coated microcapsules prepared by an interfacial polymer deposition procedure were composed of pores. Vidmar et al. [5] reported a value of 0.55 to 2.5% for the volume fraction of the water-filled pores in ethylcellulose microcapsule membranes. Chang and Poznansky [42] studied the penetration of drug molecules of differing solute radius into nylon 6-10 microcapsules under experimental conditions where it seemed reasonable to conclude that solute permeation occurred only through aqueous pores.

14.13 BOUNDARY-LAYER EFFECTS

With the exception of the section on dissolution, the equations presented so far assume that the surface drug concentration of the microcapsules or microparticles is effectively zero. This is incorrect, however, because there tends to be a stagnant boundary layer of appreciable drug concentration in contact with the surface, which can hinder drug release by diffusion. The effect of the layer is more marked with drugs of low solubility, where their concentration in the unstirred layer can tend toward the drug's solubility with resultant loss in driving force for diffusion. Corrugations on the surface of microcapsules and microparticles increase the effective thickness of the unstirred layer, whose typical thickness is approximately 500 μm in an unstirred system. Microcapsules with thin coatings may show release rates that depend on coating area, indicating the importance of the boundary-layer effect.

The total diffusional resistance R_{tot} for solute diffusion from the core of a microcapsule is equal to the sum of the resistance of the coating membrane R_m and the boundary layer R_b such that

$$R_{tot} = R_m + R_b \tag{14.47}$$

or

$$R_{tot} = \frac{l_m}{D_m K} + \frac{l_b}{D_b} \tag{14.48}$$

where D_b is the apparent diffusion coefficient in the boundary layer [43]. As the thickness of boundary layer decreases approximately as the square root of the stirrer speed employed, its effect should be theoretically eliminated at infinite stirrer speed. Therefore a plot of R_{tot} versus $1/\sqrt{\text{stirrer speed}}$ will give an extrapolated intercepts on the R_{tot} axis corresponding to infinite stirrer speed of $1/\sqrt{\text{stirrer speed}} = 0$, from which the permeability coefficient of the membrane may be determined provided that its thickness is known. Senjkovic and Jalsenjak [14] have used this approach to determine the permeability coefficient of ethylcellulose microcapsules.

Chien [44] presented the following equation to describe cumulative drug release rate from reservoir-type systems containing solid drug particles and having a boundary-layer effect, sink conditions being maintained throughout the experiment:

$$\frac{Q}{t} = \frac{C_m K D_s D_m}{K D_s l_m + D_m l_b} \tag{14.49}$$

If $KD_s l_m$ is $\gg D_m l_b$, the above equation simplifies to

$$\frac{Q}{t} = \frac{C_m D_m}{l_m} \tag{14.50}$$

or if $D_m l_b \gg KD_s l_m$, to

$$\frac{Q}{t} = \frac{C_m K D_s}{l_b} \tag{14.51}$$

Under both these extreme conditions drug release will be zero-order. In the case of the membrane-limited diffusion process (Eq. 14.50), the rate of drug release is a linear function of membrane solubility and membrane diffusivity and is inversely proportional to membrane thickness. In the case of boundary-layer-limited diffusion (Eq. 14.51), the release is a linear function of the solute solubility in the sink fluid, the diffusivity in same, and is inversely proportional to the boundary-layer thickness.

Satyanarayana Gupta and Sparks [45] have presented a mathematical model for progesterone release from spherical injectable polylactic acid microcapsules. No details of the microencapsulation procedure were given, but the model was based on a matrix device where a boundary layer and a time-variant drug concentration on the outside of the particles were considered. Though the model did not exactly predict the in vitro rate of drug release, it was found to give better agreement than the Higuchi model [36], where perfect sink conditions were assumed. It may be expected that the model would also be more applicable in the in vivo situation, where greater boundary layers are assumed to exist. The final equation for the rate of drug release was

$$\frac{dQ}{dt} = \frac{4\pi r'}{Kr_o(r_o - r')}\left[KC_sDr_o^2 + \frac{DW'C_{tot}}{\rho V_o r_o}(r_o^3 - r')\right] \qquad (14.52)$$

where r' is the radius of the shrinking core, W' is the weight of microcapsules used in release experiments, and ρ is the microcapsule density.

14.14 MOVING BOUNDARIES

Drug release from microcapsules and microparticles has been so far considered as not involving any physical change in the dimensions of the device during the course of release studies. However, many polymers and other coating materials undergo erosion or swelling upon immersion into aqueous media, and their release kinetics is further complicated by the moving boundary of the eroding or swelling coating front. The presence of moving boundaries usually introduces a nonlinear release dependence with time for which exact solutions are known only in special circumstances [7].

Lee [46] has reported a number of equations applicable to drug release from eroding or swelling polymer surfaces. An equation relating to erosion indicated that if this process was more important than the diffusional process and if the total loading was considerably greater than the concentration in the polymer, then release started with a first-order rate that shifted toward zero-order release. Provided that available surface area remained constant, this occurred because at longer times there was a identical rate of movement of the diffusional and eroding fronts. Hopfenberg [47] has also presented equations describing the idealized kinetics for erodible slabs, cylinders, and spheres, indicating that constant delivery rate is provided only by the slab geometry.

The equation of Lee relating to increasing polymer surface predicted a square root of time dependence for drug release from a swellable matrix. Drug release from swellable microcapsules produced by coacervation procedures using hydrophilic polymers such as gelatin and acacia has frequently been reported to be $t^{1/2}$-dependent.

Swelling of the coating may weaken the cohesive strength of the coating until eventually either the outer layer of coating or a larger portion thereof may slough off to permit more rapid and irregular drug release. Another possible mechanism of drug release from microcapsules and microparticles involves mechanical rupture, such as may unintentionally occur during careless handling or if chewed. If such products are tableted, their coatings usually fracture during compression.

14.15 SUMMARY

In summary, it may be stated that drug release rate from microcapsules conforming to reservoir devices is zero-order, provided that constant thermodynamic activity is maintained immediately inside the coating material. Microcapsules and microparticles conforming to monolithic

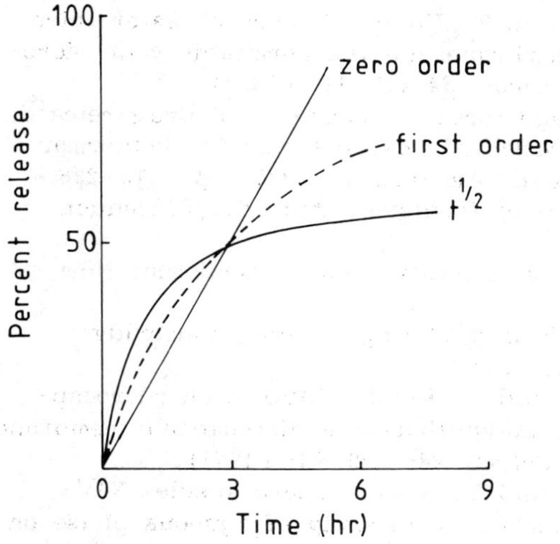

Figure 14.8 Zero-order, first-order, and $t^{1/2}$-dependent release for a drug with a half-life of 3 hr.

devices and containing dissolved drug have release rates that are $t^{-1/2}$-dependent for the first half of the total drug release and thereafter decline exponentially. However, if the monolithic device contains a large excess of dispersed drug, the release rate is essentially $t^{1/2}$-dependent throughout almost the entire drug release. Figure 14.8 shows the typical form of some of these release rates. Zero-order is not necessarily better than first-order or $t^{1/2}$-dependent release rate, as with many products an initial greater release of active ingredient followed by a slower release for maintenance is desirable. Also, in general, reservoir-type systems are more expensive and difficult to manufacture than monolithic systems. Finally, other influences, such as the presence of pores and moving boundaries, can significantly affect drug release.

REFERENCES

1. J. R. Nixon and S. E. Walker, The in vitro evaluation of gelatin coacervate microcapsules, *J. Pharm. Pharmacol. 23*:147S–155S (1971).
2. P. L. Madan, Clofibrate microcapsules II: effect of wall thickness on release characteristics, *J. Pharm. Sci. 70*:430–433 (1981).
3. L. Si-Nang, P. F. Carlier, P. Delort, J. Gazzola, and D. Lafont, Determination of coating thickness of microcapsules and influence upon diffusion, *J. Pharm. Sci. 62*:452–455 (1973).
4. M. S. Harris, Preparation and release characteristics of potassium chloride microcapsules, *J. Pharm. Sci. 70*:391–394 (1981).
5. V. Vidmar, I. Jalsenjak, and T. Kondo, Volume of water-filled pores in ethyl cellulose membrane and the permeability of microcapsules, *J. Pharm. Pharmacol. 34*:411–414 (1982).
6. B. C. Lippold, Solid dosage forms: mechanism of drug release, in *Formulation and Preparation of Dosage Forms* (J. Polderman, ed.), Elsevier/North Holland, Amsterdam, 1977, pp. 215–236.
7. J. Crank, *The Mathematics of Diffusion*, 2nd ed., Clarendon Press, Oxford, 1975.
8. W. Jost, *Diffusion in Solids, Liquids, Gases*, Academic, New York, 1960.
9. R. M. Barrer, *Diffusion in and Through Solids*, Cambridge University Press, London, 1941.
10. K. Takamura, M. Koishi, and T. Kondo, Studies on microcapsules IX: Permeability of polyphthalamide microcapsule membranes to electrolyte, *Kolloid-Z. Polym. 248*:929–933 (1971).
11. K. Takamura, M. Koishi, and T. Kondo, Microcapsules XIV: Effects of membrane materials and viscosity of aqueous phase on permeability of polyamide microcapsules towards electrolytes, *J. Pharm. Sci. 62*:610–612 (1973).

12. P. B. Deasy, M. R. Brophy, B. Ecanow, and M. M. Joy, Effect of ethylcellulose grade and sealant treatments on the production and in vitro release of microencapsulated sodium salicylate, *J. Pharm. Pharmacol.* *32*:15–20 (1980).
13. H. Takenaka, Y. Kawashima, and S. Y. Lin, The effects of wall thickness and amount of hardening agent on the release characteristics of sulfamethoxazole microcapsules prepared by gelatin-acacia complex coacervation, *Chem. Pharm. Bull.* *27*:3054–3060 (1979).
14. R. Senjkovic and I. Jalsenjak, Effect of capsule size and membrane density on permeability of ethyl cellulose microcapsules, *Pharm. Acta Helv.* *57*:16–19 (1982).
15. I. Jalsenjak and T. Kondo, Effect of capsule size on permeability of gelatin-acacia microcapsules towards sodium chloride, *J. Pharm. Sci.* *70*:456–457 (1981).
16. R. Senjkovic and I. Jalsenjak, Apparent diffusion coefficient of sodium phenobarbitone in ethylcellulose microcapsules: effects of capsule size, *J. Pharm. Pharmacol.* *33*:279–282 (1981).
17. T. Fujita, J. Iwasa, and C. Hansch, A new substituent constant, ¶, derived from partition coefficients, *J. Amer. Chem. Soc.* *86*: 5175–5180 (1964).
18. C. Hansch and W. J. Dunn, Linear relationships between lipophilic character and biological activity of drugs, *J. Pharm. Sci.* *61*:1–19 (1972).
19. A. A. Noyes and W. R. Whitney, The rate of solution of solid substances in their own solutions, *J. Amer. Chem. Sco.* *19*: 930–934 (1897).
20. E. Brunner, Reaktionsgeschwindigkeit in heterogenen systemen, *Z. Phys. Chem.* *47*:56–102 (1904).
21. L. Si-Nang and P. F. Carlier, Some physical chemical aspects of diffusion from microcapsules, in *Microencapsulation* (J. R. Nixon, ed.), Dekker, New York, 1976, pp. 185–192.
22. S. C. Khanna, M. Soliva, and P. Speiser, Epoxy resin beads as a pharmaceutical dosage form II: Dissolution studies of epoxy-amine beads and release of drug, *J. Pharm. Sci.* *58*: 1385–1388 (1969).
23. A. T. Florence and D. Attwood, *Physicochemical Principles of Pharmacy*, Macmillan, London, 1981.
24. A. N. Martin, J. Swarbrick, and A. Cammarata, *Physical Pharmacy*, Lea & Febiger, Philadelphia, 1969.
25. R. W. Baker and H. K. Lonsdale, Controlled release:mechanisms and rates, in *Controlled Release of Biologically Active Agents* (A. C. Tanquary and R. E. Lacey, eds.), Plenum, New York, 1974, pp. 15–71.
26. W. G. Chambliss, R. W. Cleary, R. Fischer, A. B. Jones, P. Skierkowski, W. Nicholes, and A. H. Kibbe, Effect of

docusate sodium on drug release from a controlled-release dosage form, *J. Pharm. Sci. 70*:1248–1251 (1981).
27. F. N. Christensen, F. Y. Hansen and H. Bechgaard, Mathematical model for in vitro drug release from controlled release dosage forms applied to propoxyphene hydrochloride pellets, *J. Pharm, Sci. 71*:694–699 (1982).
28. M. Francois and R. De Neve, Study of the influence of polyethyleneglycol 20000 and ureum on the permeability of Eudragit E30D coatings for caffeine, in *F.I.P. Abstracts, 39th International Congress of Pharmaceutical Sciences, Brighton, September,* 1979, p. 27.
29. W. D. Rhine, D. S. T. Hsieh, and R. Langer, Polymers for sustained macromolecular release: Procedures to fabricate reproducible delivery systems and control release kinetics, *J. Pharm. Sci. 69*:265–270 (1980).
30. A. D. Schwope, D. L. Wise, and J. F. Howes, Development of polylactic/glycolic acid delivery systems for use in treatment of narcotic addiction, *Natl. Inst. Drug Abuse Res. Monogr. Ser. 4*:13–18 (1976).
31. R. H. Guy, J. Hadgraft, I. W. Kellaway, and M. J. Taylor, Calculations of drug release rates from spherical particles, *Int. J. Pharm. 11*:199–207 (1982).
32. L. A. Luzzi, M. A. Zoglio, and H. V. Maulding, Preparation and evaluation of the prolonged release properties of nylon microcapsules, *J. Pharm. Sci. 59*:338–341 (1970).
33. T. Higuchi, Rate of release of medicaments from ointment bases containing drugs in suspension, *J. Pharm. Sci. 50*:874–875 (1961).
34. I. Jalsenjak, C. F. Nicolaidou, and J. R. Nixon, The in vitro dissolution of phenobarbitone sodium from ethyl cellulose microcapsules, *J. Pharm. Pharmacol. 28*:912–914 (1976).
35. P. L. Madan, Clofibrate microcapsules: III. Mechanism of release, *Drug Develop. Indust. Pharm. 6*:629–644 (1980).
36. T. Higuchi, Mechanism of sustained-action medication: theoretical analysis of rate of release of solid drugs dispersed in solid matrices, *J. Pharm. Sci. 52*:1145–1149 (1963).
37. H. Fessi, J. P. Marty, F. Puisieux, and J. T. Carstensen, Square root of time dependence of matrix formulations with low drug content, *J. Pharm. Sci. 71*:749–752 (1982).
38. T. J. Roseman, Monolithic polymer devices, section I, in *Controlled Release Technologies: Methods, Theory and Applications, Vol. I* (A. F. Kydonieus, ed.), CRC Press, Boca Raton, Fla. 1980, pp. 21–54.
39. G. L. Flynn, S. H. Yalkowsky, and T. J. Roseman, Mass transport phenomena and models: theoretical concepts, *J. Pharm. Sci. 63*:479–510 (1974).

40. T. M. S. Chang, *Artificial Cells*, Thomas, Springfield, Ill., 1972.
41. A. Tateno, M. Shiba, and T. Kondo, Electrophoretic behavior of ethyl cellulose and polystyrene microcapsules containing aqueous solutions of polyelectrolytes, in *Emulsions, Latices and Dispersions* (P. Becker and M. N. Yudenfreund, eds.), Dekker, New York, 1978, pp. 279–288.
42. T. M. S. Chang and M. J. Poznansky, Semipermeable aqueous microcapsules (artificial cells). V. Permeability characteristics, *J. Biomed. Mater. Res. 2*:187–199 (1968).
43. G. L. Flynn, Influence of physico-chemical properties of drug and system on release of drugs from inert matrices, in *Controlled Release of Biologically Active Agents* (A. C. Tanquary and R. E. Lacey, eds.), Plenum, New York, 1974, pp. 73–98.
44. Y. W. Chien, Methods to achieve sustained drug delivery. The physical approach: implants, in *Sustained and Controlled Release Drug Delivery Systems* (J. R. Robinson, ed.), Dekker, New York, 1978, pp. 211–349.
45. D. V. Satyanarayana Gupta and R. E. Sparks, Mathematical model for progesterone release from injectable poly(lactic acid) microcapsules in vitro, in *Controlled Release of Bioactive Materials* (R. Baker, ed.), Academic, New York, 1980, pp. 189–212.
46. P. L. Lee, Controlled drug release from polymeric matrices involving moving boundaries, in *Controlled Release of Pesticides and Pharmaceuticals* (D. H. Lewis, ed.) Plenum, New York, 1981, pp. 39–48.
47. H. B. Hopfenberg, Controlled release from erodible slabs, cylinders and spheres, in *Controlled Release Polymeric Formulations* (D. R. Paul and F. W. Harris, eds.), ACS Symposium Series 33, American Chemical Society, Washington, D.C., 1976, pp. 26–32.

Appendix

LIST OF NONCHEMICAL SYMBOLS

Symbol	Usual units	Definition
A	m^2	Area of reference plane
$'A$	m^2	Surface area of particles
A'	m^2	Internal surface area of coating
Å	10^{-10} m	Angstrom unit
A_1	m^2	Cross-sectional area of partition
A_2	m^2	Cross-sectional area of annular space
A_p	m^2	Effective particle cross-sectional area
a	kg	Surface weight fraction
a'	—	Exponent or index
B	—	Integration constant
C	$kg\ m^{-3}$	Concentration of solute
°C	—	Degrees centigrade
C_1	$kg\ m^{-3}$	Concentration of solids in coating solution
C_d	—	Drag Coefficient
C_h	$kg\ m^{-3}$	High concentration solution side
C_{it}	$kg\ m^{-3}$	Concentration in inner compartment after time t
C_l	$kg\ m^{-3}$	Low concentration solution side
C_m	$kg\ m^{-3}$	Concentration in membrane/matrix

Symbol	Usual units	Definition
$C_{m(h)}$	$kg\ m^{-3}$	Concentration in membrane at high side
$C_{m(l)}$	$kg\ m^{-3}$	Concentration in membrane at low side
C_{ot}	$kg\ m^{-3}$	Concentration in outer compartment after time t
C_s	$kg\ m^{-3}$	Saturated solubility of solute
C_{tot}	$kg\ m^{-3}$	Total concentration as loading
ΔC	$kg\ m^{-3}$	Difference in concentration in solution on either side of membrane
ΔC_m	$kg\ m^{-3}$	Difference in concentration across the two faces of membrane
CFM	$ft^3\ min^{-1}$	Cubic feet per minute
cP	$N\ s\ m^{-2}$	Centipoise
cSt	$cm^2\ s^{-1}$	Centistoke
D	$m^2\ s^{-1}$	Diffusion coefficient (diffusivity)
D_b	$m^2\ s^{-1}$	Diffusion coefficient in boundary layer
D_m	$m^2\ s^{-1}$	Diffusion coefficient in membrane
D_o	$m^2\ s^{-1}$	Hypothetical diffusion coefficient
D_p	$m^2\ s^{-1}$	Diffusion coefficient in fine pores
D_s	$m^2\ s^{-1}$	Diffusion coefficient in solvent
d_s	m	Mean equivalent diameter of particles
E_a	$kJ\ mole^{-1}$	Activation energy
e_1	—	Void fraction in partition
e_2	—	Void fraction in annular space
°F	—	Degrees Fahrenheit
F_d	N	Drag force
f	—	Frequency of particle circulation
ft	—	Foot
g	—	Gram
g'	$6.67 \times 10^{-11}\ N\ m^2\ kg^{-2}$	Gravitational constant

Symbol	Usual units	Definition
ΔH	kJ $mole^{-1}$	Molar heat of vaporization
hr	—	Hour
J	kg m^{-2} s^{-1}	Flux
K	—	Partition coefficient
K'	—	A constant
°K	—	Degrees Kelvin
K_a	—	Partition coefficient of parent compound
K_{appar}	—	Apparent partition coefficient
K_x	—	Partition coefficient of derivative
k	—	Dissolution rate constant
k'	—	Coefficient expressing porosity and tortuosity of coating
k_1	—	A constant
$\underline{k}$	—	Selectivity coefficient
L_1	m	Height of annular space
L_2	m	Height of coating partition
L_{max}	m	Maximum dimension of particles
L_{min}	m	Minimum dimension of particles
l_b	m	Boundary diffusion layer thickness
l_m	m	Membrane thickness
M	kg	Mass of drug
M_∞	kg	Total mass of drug dissolved in polymer
M_{it}	kg	Mass of drug remaining at any time in reservoir
M_{ot}	kg	Mass of drug remaining at any time in sink
M_t	kg	Mass of drug release at time t
M_{tot}	kg	Total mass of drug
$\overline{M}_n$	g $mole^{-1}$	Number average molecular weight
$\overline{M}_w$	g $mole^{-1}$	Weight average molecular weight

Symbol	Usual units	Definition
m	—	Meter
min	—	Minute
mm	10^{-3} m	Millimeter
N	m kg s^{-2}	Newton
'N	—	Total number of particle circulations
N_a	6.0226×10^{23} $mole^{-1}$	Avogadro's number
Oe	$cm^{-1/2}$ $g^{1/2}$ s^{-1}	Oersted
P	m^2 s^{-1}	Permeability coefficient (DK)
Q	kg m^{-2}	Mass of drug released per unit area at time t
Q_1	m^3 s^{-1}	Solution delivery rate
R	8.3143 J $mole^{-1}$	Gas constant
R_b	m s	Diffusional resistance of boundary layer
R_m		Diffusional resistance of membrane
R_{tot}	m s	Total diffusional resistance
r	m	Radius
r'	m	Radius of shrinking core
r_i	m	Inside radius
r_o	m	Outside radius
r_p	m	Radius of pore
r_s	m	Radius of solute
S	kg m^{-3}	Saturation solubility
S_0	kg m^{-3}	Solubility of undissociated species
s	—	Second
T	°K	Absolute temperature
T_g	°K or °C	Glass transition temperature
t	s	Time
t_b	s	Burst time

Symbol	Usual units	Definition
t_l	s	Lag time
t_m	m	Thickness of coating
U	kJ $mole^{-1}$	Lattice energy
V	m^3	Volume
V'	m^3 $mole^{-1}$	Molar volume
V_1	m s^{-1}	Mean particle velocity in partition
V_2	m s^{-1}	Mean particle velocity in annular space
V_3	m s^{-1}	Superficial velocity of suspending medium in partition
V_4	m s^{-1}	Superficial air velocity in annular space
V_f	m s^{-1}	Superficial gas velocity
V_i	m^3	Volume of internal reservior
V_o	m^3	Volume of outside reservoir
$\bar{V}_b$	m^3 kg^{-1}	Bulk volume of particles per unit mass
W	kg m^{-3}	Total weight of particles
W'	kg m^{-3}	Weight of microparticles used
W_1	—	Particle circulation rate in partition
W_2	—	Particle circulation rate in annular space
W_0	kg m^{-3}	Initial weight
W_p	kg m^{-3}	Particle weight
W_t	kg m^{-3}	Weight at time t
x	m	Distance
α	—	A constant
γ	—	Particle shape factor
ε	%	Porosity of leached portion of matrix
η	cP	Viscosity of medium
η_a	cP	Viscosity of air

Symbol	Usual units	Definition
η_0	$m^3\ kg^{-1}$	Intrinsic viscosity (limiting viscosity number)
μm	10^{-6} m	Micron
¶	3.142	A constant
¶'	—	A substitution constant (pi value)
ρ	$kg\ m^{-3}$	Density
ρ_a	$kg\ m^{-3}$	Density of air
ρ_g	$kg\ m^{-3}$	Density of gas
ρ_m	$kg\ m^{-3}$	Density of dry coating
ρ_s	$kg\ m^{-3}$	Mean particle density
τ	—	Tortuosity of matrix
ψ	N	Interparticulate adhesive forces
$\frac{dM}{dt}$	$kg\ s^{-1}$	Release rate
$\frac{dC}{dx}$	$kg\ m^{-3}\ m^{-1}$	Concentration gradient

Author Index

Numbers in brackets are reference numbers and indicate that an author's work is referred to although the name is not cited in text. Italic numbers give the page on which the complete reference is listed.

A

Abdel-Khalek, M. M., 260[24], *263*
Abdul Razzak, M. H., 37[51], 44[72], *56, 57*
Abe, J., 107[38,39], 112[38], *116*
Abend, P. G., 135[49], *142*
Abrahams, A., 245[11], *250*
Adir, J., 13[60], *18*
Agren, A., 42[68], *57*, 204[10], *215*
Agryilirah, G. A., 74[40], *91*
Ahmed, M., 278[18], *287*
Akimoto, M., 227[40,41], *238*
Alhaigue, F., 27[12], *53*
Allison, A. C., 278[14], *286*
Alpar, H. O., 101[11], *114*
Alpsten, M., 52[104], *59*
Amann, A. H., 146[1], *159*
Amberg, J. H., 261[26], *263*
Amsel, L., 8[22], *15*, 23[7], *53*, 177[40], *179*, 249[29], *252*
Anderson, J. L., 7[7], 13[7], *14*, 62[6], *89*, 102[17,18], *115*
Anderson, J. M., 235[67], *240*
Anderson, L. C., 190[30], *193*, 222[8], 225[25], *236, 237*
Anderson, W., 40[61], *56*
Ando, Y., 85[91], *94*, 191[33], *193*
Andrade, J. D., 272[12], *286*
Ansel, H. C., 52[101], *59*
Antonsen, O., 22[3], *53*
Arai, F., 104[31], *115*, 262[30], *264*
Arakawa, M., 134[41, 42, 43, 45], 135[47], *142*
Arens, R. P., 270[7], *286*
Arita, T., 109[42], *116*, 227[39], *238*, 257[18], *263*
Arkins, G. M., 225[26], *237*
Arnold, D. L., 235[66], *240*
Arnold, J., 7[10], *14*
Arrowsmith, M., 278[18], *287*
Asker, A. F., 188[27], 189[27], *193*
Aström, G., 42[68], *57*
Athamas, N. G., 185[14,15], *192*
Atkins, H. L., 226[32], *238*
Attwood, D., 37[56], *56*, 245[6], *250*, 302[23], *317*
Aulton, M. E., 37[51], 44[72], *56, 57*

Aupoix, M., 213[53], *218*
Autian, J., 29[28], 40[60], *54*, *56*, 210[34], *217*
Axelrod, J., 11[41], *16*

B

Baatz, G., 139[67], *143*
Babu, S., 170[18], *178*
Badawi, A. A., 110[49], *117*, 247[20], *251*
Baerns, M., 167[13], *178*
Baier, R. E., 29[26], *54*
Bailey, A., 148[11], *160*
Bajpai, H. S., 245[10], *251*
Bakan, J. A., 7[5,7], 13[5,7, 76], *14*, *19*, 51[96], *59*, 62 [6,9], 63[10], 82[75], 83[79], *89*, *93*, 103[19], *115*
Baker, R. W., 303–304[25], 308 [25], 309[25], *317*
Bala, K., 137[55], *143*
Ballard, B. E., 13[64], *18*
Banker, G. S., 34[42], 36[44, 48], 41[64], 44[42], 45[76], 47 [42,76,83], 48[76], *55*, *56*, *57*, *58*, 283[28,29,30], 284[31,32, 33,34,35], *287*, *288*
Barr, W. H., 13[60], *18*
Barrer, R. M., 293[9], *316*
Bartilucci, A. J., 256[10], *262*
Bartlett, C., 12[59], *18*
Bateman, N. E., 52[105], *59*, 105[35], *116*
Battista, J. V., 162[5], *177*
Bauduin, P., 232[57], 233[60], *240*
Baxter, G., 81[69], *93*
Bayless, R. G., 110[50,51], *117*, 139[61], *143*
Baytos, W. C., 108[40], *116*
Bechgaard, H., 22[3,4,5], *53*, 306[27], *318*
Becker, C. H., 73[34], *91*, 186 [18,19,20], 188[27], 189[27], *192*, *193*, 213[54], *218*, 261[27], *264*
Beerman, B., 11[47], *17*
Belcher, D. W., 183[2], *191*
Belden, D. H., 169[17], *178*
Bell, S. A., 101[12,13], *114*
Bellard, J. Q., 51[99], *59*
Benedikt, G., 174[33], *179*
Benita, S., 103[20,21,22,23], *115*
Bentele, V., 212[47], *218*
Berdick, M., 101[12,13], *114*
Berg, U. H., 212[47], *218*
Bergmann, R., 207[21], *216*
Bernfeld, P., 203[3], *215*
Berry, C. C., 10[26], *15*
Bingle, J. P., 10[31], *16*
Birrenbach, G., 208[30], 210[31], *217*
Bisson, H. J., 212[41,42], *217*
Bixler, H. J., 84[85], *93*
Blair, M. R., 11[43], *16*
Blake, D. A., 223[14], *237*
Blehm, D. J., 11[49], *17*
Blichare, M. S., 157[29], *160*
Blythe, R. H., 146[3], *159*
Boettger, R. M., 184[10], *192*
Boettner, F., 221[4], *236*
Bogentoft, C., 10[24], *15*, 52[104], *59*, 177[41], *179*, 282[25], *287*
Bolles, T. F., 271[9], *286*
Bolton, F. H., 205[15,16], *216*
Borodkin, S., 28[21,22], *54*, 247 [21,22], *251*
Boroshok, M. J., 106[36], *116*
Bossler, H. M., 175[35], *179*
Botham, R. A. 110[50], *117*
Bothwell, T. H., 11[40], *16*
Boxenbaum, H. G., 13[67], *18*
Boylan, J. C., 284[33], *288*
Braeckman, P., 244[1], *250*
Brandes, G., 232[51,52], *239*
Brodie, B. B., 11[41], *16*
Brolin, S. E., 204[10], *215*

Brophy, M. R., 34[41], 42[41, 71], *55*, *57*, 99[7], 100[7], 101[7], *114*, 150[15], *160*, 297 [12], *317*
Brown, E. A., 174[30], *179*
Brown, E. B., 11[38], *16*
Brown, R. W., 139[64,65,66], *143*
Brudney, N., 162[9], *178*, 246 [12], *251*
Brunner, E., 300[20], *317*
Bryant, W., 8[22], *15*, 23[7], *53*, 177[40], *179*, 249 [29], *252*
Brynko, C., 81[70], *93*
Buchanan, J. W., 226[29], *238*
Bueche, P., 33[35], *55*
Bungenburg de Jong, H. G., 1 [1], *14*, 62[11,12], *89*
Buri, P., 40[62], 41[62], *56*, 282[26], *287*
Burnstein, T., 227[37], *238*

C

Cadwallader, D. E., 52[101], *59*
Cain, V. A., 42[70], *57*
Caldwell, H. C., 173[29], *179*
Calvert, R. T., 52[102], *59*
Cameron, J. L., 232[51], *239*
Cammarata, A., 302[24], *317*
Campbell, J. A., 11[37], *16*
Cardarelli, N. F., 32[34], *55*
Carless, J. E., 31[33], *55*, 66 [15], 70[17,18], *90*
Carlier, P. F., 7[2], *14*, 290[3], 301[21], *316*, *317*
Carroll, D. G., 7[12], *14*
Carstensen, J. T., 311[37], *318*
Chambliss, W. G., 306[26], *317*
Chang, T. M. S., 7[18,19,20], *15*, 109[43], 110[44,45,46,47, 48], *116*, 122[7,8,9], 123[10], 124[12], 125[13,14,15,16,17, 18,19], *139*, *140*, 214[56], *218*, 312[40,42], *319*
Chase, T. N., 11[46], *17*
Chaudhry, N. C., 245[9], *251*
Chawla, R. S., 278[18], *287*
Chibata, I., 125[21], *140*
Chien, Y. W., 29[29], *54*, 313[44], *319*
Chikamatsu, Y., 85[91], *94*, 191 [33], *193*
Chiou, W. L., 261[26], *263*
Christensen, F. N., 306[27], *318*
Chu, W. H., 32[37], *55*
Cicero, T. J., 224[21], *237*
Claeys, D. A., 85[95,96,97], *94*
Clark, B. B., 11[43], *16*
Clarke, C. A., 248[25], *251*
Clarke, R. C., 79[60], 80[60], *92*
Cleary, R. W., 306[26], *317*
Cleland, L. G., 284[36], *288*
Coffey, J. F., 110[47], *116*
Coletta, V., 173[27], *179*
Collom, W. D., 10[28], *16*
Combs, A. B., 22[1], *53*, 128 [24,25], 130[25], *141*
Cook, E. M., 183[2], *191*
Cooper, J., 148[5], *159*
Cooper, J. R., 11[41], *16*
Cooper, J. W., 52[101], *59*
Corn, M. E., 156[26], *160*
Coulon, W. A., 52[107], *60*
Courts, A., 79[60], 80[60], *92*
Couvreur, P., 211[40], *217*, 232 [56,57,58,59], 233[60,61,62], *240*
Cowsar, D. R., 7[15], *14*, 113 [56], *117*
Crabtree, R. E., 247[18], *251*
Crainich, V. A., 77[56], *92*
Crank, J., 293[7], 307–308[7], 314[7], *316*
Crosley, A. P., 156[25], *160*
Crosswell, R. W., 213[54], *218*
Crowley, J. D., 36[43], *55*

Cuff, G. W., 22[1], *53*, 128[25], 130[25], *141*
Curry, S. H., 12[58], *17*
Curzon, G., 11[45], *17*
Cusimano, A. G., 186[19], *192*

D

Daetsu, I., 207[24], *216*
Daoust, R. G., 185[14,15], *192*
Dahm, M., 139[67], *143*
Dana, R. L., 185[14,15], *192*
Dannelly, C. C., 269[5], *286*
Davis, S. S., 250[34], *252*
Deasy, P. B., 34[41], 42[41,71], *55*, *57*, 99[7], 100[7], 101[7], *114*, 150[15], *160*, 297[12], *317*
Deer, J. E., 12[58], *17*
DeGennaro, M. D., 129[26], *141*
DeKay, G., 272[11], *286*
Dekay, H. G., 45[76], 47[76], 48[76], *57*
Delort, P., 7[2], *14*, 290[3], *316*
Delport, J. P., 36[49,50], 47 [79], *56*, *58*
De Meester, C., 233[61], *240*
De Neve, R., 307[28], *318*
DeSavigny, C. B., 139[62], *143*
De Seille, J. M., 36[49], *56*
Dillingham, E. O., 210[34], *217*
Dingeldein, E., 207[21], *216*
Di Roberto, F. M., 7[16], *15*
Dittgen, M., 206[19], *216*
Doluisio, J. T., 267[3], *286*
Donbrow, M., 28[16,17,20], 42 [67], *53*, *54*, *57*, 103[20,21, 22,23], *115*, 173[21], 175[36, 37], *178*, *179*
D'onofrio, G. P., 52[105], *59*, 105[35], *116*
Dopper, J. H., 41[65], *57*
Douglass, C. C., 7[11], *14*
Down, G. R. B., 50[93], *59*
Dramer, P. A., 36[44], *55*, 227 [36], *238*
Draper, E. B., 261[27], *264*
Dreher, D. K., 37[52], *56*, 175 [35], *179*
Dreyer, W. J., 208[26, 27, 29], *216* *217*
Dubois-Krack, G., 233[61], *240*
Dunn, W. J., 299[18], *317*

E

Eareckson, W. M., 129[29], 137 [57], *141*, *143*
Eastoe, J. E., 64[13], *89*
Eaves, T., 52[106], *60*
Ecanow, B., 34[41], 42[41], *55*, 99[7], 100[7], 101[7], *114*, 297[12], *317*
Eckenhoff, J. B., 168[15], 172 [15], *178*
Edman, P., 205[14], 212[48], *216*, *218*
Egawa, S., 79[63], *92*, 257[13], *263*
Eisenberg, S., 7[11], *14*
Ekenved, G., 10[24], *15*, 52[104], *59*, 177[41], *179*
Ekman, B., 204[8,9,10,11,12], 205[13], 212[48], *215*, *216*, *218*
Elegakey, M. A., 84[86], *93*, 260[24], *263*
Elgindy, N. A., 84[86], *93*
El-Khawas, F., 260[24], *263*
Ellis, J. R., 146[1], *159*
Ellison, T., 156[25], *160*
El-Menshawy, M. E., 83[77], *93*, 98[4], *114*
El-Samaligy, M., 208[25], *216*
El-Sayed, A. A., 110[49], *117* 151[16], *160*
Elwood, P. C., 7[9], *14*
Elworthy, P. H. 48[87], *58*, 148 [8], *159*

F

G

Goldzieher, J. W., 267[3], *286*
Gomez, G., 12[55], *17*
Gomez, J. R., 12[55], *17*
Gonczy, C., 231[47], *239*
Goodman, H., 45[75], 48[75], *57*, 283[30], *287*
Goodpart, F. W., 157[27], *160*
Goodwin, J. T., 265[1], 266[2], 270[6], *286*
Gore, A. Y., 47[83], *58*
Gough, D., 272[12], *286*
Grabner, R. W., 106[36], *116*
Granatek, A. P., 185[14,15], *192*
Granatek, E. W. 185[14,15], *192*
Granchelli, F. E., 235[66], *240*
Grass, G. M., 167[14], 172[14], 174[14], *178*, 279[22], *287*
Gratzi, M. M., 223[11], *236*
Green, B. K., 61[1,2,3], 82[2], *89*
Greenfield, J. S. B., 9[23], *15*
Greenland, T., 213[53], *218*
Gregoriadis, G., 278[14,16], *286, 287*
Gregory, J. B., 221[7], *236*
Grief, M., 155[22], *160*
Grier, L., 11[45], *17*
Grislain, L., 233[62], *240*
Gruschkau, H., 211[39], *217*
Guiot, P., 233[57], 233[60,61], *240*
Guitard, P., 40[62], 41[62], *56*, 282[26], *287*
Gung, P. A., 221[6], *236*
Gupta, J. P., 245[10], *251*
Gupta, R. C., 245[10], *251*
Gurny, R., 40[62], 41[62], *56*, 282[26], 284[34,35], *287, 288*
Gutcho, M. H., 13[82], *19*
Guy, R. H., 309[31], *318*
Gyselinck, P., 244[1], *250*

H

Hadgraft, J., 278[18], *287*, 309 [31], *318*
Haenzel, I., 212[44], *218*
Haider, I., 12[56], *17*
Hall, H. S., 37[54], *56*, 157[30], *160*, 166[10], 167[10], 175[38], *178, 179*
Hamacher, H., 249[26], *251*
Hamid, I. S., 186[20], *192*
Hammes, P. A., 102[15], 107[37], *114, 116*
Hansch, C., 299[17,18], *317*
Hansen, F. Y., 306[27], *318*
Haque, R., 73[30], *91*
Harkey, A. B., 11[49], *17*
Harrigan, S. E., 223[18], 224[22], *237*
Harrington, D. D., 284[35], *288*
Harris, J. F., 225[25], *237*
Harris, M., 99[6], 101[6], *114*
Harris, M. S., 29[24], *54*, 291[4], *316*, 158[31], *160*
Harris, R., 12[53], *17*
Hart, R. L., 139[61], *143*
Hasegawa, A., 227[39], *238*
Hashida, M., 250[33], *252*, 257 [13,14,15,16,17], 258[20,21], *263*
Hassan, M., 74[38,39], *91*
Hassler, C. R., 7[17], *15*, 108[40], *116*
Hattori, Y., 85[87], *93*
Hauser, W., 226[32], *238*
Hausheer, W., 7[8], *14*
Hawks, O. C., 187[24], *193*
Hayes, D. K., 221[6], *236*
Haynes, R. B., 10[27], *15*
Hecker, J., 187[24], *193*
Heimlich, K. R., 149[14], *160*
Hellstrom, K., 11[47], *17*
Helrich, M., 223[14], *237*

Helsby, G. H., 235[65], *240*
Herbig, J. A., 13[70], *18*
Herbst, W., 259[22], *263*
Heyd, A., 148[6,10], *159*
Higashide, F., 85[92], *94*, 172 [22], *178*
Higashitsuji, K., 135[48], *142*
Higuchi, W. I., 11[42], *16*, 309 [33], 311[36], 314[36], *318*
Hikasa, Y., 257[16], *263*
Himmel, R. K., 82[76], *93*
Hinkle, B. L., 169[16], *178*
Hinsvark, O., 8[22], *15*, 23[7], *53*, 177[40], *179*, 246[16], 249 [29], *251*, *252*
Hirata, G., 103[24], *115*
Hirscher, D. A., 246[14], *251*
Hjerten, S., 203[2], *215*
Ho, N. F. H., 11[42], *16*
Hogan, J. E., 37[51], 44[72], *56*, *57*
Hohn, P. M., 102[16], *115*
Holliday, W. M., 101[12,13], *114*
Hollister, L. E., 11[35], 12[57, 58], *16*, *17*
Hoon, J. R. 7[13], *14*
Hopfenberg, H. B., 47[84,85], *58*, 315[47], *319*
Hörger, G., 82[73,74], *93*
Hostetler, V., 51[99], *59*
Howes, J. F., 222[8], 223[16, 17,18], *236*, *237*, 308[30], *318*
Hsieh, D. S. T., 307[29], *318*
Hunke, W. A., 28[19], *54*
Hunter, E., 52[102, 103], *59*
Huschke, G., 7[8], *14*
Hussein, A. M., 247[20], *251*

I

Ihler, G. M., 285[39], *288*
Illi, V., 211[38], *217*
Illum, L., 250[34], *252*
Inaba, Y., 85[87], *93*, 104[32], 107[39], 109[41], *115*, *116*
Ishizaka, T., 229[42], *238*
Iskandar, A. G., 260[24], *263*
Ismail, A. A., 83[77], *93*, 98[4], *114*
Itoh, M., 105[34], 106[34], 109 [42], *116*
Iwasa, J., 299[17], *317*
Iwata, H., 229[42], *239*

J

Jack, D., 245[3], *250*
Jackanicz, T. M., 221[7], *236*
Jackson, G. J., 157[29], *160*
Jackson, R. D., 7[3], *14*
Jacob, J. T., 7[10], *14*
Jacobsen, H., 279[21], *287*
Jaffe, H., 221[6], *236*
Jalsenjak, I., 81[72], *93*, 98[5], 99[6], 101[6,9,10], *114*, 291[5], 298[14,15,16], 299[5], 311[34], 312[15], 313[14], *316*, *317*, *318*
James, M. J., 278[18], *287*
Jaminet, F., 36[49], *56*
Jan, S., 36[48], *56*
Jani, R. K., 256[10], *262*
Javidan, S., 73[30], *91*
Jayaratne, O. W., 254[4], *262*
Jecklin, T., 205[17], *216*
Jeffcoat, R., 222[10], 223[11], *236*
Jendon, D. J., 246[16], *251*
Jenkins, A. W., 128[22,23], 132 [23], *140*, *141*
John, P. M., 186[18], *192*
Johnson, D. E. 266[2], *286*
Johnson, E. M., 11[48,49], *17*
Johnsson, A. C., 204[7], *215*
Johnston, G. W., 28[13], *53*

Jones, A. B., 306[26], *317*
Jonsson, U. E., 10[24], *15*, 177[41], *179*
Joshi, R., 210[33], 211[36], *217*
Jost, W., 293[8], *316*
Joy, M. M., 34[41], 42[41], *55*, 99[7], 100[7], 101[7], *114*, 297[12], *317*
Juliano, R. L., 278[17], *287*
Juni, K., 85[89,90], *93*, 109 [42], *116*, 258[19], *263*
Jusko, W. J., 11[36], *16*
Juslin, M., 175[34], *179*, 282 [27], *287*
Justus, B. W., 11[39], *16*

K

Kaas, A. J., 1[1], *14*
Kagami, I., 11[51], *17*
Kala, H., 190–191[31], *193*, 206[19,20], *216*
Kalkwarf, D. R., 261[28], *264*
Kamani, A., 11[42], *16*
Kamashima, Y., 74[41], 75[42], *91*
Kanig, J. L., 45[75], 48[75], *57*, 148[10], *159*
Kante, B., 232[56,57,58,59], 233[60,61,62], *240*
Kanter, S. L., 12[58], *17*
Kasai, S., 105[33], *115*
Kassel, L. S., 169[17], *178*
Kassem, A. A., 247[20], *251*
Kastigir, B., 272[12], *286*
Katayama, S., 77[48,50,54], *92*
Kato, Y., 227[38,39,40,41], 239 [49,50], *238*, *239*
Katz, M., 129[28], 138[59], *141*, *143*
Kawano, Y., 132[33], *141*
Kawashima, Y., 80[65], 85[91], *92*, *94*, 189[28], 190[28,29], 191[33], *193*, 278[19], *287*, 297[13], 311[13], *317*
Kellaway, I. W., 46[77,78], *57*, 278[18], *287*, 309[31], *318*
Kent, D. J., 36[45], *55*
Khaja, G., 9[23], *15*
Khalil, S. A. H., 66[15], 70[17, 18], *90*
Khanna, S. C., 205[17], 206[18], *216*, 233[63], 235[64], *240*, 301[22], *317*
Kibbe, A. H., 306[26], *317*
Kildsig, D. O., 36[44], *55*
Kimmel, G. L., 222[10], *236*
King, T. E., 149[13], *160*
Kinget, R., 27[11], 45[74], *53*, *57*
Kircher, W., 215[57], *218*
Kiritani, M., 77[54], *92*
Kiritsis, G. C., 101[12,13], *114*
Kishizaki, A., 203[4], *215*
Kitajima, M., 104[28,30,31], 107 [28, 38], 112[38], *115*, *116*, 262[30], *264*
Klaui, H. M., 7[8], *14*
Klopper, J., 226[32], *238*
Knight, C. G., 278[15], *287*
Knowles, J. R., 79[64], *92*
Kobari, S., 85[87], *93*, 104[32], 107[39], 109[41], *115*, *116*
Koestler, R. C., 120[5], *139*
Koff, A., 186[21], *192*, 248[24], *251*
Koida, Y., 103[24], *115*
Koishi, M., 105[33], *115*, 129[27], 131[27,30,31], 132[32,33,34,35], 133[38,39,40], 138[58], *141*, *142*, *143*, 229[42], *238*, 295[11], *316*
Kojima, T., 250[33], *252*, 258 [20,21], *263*
Kolff, W. J., 272[12], *286*
Komiya, I., 11[42], *16*
Kondo, A., 13[84], *19*, 62[8], 68[8], 81[8], *89*, 104[27,28,30, 31], 107[28,38], 108[27], 111 [53], 112[38], 113[27], *115*, *116*, *117*, 120[4], *139*, 172[20], *178*, 183[7], *192*, 262[29,30],

[Kondo, A.], *264*, 273[13], *286*
Kondo, S., 111[54], *117*
Kondo, T., 13[80], *19*, 81[72], 85[94], *93*, *94*, 129[27], 131 [27,30,31], 132[32,33,34,35,36], 133[38,39,40], 134[41,42,43, 44, 45, 46], 135[47], 137[56] 138[56,58], 139[60], *141*, *142*, *143*, 291[5], 295[10,11], 298 [15], 299[5], 312[5,41], *316*, *317*, *319*
Kool, L. B., 183[8], *192*, 254 [6], *262*
Kopf, H., 210[33], 211[36], *217*
Kornblum, S. S., 278[20], *287*
Kosar, J., 225[36], *237*
Kouketsu, M., 257[18], 258[19], *263*
Kouzuki, K., 48[89], *58*
Kowarski, A., 261[25], *263*
Kowarski, C. R., 261[25], *263*
Kramer, H., 215[57], *218*, 227 [37], *238*
Kreuter, J., 86[104], 87[104], *94*, 210[35], 211[37,38,39], 212[35,43,44,45,46,47], *217*, *218*
Kruyt, H. R., 62[11], *89*
Kulkarni, R. K., 29[27], *54*, 232[51,52], *239*
Kumakara, M., 207[24], *216*
Kunii, D., 170[19], *178*
Kunitomo, K., 272[12], *286*
Kwolek, S. L., 123[11], *140*
Kydonieus, A. F., 28[23], *54*

L

Lachman, L., 51[97], *59*, 146[2], 148[5], *159*
Ladefoged, K., 22[4,5], *53*
Lafont, D., 7[2], *14*, 290[3], *316*
Lambrou, A., 186[22], *192*
Landolt, R. R., 113[55], *117*
Lang, S., 11[48], *17*
Langer, G., 272[10], *286*, 307 [29], *318*
Lantz, R. J., 148[11], *160*, 167 [14], 172[14], 174[14], *178*
Larsen, H., 41[65], *57*
Latiolais, C. J., 10[26], *15*
Lawrence, W. H., 40[63], *56* 210[34], *217*
Layton, D., 278[17], *287*
Lazarus, J., 51[97], *59*
Leach, A. A., 64[13], *89*
Leafe, T. D., 221[4], 223[15], 224[23], *236*, *237*
Lee, C. O., 272[11], *286*
Lee, P. L., 314[46], *319*
Lee, V. H., 13[65,66], *18*
Leeson, G. A., 10[29], *16*
Lehmann, K., 23[10], 37[52,53], *53*, *56*, 175[35], *179*
Leiberman, H. A., 148[9], *159*
Lenaerts, V., 232[59], 233[60], *240*
Lentz, R. J., 185[17], *192*
Leonard, F., 232[51,52,53], *239*
Lesko, L. J., 137[52], *143*
Leuenberger, H., 40[62], 41[62], *56*
Levine, R. M., 11[43], *16*
Levine, R. R., 11[44], *17*
Levy, G., 11[35,36], *16*
Levy, J., 137[53,54], *143*
Levy, M. C., 137[53,54], *143*
Lewis, D. H., 7[15], *14*, 113[56], *117*
Lieberman, H. A., 146[2], 157 [27], *159*, *160*, 165[12], *178*, 183[5], *192*
Liehl, E., 212[45,46], *218*
Lillie, K. D., 37[54], *56*, 157[30], *160*, 175[38], *179*
Lim, F., 259[23], *263*
Lin, S. Y., 74[41], 75[42], 80 [65], *91*, *92*, 189[28], 190[28, 29], *193*, 297[13], 311[13], *317*
Lindberg, B., 284[38], *288*

Lindblom, R., 284[38], *288*
Lindheimer, T. A., 229[43], *239*
Lindlof, J. A., 162[4,5], *177*
Linnell, W. H., 245[11], *251*
Linsell, W. D., 9[23], *15*
Lippold, B. C., 291[6], *316*
Lister, C., 110[47], *116*
Littenberg, R. L., 226[31], *238*
Ljungstedt, I., 204[11], 205[13], *215, 216*
Lofter, C., 204[9], *215*
Loishi, M., 295[10], *316*
Long, S., 155[21], *160*
Longo, W. H., 229[43], 239
Lonsdale, H. K., 303-304[25], 308[25], 309[25], *317*
Lovgren, K., 282[25], *287*
Lowe, J. W., 36[43], *55*
Lowey, H., 154[19], *160*
Lubbers, D. W., 212[41,42], *217*
Ludwig, A., 51[100], *59*
Luzzi, L. A., 8[21], 13[72], *15, 18*, 62[4], 67[4], 71[26], 72[27], 73[28,31,32,33], *89, 90, 91*, 125[20], 126[20], 129[26], *140, 141*, 254[2], *262*, 309[32], *318*
Lysher, W. M., 254[5], *262*

M

McAfee, J. G., 226[28], *237*
McAinsh, J., 154[18], *160*
McCall, M. S., 7[11], *14*
McCarthy, D. A., 223[18], 224[22], *237*
McCormick, G. J., 190[30], *193*, 225[25], *237*
MacDonnell, D. R., 149[14], *160*
Macek, T. J., 248[23], *251*
McGinity, J. W., 22[1], *53*, 128[24,25], 139[25], *141*
MacIntosh, F. C., 109[43], *116*, 122[8], *140*
McKinlay, I., 10[30], *16*
McMullen, J. N., 73[34], *91*
McNiff, R. G., 98[1], *114*
Madan, P. L., 62[7], 71[24,25], 73[31,32,33], *89, 90, 91*, 129[3], *139*, 254[2,8], 256[10], *262*, 289[2], 307[2], 311[35], *316, 318*
Maekawa, Y., 111[53], *117*
Maggi, G. C., 7[16], *15*
Maierson, T., 41[66], *57*, 77[55,56], *92*
Malani, R. I., 28[13], *53*
Malave, N., 7[19], *15*, 110[46], *116*
Malson, G., 284[38], *288*
Malson, T., 284[38], *288*
Malspeis, L., 224[19], *237*
Mangold, D. J., 267[3], *286*
Mank, R., 206[20], *216*
Marchetti, M., 27[12], *53*
Marinelli, N., 81[68], *92*
Marks, T. A., 223[12], *236*
Marotta, N. G., 184[10], *192*
Marriott, C., 46[77,78], *57*
Marsden, C. D., 11[45], *17*
Marston, H. R., 22[2], *53*
Martin, A., 22[1], *53*
Martin, A. N., 302[24], *317*
Martin, A. W., 128[24,25], 130[25], *141*
Martindale, 29[30], *54*
Marty, J. J., 86[103], 88[107, 108,109], *94, 95*, 311[37], *318*
Mason, B. J., 254[4], *262*
Mason, N., 224[21], *237*
Mason, S. G., 109[43], *116*, 122[8], 137[56], 138[56], *140, 143*
Masters, K., 183[1], *191*
Matheson, L. E., 28[19], *54*
Matsuda, Y., 48[89], *58*, 85[87], *93*
Matsukawa, H., 76[46], 77[48,49, 50,51,52,53,54], 79[62], 84[84], *92, 93*
Matsuo, Y., 125[21], *140*
Matsushita, T., 79[63], *92*

Matsuyama, J., 77[48], *92*
Matthews, B. R., 71[22,23], *90*
Mattson, H. W., 13[68], *18*
Maulding, H. V., 8[21], *15*, 125[20], 126[20], *140*, 309 [32], *318*
Mauler, R., 211[39], *217*
Meadow, S. R., 10[29], *16*
Meeker, R., 284[34], *288*
Meier, F. G., 137[51], 139[69], *143*
Mellan, I., 37[57], *56*
Mercier, M., 233[61], *240*
Merkle, H. P., 83[78], *93*
Merory, J., 184[9], *192*
Mesiha, M. S., 101[8], *114*
Meyer, F. J., 223[14,15], 224 [23], *237*
Meyers, W. E., 7[15], *14*, 113 [56], *117*
Michaels, A. S., 84[85], *93*, 174[32], *179*
Middelbeek, E. M., 41[65], *57*
Migchelsen, M., 110[48], *116*
Miller, J. A., 221[6], *236*
Miller, O. H., 246[14], *251*
Miller, R. E., 98[1], *114*
Milosovich, G., 174[31], *179*, 256[9], *262*
Minatoya, H., 102[16], *115*
Miyagishima, A., 172[22], *178*
Miyamoto, E., 11[51], *17*
Miyano, S., 11[53], *117*
Mody, D. S., 148[9], *159*
Molday, R. S., 208[26,27,28], *216*, *217*
Moldenhauer, H., 190–191[31], *193*
Morehouse, D. W., 205[15,16], *216*
Morgan, P. W., 121[6], 123[11], *139*, *140*
Mori, T., 125[21], *140*
Morimoto, Y., 227[38,39,40,41], 231[49,50], *238*, *239*
Morishita, M., 85[87], *93*, 104 [32], 107[38,39], 109[41], 112 [38], *115*, *116*
Morris, N. J., 29[25], *54*, 86 [98,99,100], *94*
Morrisson, A. B., 11[37], *16*
Morse, L. D., 73[29], *90*, 102 [14,15], 106[36], 107[37], *114*, *116*
Mortada, S. A., 83[82], *93*
Mortada, S. M., 85[88], *93*
Morton, J. F., 224[24], *237*
Mosbach, K., 203[6], 204[7], *215*
Mosbach, R., 203[6], *215*
Moss, R. D., 259[23], *263*
Motycka, S., 42[69], *57*, 249[27, 28], *252*
Mrteck, R. G., 73[30], *91*
Munden, B. J., 45[76], 47[76], 48[76], *57*
Muramatsu, N., 132[36], 134[46], *142*
Muranishi, S., 250[33], *252*, 257 [13,14,15,16,17], 258[20,21], *263*
Muroya, N., 104[28], 107[28], *115*
Myers, R. D., 284[34], *288*

N

Nack, H., 13[69,71], *18*, 120[1], *139*
Nadai, T., 227[38,39,40,41], *238*
Nadkarni, P. D., 36[44], *55*
Nagai, T., 75[44], *91*
Nagata, A., 107[39], *116*
Nairn, J. G., 42[69], *57*, 249 [27,28], *252*
Nakamura, T., 134[45], *142*
Nakamura, Y., 257[18], 258[19], *263*
Nakano, M., 85[89,90], *93*, 105 [34], 106[34], 109[42], 111[54],

[Nakano, M.], *116, 117*, 257[18], 258[19], *263*
Nambu, N., 75[44], *91*
Nappen, B. H., 184[10], *192*
Nash, H. A., 221[7], *236*
Nash, J. F., 247[18], *251*
Natarajan, T. K., 227[33], *238*
Navari, R. M., 70[20], *90*
Needleman, P., 11[48,49], *17*
Nelsen, L., 235[66], *240*
Nelson, J., 232[51], *239*
Newton, D. W., 73[34], *91*
Newton, J. M., 47[80,81], *58*, 183[3], *191*
Niazi, S., 13[61], *18*
Nicholes, W., 306[26], *317*
Nicholson, A. E., 151[17], *160*
Nicolaidou, C. F., 98[5], 99[6], 101[6], *114*, 311[34], *318*
Nielsen, F., 183[6], *192*
Nilsson, H., 203[6], *215*
Nishizawa, K., 135[48], *142*
Niveleau, A., 213[53], *218*
Nixon, J. R., 13[79], *19*, 31[33], 47[86], *55, 58*, 66[15], 70[17, 18], 71[21,22,23], 74[38,39, 40], 75[43], *90, 91*, 98[5], 99 [6], 101[6,9], *114*, 289[1], 311[34], *316, 318*
Nolen, R. L., 183[8], *192*, 254[6], *262*
Norhdurft, H., 86[101], *94*
Notari, R. H., 224[19], *237*
Nouh, A., 47[86], *58*
Noyes, A. A., 300[19], *317*
Nozawa, Y., 85[92,93], *94*
Noznick, P. P., 184[11], *192*
Nuessle, N. O., 281[24], *287*
Nunning, B. C., 185[14,15], *192*
Nygren, A., 284[38], *288*
Nystrom, B., 250[30], *252*

O

O'Brien, P., 250[31], *252*
Ogawa, T., 131[31], *141*
Ohno, H., 278[19], *287*
Okano, T., 250[32], *252*
Okor, R. S., 28[18], 40[61], *54, 56*
Okumaura, M., 231[49,50], *239*
Olderman, G. M., 81[70], *93*
Omura, Y., 172[22], *178*
Opferman, L. P., 78[57], *92*
Opitz, N., 212[41,42], *217*
Oppenheim, R. C., 52[105], *59*, 86[103,105,106], 88[107,108,109], 89[110], *94, 95*, 105[35], *116*
Orr, N., 51[98], *59*
Ovadia, H., 231[48], *239*

P

Pagliery, M., 51[97], *59*
Palmer, E., 188[26], *193*
Palmieri, A., 73[35], 74[36,37], *91*
Pampus, G., 83[83], *93*
Pani, K. C., 232[52], *239*
Paradissis, G. N., 70[19], *90*, 256[12], *263*
Parks, J. D., 11[45], *17*
Parks, J. Y., 11[42], *16*
Parrott, E. L., 33[38], *55*, 70[19], *90*, 256[12], *263*
Pasin, J. Z., 188[25], *193*
Patanus, A. J., 108[40], *116*
Paterson, P. Y., 231[48], *239*
Pauls, J. F., 185[17], *192*
Peck, G., 36[48], *56*
Peck, G. E., 113[55], *117*
Pemberton, J., 11[34], *16*
Peppas, N. A., 284[35], *288*
Perusse, C. B., 11[37], *16*
Peters, D., 157[27], *160*
Phares, R. E., 66[14], 70[16], *90*, 247[19], *251*
Philipps, P. S., 139[64, 65, 66] *143*
Pickard, J. F., 48[87], *58*, 148 [8,12], *159, 160*
Pilkington, J. R. E., 11[50], *17*

Pines, A., 9[23], *15*
Pirakitikuir, P., 36[48], *56*
Pirzio-Biroli, G., 11[40], *16*
Pitt, C. G., 222[10], 223[11, 12,13], *236*
Pomare, E. W., 7[12], *14*
Pondell, R. E., 37[54], *56*, 157[30], *160*, 162[10], 166[10], 167[10], 175[38], *178*, *179*
Porter, A. M. W., 10[25], *15*
Porter, S. C., 36[46], 43[46], *55*
Potron, G., 137[54], *143*
Powell, T. C., 98[2], 102[17, 18], 103[19,25], *114*, *115*
Poznansky, M. J., 110[44], *116*, 123[10], 125[16], *140*, 284[36], *288*, 312[42], *319*
Price, J. C., 73[31,32,33], *91*, 254[2], *262*
Prillig, E. B., 146[1], *159*
Puisieux, F., 311[37], *318*
Purcell, A. M., 110[52], *117*
Purves, M. J., 11[50], *17*
Puumalainen, P., 175[34], *179*, 282[27], *287*

Q

Quash, G. A., 213[53], *218*
Quisumbing, A. R., 28[23], *54*

R

Raafat, H., 9[23], *15*
Raghunathan, Y., 8[22], *15*, 23[7], *53*, 177[40], *179*, 249 [29], *252*
Rambourg, P., 137[53,54], *143*
Rankell, A., 165[12], *178*, 183[5], *192*
Ranney, D. F., 229[45], *239*
Ranney, M. W., 13[81], *19*
Raun, E. S., 7[3], *14*
Rawlins, M. D., 13[63], *18*
Rednick, A. B., 151[17], *160*
Rees, J. E., 48[87], *58*, 148[8, 12], *159*, *160*
Rees, J. T., 247[17], *251*
Regalado, R. G., 12[53], *17*
Reich, S. D., 231[47], *239*
Rembaum, A., 208[26,27,29], *216*, *217*
Renbaum, A., 208[28], *217*
Reul, B., 186[23], *193*
Reuning, R. H., 224[19,22], *237*
Reuss, K., 207[21], *216*
Reyes, Z., 103[26], *115*, 214[55], *218*, 281[23], *287*
Rhine, W. D., 307[29], *318*
Rhodes, E. A., 220[29,30], 227 [33], *238*
Rhodes, C. T., 284[31,32], *288*
Riccieri, F. M., 27[12], *53*
Richards, F. M., 79[64], *92*
Richards, P., 226[32], *238*
Ridgway, K., 36[46], 43[46], *55*
Riegelman, S., 261[26], *263*
Riley, B., 7[10], *14*
Robinson, J. A. J., 46[77,78], *57*
Robinson, J. R., 13[65,66], *18*
Robinson, M. J., 148[11], *160*, 167[14], 172[14], 174[14], *178* 185[13,16,17], *192*, 279[22], *287*
Roe, H. L., 250[31], *252*
Rohdewald, P., 208[25], *216*
Roland, M., 39[58], 48[88], *56*, *58*, 211[40], *217*, 232[56,57,58, 59], 233[60,61,62], *240*
Roseman, T. J., 311[38], 312[39], *318*
Rosen, A., 11[47], *17*
Rosen, E., 154[20], 156[25], *160*, 173[29], *179*
Rosenberg, B., 224[24], *237*
Rosenbergy, F. J., 102[16], *115*
Ross, G., 186[23], *193*
Rowe, E. L., 44[73], *57*

Rowe, R. C., 23[8,9], 30[31], 31[32], 33[36,39], 34[40], 36 [45], 39[39], 47[80,81], 49 [91], 50[92,94,95], *53, 54, 55, 56, 58, 59,* 154[18], *160*
Rubin, H., 173[27], *179*
Rupprecht, H., 215[57], *218*
Ryede, M., 250[30], *252*

S

Sadek, 22[9]
Saeki, K., 76[46], 77[49, 50, 51, 52,53], 79[62], 84[84], *92, 93*
Said, S. A., 151[16], *160*
Sakaguchi, H., 11[51], *17*
Sakamoto, M., 79[63], *92*
Salib, N. N., 83[77], *93,* 98 [3,4], *114*
Samejima, M., 103[24], *115*
Samuelov, Y., 28[17,20], *53, 54,* 175[36,37], *179*
Sanderson, J. E., 221[2], 222[2], *236*
Sandine, W. E., 254[7], *262*
Santo, J. E., 135[49,50], *142, 143*
Santucci, E., 27[12], *53*
Sarnotsky, A. A., 165[11], *178*
Sartori, M. F., 221[3], *236*
Sato, T., 125[21], *140*
Satomura, K., 257[16], *263*
Satyanarayana Gupta, D. V., 314[45], *319*
Saunders, K. J., 213[49], *218*
Saunders, L., 245[2,9], *250, 251*
Scailteur, V., 232[59], *240*
Scarpelli, D. G., 229[44], *239*
Schact, E., 244[1], *250*
Schäfer, W., 139[67], *143*
Scheffel, U., 227[33], *238*
Scher, H. B., 139[68], *143*
Scheu, J. D., 113[55], *117*
Schindler, A., 222[10], 223[11,12, 13], *236*
Schlameus, H. W., 267[3], *286*
Schleicher, L., 61[1,2,3], 82[2], *89*
Schlichting, D. A., 246[15], *251*
Schmollack, W., 206[19], *216*
Schnoring, H., 80[66], 83[83], *92, 93*
Schon, N., 80[66], 83[83], *92, 93*
Schroeter, L. C., 156[24], *160*
Schwope, A. D., 223[16,17,18], *237,* 308[30], *318*
Scott, M. W., 28[13], *53,* 148[9], *159,* 185[17], *192*
Seager, H., 7[4], *14,* 184[12], *192*
Sekikawa, H., 109[42], *116*
Senjkovic, R., 101[9,10], *114,* 298[14,16], 313[14], *317*
Senyei, A. E., 229[44,45], 231 [46,47,48], *239*
Sezaki, H., 250[33], *252,* 257 [13,14,15,16,17], 258[20,21], *263*
Shah, B., 170[18], *178*
Shanbhag, S. R., 254[8], *262*
Shank, C. P., 110[50], *117*
Shank, J. L., 285[40], *288*
Shapiro, S., 11[41], *16*
Sharma, H., 52[102,103], *59*
Shashoua, V. E., 129[29], *141*
Shaw, S. M., 113[55], *117*
Shepard, M, 156[23], *160*
Sheth, S. G., 235[66], *240*
Shiba, M., 85[94], *94,* 131[30], 132[32,33], *141,* 312[41], *319*
Shigeri, Y., 131[30], 133[40], 139[6], *141, 142, 143*
Shimada, T., 77[49,53], *92*
Shimosaka, Y., 79[61], *92*
Shiota, H., 213[50], *218*
Shipley, M., 11[45], *17*
Shoop, C. E., 248[23], *251*
Shovers, J., 254[7], *262*

Sidhom, M. B., 101[8], *114*
Sidman, K. R., 235[66], *240*
Signorino, C. A., 28[14,15], *53*
Simmons, S. P., 284[34], *288*
Simon, E., 173[26], *178*
Simpson, H., 10[30], *16*
Sina, A., 247[20], *251*
Si-Nang, L., 7[2], *14*, 290[3], 301[21], *316, 317*
Sinclair, R. G., 221[5], *236*
Singer, J. M., 213[52], *218*
Sinkula, A. A., 284[37], *288*
Sirine, G. F., 81[67], *92*
Siu Chong, E. D., 7[20], *15*
Sjoholm, I., 204[8,9,10,11,12], 205[13,14], 212[48], *215, 216, 218*
Sjökvist, R., 42[68], *57*
Skierkowski, P., 306[26], *317*
Sliwka, W., 13[75], *18*
Smith, D. A., 183[2], *191*
Smith, H. A., 246[13], *251*
Smith, R. A., 186[22], *192*
Smith, T. L., 32[37], *55*
Smith, W. V., 197[1], *215*
Smolen, V. F., 41[64], *57*
Soliva, M., 191[32], *193*, 210[33], 211[36], *217*, 235[64], *240*, 301[22], *317*
Sölvell, L., 52[104], *59*
Somerville, G. R., 265[1], 266[2], 270[6], *286*
Sparks, R. E., 314[45], *319*
Speaker, T. J., 137[52], *143*
Speiser, P., 13[62], *18*, 83[78], 88[107,108], *93, 95*, 191[32], *193*, 205[17], 206[18], 208[30], 210[31,32,33], 211[36,37,39,40], 212[41,42, 43], *216, 217, 218*, 232[57, 58,59], 233[60,61,62,63], 235[64], *240*, 301[22], *317*
Sperandio, G. J., 66[14], 70[16], *90*, 113[55], *117*, 245[7,8], 246[13], 247[19], *250, 251*
Spitael, J., 27[11], 45[74], *53, 57*
Spross, B., 250[30], *252*
Sreedharan, K. S., 9[23], *15*
Stafford, J. W., 37[55], *56*
Stanley, P., 47[80], *58*
Stannett, V., 47[84], *58*, 213[50], *218*
Stark, A., 110[47], *116*
Stauffer, D. R., 248[23], *251*
Steber, W. D., 235[66], *240*
Stein, H. S., 226[28], *237*
Steinborn, J. A., 246[16], *251*
Steinle, M. E., 103[25], *115*
Stern, P. W., 148[7], *159*
Sternberg, S., 84[85], *93*
Stewart, N. F., 88[109], 89[110], *95*
Stinson, N. E., 86[102], *94*
Stivic, I., 101[9], *114*
Stoopak, S. B., 278[20], *287*
Storey, G. W., 10[31], *16*
Striley, D. J., 81[71], *93*
Strong, P., 235[66], *240*
Sucker, H., 40[62], 41[62], *56*, 282[26], *287*
Sudekum, R. H., 7[14], *14*, 270[8], *286*
Sugibayashi, K., 227[38,39,40,41], 231[49,50], *238, 239*
Sui Chang, E. D., 125[15], *140*
Sullivan, M. F., 261[28], *264*
Sundberg, D. P., 247[22], *251*
Sutaria, R. H., 148[4], *159*
Suzuki, H., 79[61], *92*, 134[44, 45], 137[56], 138[56], *142, 143*
Svedres, E. V., 185[13], *192*
Swarbrick, J., 47[83], *58*, 302[24], *317*
Sweeney, N. P., 270[7], *286*
Swintoskey, J. V., 154[20], *160*, 185[16], *192*
Syarto, J., 173[28], *179*
Szymanski, C. D., 184[10], *192*

T

Takahashi, K., 132[34], 133[39, 40], *141, 142*

Takahashi, Y., 257[14], 258[20], *263*
Takamura, K., 131[31], 132[35], *141*, 295[10,11], *316*
Takeda, Y., 75[44], *91*
Takenaka, H., 74[41], 75[42], 80[65], 85[91], *91, 92, 94*, 189[28], 190[28,29], 191[33], *193*, 278[19], *287*, 297[13], 311[13], *317*
Takikawa, K., 257[18], *263*
Takino, S., 78[58], *92*, 256[11], *262*
Talwalker, A., 170[18], *178*
Tamamushi, B., 134[41,42], *142*
Tanaka, M., 48[89], *58*
Tanaka, N., 78[58], *92*, 256 [11], *262*
Tanaka, Y., 48[89], *58*
Tani, N., 235[62], *240*
Tanigaki, J., 48[89], *58*
Tanigawa, N., 257[16], *263*
Tannenbaum, P., 156[25], *160*
Taroy, E., 110[47], *116*
Tateno, A., 85[94], *94*, 312[41], *319*
Tatter, C. W., 184[11], *192*
Tauber, U., 211[38], *217*
Taylor, D., 36[48], *56*
Taylor, L. D., 75[45], 76[47], *92*
Taylor, M. J., 278[18], *287*, 309 [31], *318*
Teaque, G. S., 36[43], *55*
Thely, M. H., 52[107], *60*
Thies, C., 49[90], *58*, 224[20, 21,22], *237*
Thompson, B. B., 129[26], *141*
Tice, T. R., 7[15], *14*, 113[56], *117*
Timreck, A. E., 268[4], *286*
Tomioka, S., 131[30], 132[32, 33], *141*
Tomizawa, M., 133[40], *142*
Tosa, T., 125[21], *140*
Toupin, P. Y., 162[9], *178*
Traue, J., 190–191[31], *193*
Treadwell, B. L. J., 7[12], *14*
Trouet, A., 211[40], *217*
Truant, A. P., 246[16], *251*
Trubiano, P. C., 259[22], *263*
Tsuji, A., 11[51], *17*
Tsukinaka, Y., 11[51], *17*
Tsuneoka, Y., 104[30], *115*
Tucker, F. E., 28[21,22], *54*
Tucker, S. J., 151[17], *160*
Tulkens, P., 211[40], *217*

U

Umezawa, H., 22[6], *53*, 157[28], *160*
Unger, K., 215[57], *218*
Unsworth, M., 235[65], *240*
Updike, O. L., 70[20], *90*
Utsumi, I., 78[58], *92*, 256[11], *262*

V

Van Abbé, N. J., 247[17], *251*
Van Besauw, J. F., 85[97], *94*
Vandegaer, J. E., 13[78], *19*, 62 [5], *89*, 135[50], 137[51], 139 [69], *143*
Vanderhoff, J. W., 213[51], *218*
Van Dress, M., 235[67], *240*
Van Ooteghem, M., 51[100], *59*
Vanseveren, R., 244[1], *250*
Van Wagenon, R., 272[12], *286*
Vassiliades, A. E., 139[63], *143*
Vasudevan, P., 137[55], *143*
Veldkamp, W., 155[21], *160*
Vemba, T., 39[58], 48[88], *56*, *58*
Versanno, J. O., 261[25], *263*
Vidmar, V., 291[5], 299[5], 312 [5], *316*

Voellmy, C., 191[32], *193*
Volberger, B., 261[25], *263*
Vrancken, M. N., 85[95,96], *94*

W

Wada, A., 203[4], *215*
Wagner, H. N., 226[29], *238*
Wagner, J. G., 155[21], *160*
Wagner, N. H., 226[30], 227 [33], *238*
Wahlig, H., 207[21], *216*
Wai, K., 284[31,32], *288*
Wakamatsu, Y., 138[58], *143*
Wakiyama, N., 85[89,90], *93*
Walker, S. E., 52[106], *60*, 71 [21], *90*, 289[1], *316*
Walker, W. G., 107[37], *116*
Wall, M. E., 222[10], *236*
Walters, V., 101[11], *114*
Wan, J., 203[3], *215*
Warburton, B., 29[25], *54*, 86[98,99,100], *94*
Ward, J. B., 245[8], *250*
Ward, L., 221[4], *236*
Watanabe, A., 135[48], *142*
Watanabe, Y., 250[32], *252*
Webb, N., 210[34], *217*
Wecht, C. H., 10[28], *16*
Weiner, M., 11[41], *16*
Wentworth, R. L., 221[2], 222 [2], *236*
Werkmeister, D. W., 110[50], *117*
Whatley, T. L., 133[37], *142*, 232[54,55], *239*
Wheby, M. S., 11[39], *16*
Whitney, W. R., 300[19], *317*
Widder, K. J., 229[44,45], 231 [46,47,48], *239*
Wiedhaup, K., 41[65], *57*
Willet, G. P., 190[30], *193*, 225[25], *237*
Williams, G., 7[9], *14*
Williams, J. E., 81[71], *93*
Williams, J. L., 47[84], *58*, 213 [50], *218*
Wilson, C. H., 253[1], 261[1], *262*
Winek, C. L., 10[28], *16*
Winter, J. M., 10[31], *16*
Wirbrant, A., 42[68], *57*
Wise, D. L., 190[30], *193*, 221 [2,7], 222[2,8], 223[16,17,18], 225[25], *236*, *237*, 308[30], *318*
Wittbrecker, E. L., 121[6], 138 [59], *139*, *143*
Witte, J., 83[83], *93*
Wolf, W. R., 254[3], *262*
Wood, D. A., 133[37], *142*, 219 [1], 232[54,55], *236*, *239*
Wood, J. H., 173[28], *179*
Woodland, J. H. R., 223[14,15], *237*
Woodruff, C. W., 281[24], *287*
Woods, A. C., 11[46], *17*
Woods, J. D., 254[4], *262*
Wurster, D. E., 162[1,2,3,4,5,6, 7,8], *177*
Wurzburg, O. B., 259[22], *263*

Y

Yalkowsky, S. H., 312[39], *318*
Yamaguchi, T., 104[28], 107[28], *115*
Yamatamoto, M., 77[48], *92*
Yamate, G., 272[10], *286*
Yang, W., 283[29], *287*
Yazawa, K., 262[30], *264*
Yen, S. P. S., 208[26,27,28,29], *216*, *217*
Ynker, M. H., 247[21], *251*
Yolle, I., 226[29], *238*
Yolles, S., 221[3,4], 223[14,15], 224[23,24], *236*, *237*

Yoncoskie, R. A., 103[25], *115*
Yoon, S. M., 170[19], *178*
Yoshida, N. H., 104[29], *115*, 207[24], *216*
Yoshimura, K., 250[32], *252*
Yoshioka, T., 134[46], *142*, 257[17], *263*
Yum, S. I., 168[15], 172[15], *178*
Yurkowitz, I. L., 83[80], *93*

Z

Zaborsky, O. R., 203[5], *215*
Zehnder, H. J., 210[35], 212[35], *217*
Zessin, G., 190–191[31], *193*
Zilkha, K. J., 11[45], *17*
Zoglio, M. A., 8[21], *15*, 125[20], 126[20], *140*, 309[32], *318*
Zolle, I., 226[28,30], 227[35], *237*, *238*
Zweidinger, R., 222[10], 223[11], *236*

Subject Index

A

Absorption of drugs, 11
Acacia:
  air suspension coating, 165
  binder, 102
  coacervation, 61–75, 82, 249, 290, 291
  composition, 64
  diffusion coefficient, 298
  gelatin films, 73–74
  interfacial polycondensation, 137
  spray drying, 184, 188
  swelling, 297–298, 315
  tortuosity, 311
  triple-walled microcapsules, 86
Acetanilide, 70
Acetazolamide, 3
Acrylic polymers and copolymers:
  air suspension coating, 165, 175
  anionic, 26
  aqueous-based coatings, 37
  cationic, 26
  chemistry, 199–203
[Acrylic polymers and copolymers]
  coacervation, 83–84, 110, 112–113
  cohesive energy density, 35
  microparticles, 203–208
  molecular-scale entrapment, 283–284
  nanoparticles, 208–213
  pan coating, 151
  permeation considerations, 291
  plasticizer effect, 40–41
  shock preventing agents, 77
Adjuvants, 209, 212
Adsorption of drugs, 293, 297
After-hardening, 27–28
Agar:
  coacervation, 65
  congealable procedures, 253, 257–259
  graft-polymerization, 214
Agarose congealable procedures, 257–259
Aggregation:
  air suspension coating, 167
  reduction by or in:
    air suspension coating, 175
    charge repulsion, 133
    ethylcellulose, 113
      microcapsules, 106–107

[Aggregation]
inert fillers, 77, 139
sonication, 227
starch, 187–188
surfactants, 77, 262
thickening agent, 262
suspension polymerization, 196
Air suspension coating, 33, 36–37, 43, 161–177
Albumin:
coating of microcapsules, 110
complex coacervation, 82
dip coating, 272
drug carrier, 284
nanoparticles, 88, 89
permeation considerations, 292
polymerization procedures, 225–231
^{99m}Tc-labeled spheres, 226-227
Alginates:
air suspension coating, 165
congealable procedures, 259–260
graft-polymerization, 214
Aluminum monostearate, 189, 279
Amaranth, 113, 256
Amino acids, 262
Aminophylline, 3
Amitriptyline, 3, 10, 12
Amphetamine (sulfate), 246–247
Ampicillin (sodium or trihydrate), 3, 7, 11, 113, 184-185
Amprotropine, 248
α-Amylase, 85
Amylobarbitone, 102
Animal models, 156, 261
Antibodies, 208–209
Anticholinesterases, 12
Antiphlogistic enzyme, 107
Antitack agents, 42–43, 44
Aqueous-based coatings, 36–37, 157–158
Arginase, 132–133
Arrhenius equation, 293–294
Ascorbic acid (*see* Vitamin C)
L-ascorbyl monostearate, 191
Asparaginase, 8, 125, 128, 205
Aspirin:
air suspension coating, 173, 174, 177
coacervation, 70, 72, 101, 103, 104-105
congealable procedures, 256–257
dip coating, 272
filling of microcapsules, 52
pharmacological considerations, 10, 11, 12
reasons for microencapsulation, 3, 7
Atomizers:
air suspension coating, 164
effect on film formation, 45
pan coating, 148
spray congealing, 182
spray drying, 182
Atropine sulfate, 246
Attapulgite, 4

B

Barbiturates, 12
Barbituric acid, 72
Barium sulfate, 155, 235
Bead polymerization (*see* Suspension polymerization)
Beclamide, 4
Beeswax:
air suspension coating, 174
centrifugal extrusion process, 267, 269
congealable procedures, 253, 261
ion-exchange resins, 249
pan coating, 150, 154, 156
spherical matrices from liquid suspension, 278
spray congealing, 186

[Beeswax]
waxy sealant, 42
Bellafoline, 278–279
Benzaldehyde, 47
Benzalkonium chloride, 128–129
Benzodiazepines, 206–207
Benzylcellulose, 113
Benzylpenicillin sodium, 156
Bitolterol, 102
"Blank" microcapsules, 107
Bleomycin, 257
Bloom rating, 31
Blowing agent, 213
Boundary-layer effects, 312–314
Brewers' yeast, 259
Bulk polymerization, 195–196, 220, 223, 233
Burst effect, 307–308, 311
Butamben, 85
Butobarbitone, 4
Butyl rubber, 103

C

Caffeine, 28, 45, 128–129, 175
Calcium alginate, 128–129, 254, 259
Calcium sulfate hemihydrate, 128–129
Camphor, 4
Capillary extrusion method, 73, 256, 261
Caprolactam, 113
Carbinoxamine, 246
Carbon (activated or black), 72, 81, 256, 272
Carbonic anhydrase, 125, 204–205
Carbonless carbon paper, 1–2, 61
Carbon tetrachloride, 7
Carboxymethylcellulose, 82, 191
Carboxymethylethylcellulose, 109, 165
Carboxyvinyl polymer, 84
Carnauba wax:
electrostatic encapsulation, 271–272
ion-exchange resins, 249
pan coating, 150, 157, 159
spray congealing, 186
waxy sealant, 42
Castor oil, 4, 70
Castor wax, 157, 189, 248
Catalase, 84, 110, 125
Cells, encapsulation of mammalian, 259–260
Cellulose (and derivatives), 34, 88, 250, 282
Cellulose acetate, 35
Cellulose acetate -N, N-di-n-butylhydroxypropyl ether, 112–113
Cellulose acetate butyrate, 108, 159, 165, 249
Cellulose acetate phthalate:
air suspension coating, 165, 174
coacervation, 83, 107, 247
congealable procedures, 254
enteric properties, 25, 26, 27
filler, 102
ion-exchange resins, 247, 248, 249, 250
mixed solvent, 36, 45
pan coating, 151, 152, 154, 156, 159
permeation considerations, 291
plasticizer effect, 40
spray embedding, 189–190
stress dissipation, 45
Cellulose nitrate, 35, 109–110, 113
Centrifugal devices, 265–269
Cetyl alcohol, 156, 186
Channeling agents, 42
Charcoal, 7, 110, 272

Chloramphenicol:
  air suspension coating, 174
  coacervation, 98
  dissolution behavior, 301
  permeation considerations, 291
  polymerization procedures, 205–206, 235
  reasons for microencapsulation, 4
Chlordiazepoxide, 12
Chlorinated rubber, 110–111
Chlorothiazide, 75, 205–206
Chlorpheniramine (maleate), 4, 7, 157, 249, 306
Chlorphentermine, 12
Chlorpromazine hydrochloride, 4, 10
Chlorpropamide, 12
Cholestyramine, 245
α-Chymotrypsin, 133
Citric acid, 4
Clofibrate, 4, 71
Cloxacillin, 4, 11
Coacervation:
  acrylates, 83–84
  aggregation reduction, 44
  carboxyvinyl polymer, 84
  cellulose acetate phthalate, 83
  charge alteration, 75–76
  complex, 29, 47
    aqueous vehicles, 61–89
    definition, 65
    effect of pH, 68–69
    pan coating, 158–159
    partitioning, 73
    sodium polyacrylate, 84
    three-phase diagram, 65–66
  core exchange, 81
  definition, 62
  enhancement of speed, 80
  ethylcellulose, 97–107
  induced by:
    incompatible polymers, 101-103
[Coacervation]
    solvent alteration, 103–106
    temperature change, 98–101
  methylcellulose, 82–83
  nonaqueous vehicles, 97–114
  poly (methyl vinyl ether/maleic anhydride), 83
  shock prevention, 76–77
  simple:
    definition, 64
    effect of pH, 70
    microencapsulation, 70–71
    modified for producing microparticles, 86–88
    recovery of product, 70
    three-phase diagrams, 64–65
  sodium polymetaphosphate, 82
  sodium silicate, 82
  solvent evaporation process, 85–86
  three-walled microcapsules, 86
Coating:
  area, 302
  defects, 49-50
  density, 101
  desolvation, 44–45
  gelation, 44–45
  other terms, 1
  properties, 23
  thickness, 49–50, 290–291, 302
Coconut oil, 71
Codeine phosphate, 4
Cod liver oil, 4, 70
Cohesive energy density, 34
Colestipol hydrochloride, 245
Colorants, 43
Congealable disperse-phase encapsulation procedures, 253–262
Co (polyether) polyurethane, 28
Core:
  Buffers, 22
  coacervation, 66–67
  density, 22
  exchange, 81
  increase in hydrophobicity, 278
  mechanical support, 22

[Core]
other terms, 1
pan coating, 149–151, 167
properties, 21-23
shape, 22, 167
size, 22, 155–156, 167
swelling, 23, 177, 249
thermal expansion, 23
Corticosteroids, 12
Cotton seed oil, 225
Cracks, 291
Crosslinked formaldehyde-urea, 273
Crosslinking (hardening) agents:
2,3-butanedione, 231
control of degree, 78–79
diethylenetriamine, 129
effect on diffusion, 296–297
epoxy resins, 235
formaldehyde, 67, 68, 78–79, 128, 231, 256, 298
glutaraldehyde, 67, 79–80, 125, 134, 137, 191, 229, 272, 290
glyoxal, 79
heat for albumin, 225
mechanism of action, 78–80
N,N^1-methylene-bis (acrylamide), 202, 203
2-methylglutaraldehyde, 79
tortuosity, 311
triethylenetetramine, 129
Cyclandelate, 4
Cyclazocine, 223, 224
Cyclophosphamide, 224
Cysteine, 4

D

Dactinomycin, 232–234
Daunomycin hydrochloride, 227
Daunorubicin, 232
D & C yellow 11, 256
DDT, 72
Dehydroemetine dihydrochloride, 235
Dexamphetamine (sulfate), 156, 173–174, 245-246
Dextran, 82, 249, 250, 257, 284
Dextrins, 184, 188
Dextromethorphan (hydrobromide), 177, 247–248, 250
Diacetyl reductase, 254, 259
2,4-Diamino-6-(2-naphthylsulfonyl)-quinazoline, 190, 224–225
Diazepam, 4, 12, 128–129
Diazoxide, 12
Dibucaine, 85
cis-Dichlorodiamineplatinum, 224
Dicloxacillin sodium, 4, 172, 185
Dicumarol, 11, 12
Differential scanning calorimetry, 299
Diffusion, 293-295
coefficient (diffusivity), 293–294, 295–299, 301, 311
Digoxin, 12, 177, 190-191
Dimethicone fluid, 4
Dip coating, 272
Diphenhydramine hydrochloride, 4, 128–129, 174, 256
Dissolution of polymers, 34–36
Disulfiram, 4
Doxorubicin (hydrochloride), 224, 229–231, 233
Doxycycline hydrochloride, 4

E

Electrocapillary emulsion technique, 135
Electrostatic encapsulation, 271–272
Emulsion polymerization, 196–197, 208-209, 212-213
Encapsulation of microcapsules, 51–52
Ephedrine (base or hydrochloride),

[Ephedrine]
245, 247–248
Epoxy resin, 233, 235
Eprazinone, 4, 7, 290
Erosion of polymers and permeation considerations, 314–315
Erythrocytes as drug carriers, 285
Erythromycin ethyl succinate, 225–226
Esterases, 26
Ethylcellulose:
air suspension coating, 23, 165, 172, 173, 174, 175, 177, 248
anticlumping agent, 113
apparent diffusion coefficient, 298–299
aqueous-based coating, 37, 157–158, 175, 177
coacervation, 52, 97–107, 249
congealable procedures, 256
effect of or on:
application method, 148
diffusion, 309
humidity, 48
plasticizer, 37, 39
viscosity grade, 297
films, 28
glass transition temperature, 33
mixed solvent, 36
molecular-scale entrapment, 284
molecular weight, 31
pan coating, 148, 151, 154, 156, 157, 159, 174, 175, 177
permeability coefficient, 313
permeation considerations, 291–292
polyethylene glycol as channeling agent, 42
pores, 312
[Ethylcellulose]
reduction of aggregation in microcapsules, 106–107
solvent evaporation process, 85–86, 104–105
solvents, 34, 36
spherical matrices from liquid suspension, 278
spray congealing, 185
spray drying, 185
triple-walled microcapsules, 86
waxy sealant treatment, 42
Ethylene-vinyl-NN-diethylglycinate, 27
Ethyl methacrylate, 165
Ethynylestradiol, 267–268
Evaporation rate of solvents, 164–166
Extrusion, 174, 253–255, 265–271

F

Fenfluramine, 4, 10
Ferrous citrate, 4
Ferrous fumarate, 5
Ferrous sulfate, 5, 7, 10
Fibrinogen as drug carrier, 284
Fick's equations, 293
Fillers:
aggregation reduction, 77, 112, 196
function, 44
permeation considerations, 291, 297
polyethylene, 101
Film formers, 23–24
adhesive forces, 29
air suspension coating, 164–165
cohesive forces, 29
diffusion, 26, 29
disintegration, 26
dissolution, 26
enteric, 25–28
enzymatic breakdown, 26
hydrophilic, 28

[Fillers]
hydrophobic, 28
molecular weight, 29
plasticizer effect, 40–41
regulatory status, 29
release mechanisms, 26
swelling, 26, 27
tack, 33
toxicity, 29
viscosity, 30
Films:
adhesiveness, 33–34
dispersed solids, 43–44
laminated, 28–29
mechanical properties, 31, 45–47
permeability to gases, 47–48
photostability, 48
First-pass effects, 11
Fluorescin (isothiocyanate), 89, 211, 232
5-Fluorouracil, 105, 227, 229, 257
Furosamide, 11, 282

G

Gastric emptying, 9–11, 177
Gastrointestinal transit time, 11, 22–23, 52
Gelatin:
acacia films, 73–74
air suspension coating, 165, 173
bloom rating, 31
coacervation, 61–82, 83, 86–89, 103, 247, 249, 290, 291
composition, 62, 64
congealable procedures, 253, 254, 256–257
core support, 22, 24, 128–129
crosslinking agents, 42, 77–80
[Gelatin]
derivatives, 75, 76, 77
diffusion, 31, 297–298, 309
extrusion, 174, 270–271
micropellets, 78
modulus of elasticity, 46
nanoparticles, 86–89
pan coating, 149, 154
pharmagel A type, 247
photostability, 48
polymerization procedures, 208
separation, 73
solvent evaporation process, 85–86
spray congealing, 186, 187–188
spray drying, 185, 186
swelling, 297–298, 315
tortuosity, 311
Gelation, 44–45
Gel permeation chromatography, 30, 31
Gentamicin (sulfate), 113, 207
Glass transition temperature of polymers, 32–33
aliphatic polyesters, 223
diffusion, 296-297
effect of:
filler, 44
plasticizer, 39
β-Glucosidase, 204
Gluten, 24, 184
Gluthemide, 205–206
Glycerides, 266–269
Glycerin, 271–272
Glyceryl dipalmitate, 186, 261–262
Glyceryl distearate, 154, 156, 174, 186, 279
Glyceryl monopalmitate, 186, 261–262
Glyceryl monostearate, 111, 154, 156, 174, 186, 189
Glyceryl stearates, 26
Glyceryl trinitrate, 5
Glyceryl tripalmitate, 261–262
Glyceryl tristearate, 186, 262

Glycowax (S-932), 188–189, 253, 261
Gold sodium thiosulfate, 113
Granulation processes, 278–279
Griseofulvin, 12
Guanethidine, 12

H

Half-life:
 biological, 11–12
 drug release, 306
Hardened oils and fats, 111–112, 253, 262
Hardening agents (*see* Cross-linking agents)
Histidase, 133
Horse anti (human lymphocyte) globulin, 205
Human immunoglobulin, 208–209
Hydrochlorthiazide, 12
Hydrocortisone, 11, 205–206
Hydroflumethiazide, 71
Hydrogenated beef tallow, 111, 253, 262
Hydrogenated castor oil:
 air suspension coating, 165, 174
 coacervation, 111–112
 congealable procedures, 253
 granulation processes, 279
 pan coating, 150, 154, 159
 spray congealing, 185, 186
  drying, 185
Hydroxypropylcellulose:
 air suspension coating, 165
 aqueous-based coating, 36
 films, 28, 33
 polymerization procedures, 235
[Hydroxypropylcellulose]
 spray drying, 185
Hydroxypropylmethylcellulose:
 adhesiveness, 34
 air suspension coating, 165, 248
 aqueous-based coating, 36, 37
 effect of:
  fillers, 44
  method of application, 148
  plasticizers, 37, 39
 film defects, 49, 50
 floatation, 22
 glass transition temperature, 33
 mechanical properties, 47
 pan coating, 157, 159
 spray drying, 185
Hydroxypropylmethylcellulose phthalate:
 air suspension coating, 165
 aqueous-based coating, 37
 coacervation, 108–109
 enteric properties, 25, 27
 pan coating, 150
12-Hydroxystearyl alcohol, 156, 174
Hyoscine methonitrate, 5

I

Immunosuppressive agents, 227
Indomethacin, 5, 73, 75, 102, 206–207
Influenza virions, 211, 212
In situ polymerization, 272-273
Instagliations, 49–50
Instron tensile tester, 46
Interfacial polycondensation, 119–139
Interfacial polymerization, 232
Intrinsic factor, 5
Invertase, 137

Ion-exchange resin:
  coated, 70, 125, 247–249
  drug complex, 23, 80, 177, 241–250
Iron (phosphate) 7, 11
Isoniazid, 101, 298

J

Japanese (synthetic) wax, 253, 261, 262

K

Kinetics of drug release, 302–306, 308–311

L

β-Lactam antibiotics, 11
Lactate dehydrogenase, 204
Lag time effect, 307–308, 311
Lectin as drug carrier, 284
Lemon oil, 76, 82, 184, 186, 188
Levodopa, 5, 11
Lipase, 26
Liposomes, 273–278
Liquid paraffin, 71, 84
Lithium carbonate, 5
Lucanthone hydrochloride, 247
Lycopodium spores, 72

M

Magnesium aluminum hydroxide hydrate, 105
Magnetite, 229–231
Marumerizer, 279–281
Mechanical properties of films, 45-47
Meclofenoxate hydrochloride, 5
Melt index, 31
Menthol, 82
Meprobamate, 5, 10, 12, 206–207
Mepyramine maleate, 256
Mercaptopurine, 227
Methantheline bromide, 128–129
Methapyrilene (hydrochloride), 246, 247–248, 283
Methaqualone, 5, 175, 176
Methionine, 6, 262
Methotrexate, 232
Methylamphetamine (hydrochloride), 5, 247
Methylcellulose:
  air suspension coating, 165, 173, 174
  coacervation, 65, 82–83
  pan coating, 154
  permeation considerations, 292
Methylene blue, 31, 150–151, 260
Methylglyoxal, 85
Methyl methacrylate, 165
Methyl salicylate, 7
β-Methylumbelliferone, 212
Micelle polymerization, 197, 199, 208
Microcapsules:
  artificial cells, 124–125
  "blank", 107
  colored, 81
  definition, 1
  dual-walled, 28–29, 81
  electrophoretic properties, 75
  encapsulation of, 51–52, 105
  number per unit weight, 51
  other terms, 1
  pearlized, 81
  radioimaging agents, 113
  reduction of aggregation, 106–107

[Microcapsules]
  size:
    fractionation, 51
    range, 1, 50
  surface area, 21
  tableting (*see* Tableting of microcapsules)
  triple-walled, 29, 86
  types, 1–2
Microcrystalline:
  cellulose, 282
  wax, 156, 173–174
Microencapsulation:
  disadvantages for sustained release, 10
  environmental protection, 3–6, 7, 259
  enzyme replacement therapy, 8
  flavoring agents, 184, 186, 188
  erythrocyte replacement therapy, 8
  historical considerations, 1–3
  gastric irritation reduction, 5,7
  general references, 13
  improved:
    flow, 7
    handling, 7
  intrauterine contraceptive device, 7, 108
  liquid solid conversion, 4, 7
  odor masking, 4–5, 7
  mammalian cells, 259–260
  nonpharmaceutical applications, 2, 285
  parenteral therapy, 8
  pharmacological considerations, 8–13
  physicochemical considerations, 12–13
  problems encountered, 3
[Microencapsulation]
  reduction of:
    hygroscopic properties, 7
    incidence of side effects, 9–13
    plasma peaks, 267–268
    sensitization to penicillins, 7, 184–185
    volatility, 7, 184, 188
  regulatory considerations, 285–286
  separation of incompatibilities, 3–6, 7
  suspensions, 8, 185, 279
  taste masking, 3–6, 7, 185, 186, 248
  thermography, 7
  3M process, 270–271
  topical products, 7, 113
Microparticles, 1, 253, 254, 256, 257, 259
Microspheres:
  albumin, 226–231
  cellulose, 250
  congealable procedures, 254, 257
  magnetic, 229-231
  solvent evaporation, 85
Mitomycin C, 250, 257 258–259
Modulus of elasticity, 45, 46
Molecular-scale entrapment, 281, 283–284
Molecular weight of:
  drugs, 12, 296
  polymers, 29–31, 297
Monoamine oxidase inhibitors, 12
Monolithic (matrix) devices, 289, 303, 308–311, 315–316
Morphine sulfate, 128–129
Moving boundaries, 314–315
Myristyl alcohol, 24

N

Naltrexone (pamoate), 223–224, 235, 261–262, 308
Nanocapsules, 1, 86, 135
Nanoparticles:
  acrylic, 208–213
  adjuvants, 209, 212
  advantages, 86, 88–89
  albumin, 89
  congealable procedures, 257
  definition, 1
  disadvantages, 88
  gelatin, 86–89
  polyalkyl cyanoacrylate, 232
Nanospheres, 257
Niacin, 7
Nicotinamide, 6
  adenine dinucleotide, 254
Nitrofurantoin, 5
Nitroglycerin, 11, 154
Noreleagnine hydrochloride, 284
Norethindrone (acetate), 222, 267–268
d-Norgestrel, 221–222
Nortriptyline, 5,10
Noscapine, 5
Novobiocin (acid or sodium), 156
Nylon, 6, 33, 42, 121–129

O

Oils flavored, 270–271
Olive oil, 81
Opaquant-extenders, 43–44
Orange oil, 188
Oxytetracycline (base or hydrochloride), 5, 101

P

Palmitic acid, 278
Pan coating, 145–159
  adhesion, 33
  antitack agents, 42–43
  aqueous-based coating, 36–37
  colorants, 43
  effect of talc, 44
Pancreatin, 107, 109
Papaverine (base or hydrochloride), 5, 205–206
Paracetamol:
  air suspension coating, 174
  coacervation, 98, 102, 103
  pan coating, 149
  polymerization procedures, 214
Paraffin (wax):
  congealable procedures, 262
  ion-exchange resins, 249
  pan coating, 157, 159
  release mechanism, 26
  sealant, 42, 98, 249
  solubility, 27
Paraminosalicylic acid, 12
Particle interactions, 167
Partition (distribution) coefficient, 13, 295, 299–300
Pearl polymerization (*see* Suspension polymerization)
Pectic acid, 77
Pectin, 65
Penicillins, 184, 185
Pentaerythritol tetranitrate, 5
Pentobarbituric acid, 72–73
Peppermint oil, 7, 184
Pepstatin, 22, 157
Pericyazine embonate, 128
Permeability:
  coefficient (apparent), 295, 298–299, 313
  considerations, 289–293
  constant, 295
  definition, 295

[Permeability]
microcapsules to:
electrolytes, 132, 139
oxygen, carbon dioxide, and water vapor, 47–48
Pharmagel A (*see* Gelatin)
Phenacetin 5, 72, 83, 229
Phenazone, 175, 260, 282
Phenethicillin potassium, 101
Phenformin (hydrochloride), 5, 10
Phenobarbitone (sodium):
apparent diffusion coefficient, 298
biological half-life, 12
coacervation, 74, 98–99
diffusion, 309
granulation processes, 278–279
ion-exchange resins, 245
molecular-scale entrapment, 284
polymerization procedures, 206–207
reasons for microencapsulation, 5
spray:
embedding, 190–191
polycondensation, 191
Phenobarbituric acid, 73
Phenothiazines, 12, 13
Phenoxymethylpenicillin, 246
Phentermine, 246–247
Phenylbutazone, 5
Phenylephrine (hydrochloride), 5, 157
Phenylpropanolamine (hydrochloride), 6, 157, 177, 211, 249
Phenytoin, 12, 128–129
Photostability, 48, 81
Pigments, 36, 43, 152, 291, 297
Placental alkaline phosphatase, 85
Plasma drug profiles, 8, 12
Plasticizers, 37–41, 296–297
Polidexide, 245
Pollen extracts, 174
Polyacrylamide:
modified, 212–213
nanoparticles, 211, 212
polymerization procedures, 203, 204–205, 208
Polyacryldextran, 212–213
Poly(adipyl L-lysine) microcapsules, 135–136
Polyalkyl cyanoacrylate:
in situ polymerization, 273
products, 231–233
Polybutadiene, 103
Poly (ε-caprolactone), 222–223
Polychloroprene, 86
Polydimethylsiloxane, 103, 214–215
Polyester, 137–138, 219–225
Polyethylene, 33, 35, 101–103, 113, 150
Polyethylene glycol:
air suspension coating, 165, 175
aqueous-based coatings, 36
congealable procedures, 260–261
ion-exchange resin swelling, 23, 249
mixed films, 28
permeation considerations, 291
Polyethyleneimine, 113
Poly (ethylene-vinyl acetate), 110–111
Polyglutamic acid and copolymers, 235
Polyglycolic acid and copolymers:
coacervation, 113
diffusion, 308
molecular-scale entrapment, 284
polymerization procedures, 219–225
spray embedding, 190

Poly (hexamethylene sebacamide) (*see* Nylon)
Polyhydroxyethyl methacrylate, 272
Polyisobutylene, 103
Polylactic acid and copolymers:
  air suspension coating, 174
  boundary-layer effects, 314
  coacervation, 109, 113
  diffusion, 308
  molecular-scale entrapment, 284
  permeation considerations, 291–292
  polymerization procedures, 219–225
  solvent evaporation process, 85
  spray embedding, 190
Polylysine, 259–260, 284
Polymerization procedures, 195–215, 219–235
Polymers:
  amorphous, 32
  annealing, 32
  crystalline structure, 31–32
  glass transition temperature, 32–33
  swelling, 292
  tacticity, 32
Polymethacrylate, 207
Polymethyl methacrylate, 183, 200–202, 207, 210–212
Poly (methyl vinyl ether/maleic anhydride), 20, 41, 83
Polypeptides, 262
Polyphthalamide, 129, 131–135, 295
Polysiloxane, 214–215
Polystyrene:
  coacervation, 110, 113
  expanded, 213
[Polystyrene]
  in situ polymerization, 273
  polymerization procedures, 207, 213–214
  solvent evaporation process, 85
Poly (styrene-maleic acid) copolymer, 155
Polyterephthalamide, 131–132
Poly (terephthaloyl L-lysine), 133–135
Polyurethane, 138–139, 273
Polyvinyl acetate, 28
  phthalate, 25, 27, 40
Polyvinyl alcohol:
  coacervation, 65, 82, 103, 112–113
  congealable procedures, 261
  polymerization procedures, 214
  spray:
    drying, 185
    embedding, 190–191
    polycondensation, 191
Polyvinyl chloride, 113
Polyvinylpyrrolidone:
  adhesive, 149–150, 151, 156–157
  after-hardening of shellac, 28
  coacervation, 82, 112–113
  permeability of core, 132
  spray:
    drying, 185
    embedding, 190–191
    polycondensation, 191
Pore effects, 291, 312
Porosity, 311
Potassium chloride:
  coacervation, 103, 158–159
  gastrointestinal irritation, 7, 11
  pan coating, 158–159
  polymerization procedures, 207
  tableting of microcapsules, 52

Potassium dichromate, 103
Prednisolone, 6, 12, 74
Procainamide (hydrochloride), 6, 12
Procaine penicillin (G), 70, 279
Progesterone, 108, 314
Propantheline (bromide), 6, 11
Propoxyphene hydrochloride, 6, 306
Propranolol (hydrochloride), 6, 154, 244
Propylthiouracil, 11
Pseudoephedrine hydrochloride, 247–248
Pyrene butyric acid, 212

Q

Quaternary ammonium salts, 11
Quinacrine hydrochloride, 137
Quinidine sulfate, 6, 250, 266–267, 269

R

Regulatory considerations, 285–286
Release of drug from microcapsules and microparticles, 289–316
Reserpine, 6
Reservoir (capsular) devices, 289, 303–306, 315–316
Riboflavin (*see* Vitamin B_2)
Ribonuclease, 204

S

Saccharin, 174
Safety precautions, 148, 173, 183
Salicylamide, 12, 103
Salicyclic acid, 40, 175, 204, 282
Scaling-up problems 285
Secretin, 137
Selectivity coefficient, 243, 247
Shellac, 28, 156–157, 165, 175, 189
Shock prevention, 76–77
Side-vented coating pan process, 153
Silica (mica), 74, 77, 81
Silicone rubber, 32
Sodium alginate:
  congealable procedures, 254, 259
  crosslinking agents, 42
  ion-exchange resins, 250
  solvent evaporation process, 85
Sodium bicarbonate, 6, 22, 174
Sodium carboxymethylcellulose:
  coacervation, 82
  ion-exchange resins, 250
  shock preventing agent, 77
  solubility, 34
  spray drying, 183, 184
Sodium carboxymethylstarch, 77
Sodium chloride, 7, 104, 113
Sodium o-iodohippurate, 257
Sodium pentobarbital, 125–127, 129
Sodium polyacrylate, 84
Sodium polymetaphosphate, 82
Sodium salicylate:
  air suspension coating, 174–175
  coacervation, 99–100, 103

[Sodium salicylate]
congealable procedures, 254, 256
effect of ethylcellulose grade, 297
interfacial polycondensation, 128–129
solvent evaporation process, 85
waxy sealants, 42
Sodium silicate, 82
Sodium thiocyanate, 79
Solubility:
drugs, 12–13, 300-302
parameter, 36, 39
polymers, 27, 34–36
Solvent evaporation process, 85–86, 104–105
Solvents, 34–37, 164–166
Spermaceti, 42, 185
Spherical matrices from liquid suspension, 278
Spheronization, 102, 175, 279–282
Spiramycin, 250
Spray:
congealing, 181–183, 185–187
drying, 181–185
embedding, 188–191
polycondensation, 191
Stagnant diffusion layer, 290–291
Starch, 253, 259, 282, 284
Stearic acid:
air suspension coating, 165
antitack agent, 42
dip coating, 272
ion-exchange resin, 249
pan coating, 159
spray congealing, 185
Stearyl alcohol, 73, 249, 254
Stokes-Einstein equation, 294
Stress-strain testing of films, 45–47
Succinimide, 6
Sugars, 165, 270–271
Sulfadiazine (sodium):
coacervation, 70, 71, 83
congealable procedures, 256–257
polymerization procedures, 225
Sulfaethylthiadiazole:
congealable procedures, 261
granulation processes, 279
spray:
congealing, 185, 186–187
embedding, 188–189
Sulfamerazine, 70, 73
Sulfamethizole:
coacervation, 105, 109
congealable procedures, 257–258
polymerization procedures, 227–228
tortuosity, 311
Sulfamethoxazole:
apparent diffusion coefficient, 298
coacervation, 74–75, 80
spherical matrices from liquid suspension, 278
spray embedding, 189-190
Sulfamethoxydiazine, 6
Sulfamethylthiadiazole, 185
Sulfanilamide, 78, 256
Sulfathiazole (sodium), 85, 128
Sulfur, 72
Surface roughness, 50
Surfactants:
anionic as shock preventing agents, 77
cationic to reduce aggregation, 77
reduction of aggregation, 41

Suspension polymerization, 196, 213–214, 233, 241
Suspensions of microcapsules, 185
Sustained release terms, 8
Swelling of polymers, 292, 297–298, 314–315

T

Tableting of microcapsules, 52
  aspirin, 101, 173
  barbiturates, 99, 125
  drug-loaded ion-exchange resins, 247, 248
  isoniazid, 101
  penicillins, 101, 156, 185
  phenazone, 175
  sulfamethoxazole, 189–190
  tridihexethyl iodide, 156
  vitamin B_2, 110
Tartaric acid, 6
Testosterone, 284
Tetanus toxin, 208–210
Tetracaine, 85
Tetracycline (base, hydrochloride or phosphate), 6, 9, 12, 156–157, 208
Theobroma oil, 111
Theobromine, 128–129
Theophylline:
  coacervation, 103
  ion-exchange resins, 245
  interfacial polycondensation, 128–129
  polymerization procedures, 208
  therapeutic index, 12
Therepeutic index, 12
Thiabendazole, 74
Thiamine (hydrochloride or mononitrate) (*see* Vitamin B_1)
Thiopropazate hydrochloride, 246
Three-phase diagram construction, 66
Tolbutamide, 12
Tortuosity, 80, 311
Tridihexethyl iodide, 156
Trifluoperazine embonate, 6, 128, 132
Trimeprazine tartrate, 6, 154
Triple-walled microcapsules, 29, 86
Trypsin, 203–204

U

Undecenovanillylamide, 74
Urea, 40
Urease, 8, 124–125, 104

V

Vacuum deposition, 285
Vibrating reed method, 73
Vinblastine, 232–233
Viscosity of polymers, 30, 31, 297
Vitamin A (palmitate), 7, 187–188, 259
  B_1 (thiamine hydrochloride or mononitrate), 6, 7, 186
  B_2 (riboflavin):
    absorption, 11
    "blank" microcapsules, 107
    coacervation, 70, 78, 110
    congealable procedures, 256

[Vitamin A]
ion-exchange resins, 246
reasons for microencapsulation, 6, 7
B_6 (pyridoxine hydrochloride), 6
B_{12} (cyanocobalamin), 6, 135–136, 246
C (ascorbic acid), 6, 103, 260
Vitamins, fat soluble, 268–269
Volatile oils, 184

W

Wall thickness, 49–50
Warfarin, 12, 204
Waxes, 165
Waxy sealants, 42, 98, 99, 111, 249, 291–292
Wurster coating (*see* Air suspension coating)

Y

Yeast alcohol dehydrogenase, 204, 208
Yeast cells as drug carriers, 285

Z

Zein, 88